Organization and Expression of Globin Genes

Organization and Expression of Globin Genes

Editors

George Stamatoyannopoulos

Professor of Medicine, Division of Medical Genetics
School of Medicine
University of Washington
Seattle, Washington

Arthur W. Nienhuis

Chief, Clinical Hematology Branch
National Heart, Lung, and Blood Institute
National Institutes of Health
Bethesda, Maryland

Alan R. Liss, Inc., New York

Address all Inquiries to the Publisher
Alan R. Liss, Inc., 150 Fifth Avenue, New York, NY 10011

Printed in the United States of America

Library of Congress Cataloging in Publication Data

Conference on Hemoglobin Switching (2nd:1980:
Airlie House)
Organization and expression of globin genes.

Includes bibliographical references and index.
1. Globin—Congresses. 2. Gene expression—Congresses. 3. Chromosome mapping—Congresses. I. Stamatoyannopoulos, George. II. Nienhuis, Arthur W. III. Title. [DNLM 1. Globin—Genetics—Congresses. 2. Genes—Congresses. W3 C676 2nd 1980e / WH 190 C7476 1980o]
QP552.G5C66 1980 612'.1111 81-3677
ISBN 0-8451-0211-7 (v. 1) AACR2

Contents

Preface

This volume, the first of two containing the Proceedings of the Second Conference on Hemoglobin Switching held at Airlie House, Virginia, June 22–26, 1980, is devoted to the structure and organization of the globin genes and the mechanism of their regulated expression. The second volume, entitled "Hemoglobins in Development and Differentiation" describes observations of hemoglobin switching in vivo and in culture and the regulation of hemopoietic cells.

The First Conference on Hemoglobin Switching was held at Seattle, Washington, in June 1978. In the two years that have passed, striking progress has been achieved in several areas related to hemoglobin switching. In organizing the Second Conference we again attempted to be comprehensive, including work related to all aspects of the phenomenon of hemoglobin switching, e.g., gene structure, the informational macromolecules, hemopoietic cell differentiation, and in vivo and clinical observations. Hence, as was true of the first conference in Seattle, this Second Conference was a gathering of molecular, cellular, and developmental biologists, and also clinical investigators who shared common interest in the question of regulation of the globin gene expression in development and differentiation.

This Conference would not have been held without the support and encouragement of Dr. Earl Chamberlayne of the Fogarty International Center, Dr. John Hercules of the National Heart, Lung, and Blood Institute, and Dr. David Badman of the National Institute of Arthritis, Metabolism, and Digestive Diseases. We would like to thank Dr. Arthur Axelrad for help in planning the program and Mrs. Linda Garza and Laura Short for assistance in the organization of the Conference. The Conference was sponsored jointly by the National Heart, Lung, and Blood Institute; The National Institute of Arthritis, Metabolism and Digestive Diseases; The Clinical Center, NIH; The National Institute of General Medical Sciences; The National Cancer Institute; The Fogarty International Center; and the Center for Advanced Studies in Biomedical Sciences, University of Washington.

George Stamatoyannopoulos
Arthur W. Nienhuis

Organization and Expression of Globin Genes: An Introduction

Arthur W. Nienhuis and George Stamatoyannopoulos

Clinical Hematology Branch, National Heart, Lung, and Blood Institute, Bethesda, Maryland (A.W.N.) and University of Washington, Seattle (G.S.)

The rapid advances in our knowledge of globin gene structure are based on several powerful experimental methodologies including restriction endonuclease mapping, molecular cloning, and rapid DNA sequencing. Based on the advent of these techniques we have, in a sense, entered the golden age of molecular anatomy. Much of this volume is devoted to an account of the progress made in defining the structure of the globin genes during the past two years since the first Hemoglobin Switching Conference was held in Seattle [1].*

Structural information ultimately is rather sterile unless functional relevance can be defined. Generally the ability to delineate the function of specific DNA sequences and genomic regions has lagged behind the ability to define structure. However, new and more powerful approaches have recently evolved. These include sophisticated techniques to detect and characterize precursor RNA molecules, in vitro transcription of isolated gene fragments, careful analysis of the naturally occurring mutants of the human globin genes, and transfer of intact genes into cells by a variety of procedures. The enthusiastic application of these has yielded data regarding gene function which are also included in this volume.

The purposes of this introductory paper are several. We wish to provide a brief summary of the contents of this volume so that the reader's attention may be directed to topically related information found in several different papers. Interpretations as to the relevance and limitation of certain data and experimental techniques are provided where we have felt that these could be helpful. Areas which appear to be ready for fruitful investigation are also indicated. Finally, we have provided a brief summary of our current understanding of the molecular basis for the differential expression of the individual globin genes during development and differentiation.

*The papers within this volume are referenced in this chapter by referring to the senior author. When data considered particularly relevant to this review are not presented or discussed within this volume, the relevant references are cited.

GENE STRUCTURE

The Canonical Globin Gene

The general structure of the mouse β-globin gene, as defined by Leder and coworkers, includes three exons or coding blocks and two introns or intervening sequences [2, 3]. It is now recognized as being characteristic of all functional globin genes and thus may be referred to as the canonical structure. The introns always occur at the same relative position in each gene; the shorter of the two at a position corresponding to amino acids 29-30 or 30-31 and the longer at a position corresponding to amino acids 103-104 or 104-105. In general the second of the two introns is considerably longer in the β-like genes (600–900 nucleotide pairs) than in the α-like genes (120–150 nucleotide pairs). The paper by Maniatis et al contains a detailed summary of the primary structure of the mammalian globin genes with specific emphasis on those found in the human genome, while the papers that follow contain information about the globin genes found in several other species.

The Globin Gene Promoter

Current consensus is that initiation of transcription of the globin genes occurs at a position referred to as the Cap site which corresponds to the 5′ end of mature messenger RNA (mRNA). As reported by Luse, Talkington, and Flavell and their coworkers in papers found in this volume, pure globin gene fragments when incubated in a crude cell extract, support the synthesis of RNA molecules which begin at the Cap site. For these reasons much attention has been focused on the DNA sequences in this region and immediately upstream in an effort to define the nature of the globin gene promoter. The analysis by Maniatis and his colleagues has led to the identification of at least two sequences upstream from the Cap site which are conserved in several species. The first, referred to as the ATA box occurs 29–30 nucleotide pairs upstream. It is an AT-rich region which is analogous to that which functions in RNA polymerase binding in bacteria. The second occurring 70–78 nucleotide pairs upstream is referred to as a CCAAT box.

Further support for the importance of the sequences upstream from the Cap site and their possible promoter function is provided by the in vitro transcription studies. Talkington and Leder inserted a 148 nucleotide pair fragment of the mouse α-globin gene, 141 nucleotide pairs of which are upstream from the Cap site, into plasmid DNA by molecular cloning. It gave a specific RNA product of defined length and thus functioned as a promoter in a crude cell-free extract of Hela cells supplemented with RNA polymerase II. Flavell and coworkers prepared a number of deletion mutants of the putative promoter region of the rabbit β-globin gene. Removal of DNA sequences 100 nucleotide pairs upstream from the Cap site had no effect but the nucleotides between 10 and 75 upstream were absolutely required for accurate transcription. Deletion of the nucleotides between 4 and 30 upstream from the Cap site also abolished transcription while deletion of the nucleotides between 8 and 16 upstream gave a template which functioned poorly compared to the "normal" promoter. Furthermore, this mutant gave a displaced site of initiation 30 nucleotide pairs downstream from the ATA box. Thus the ATA sequence may serve to bind RNA polymerase and phase the initiation of transcription.

Study of globin pseudogene fragments in the cell-free extracts also supports a role for the ATA consensus sequence (box) in promoter function. These elements are interspersed among and have considerable homology to functional globin genes but substitution, frame shift, or deletion mutations in coding regions or at splice sites makes it impossible for pseudogenes to encode for a globin product. One such pseudogene, the α_3 gene of mouse, has a change in the ATA box-ATA to AGA: appropriate restriction endonuclease fragments containing this gene are not transcribed in vitro [Talkington and Leder]. Another example is the β^X gene of goat. As reported by Lingrel et al, this gene has a sequence TCAC rather than ATAA found in normal genes. The β^X gene is not transcribed in vitro [Luse et al]. Evidence indicating that the CCAAT box is required is less substantial although the human δ-globin gene which appears to be transcribed inefficiently both in vivo [Kantor et al] and in vitro [4] differs from the other human β-like genes both with respect to the distance between the CCAAT and ATA sequences and also in having a C rather than T in the CCAAT box [Maniatis et al].

A word of caution is in order in interpreting data obtained using the crude cell-free extracts for transcription. The efficiency of these systems is low; less than 5% of added DNA fragments give an RNA product. Furthermore, no tissue specificity is evident since putative globin gene promoters function in extracts of Hela cells. Thus while analysis using these crude extracts may define the minimum sequences required for promoter function, they may provide little or no insight into quantitative aspects of transcription. Hamer et al have put the mouse α-globin gene into SV40 and used such recombinants to infect susceptible monkey kidney cells [5]. Reported by Hamer at the Airlie House Meeting was evidence to be published elsewhere, that sequences several hundred nucleotide pairs upstream from the putative mouse α-globin gene promoter may influence the transcriptional rate of the α-globin gene in this system. Thus refinement of the in vitro transcription systems and/or further development of other strategies may be necessary to define DNA sequences important for quantitative aspects of promoter function.

GLOBIN GENE ORGANIZATION

Linkage

The human α-like and β-like globin genes form clusters found on chromosomes 16 and 11, respectively [Maniatis et al and Smithies et al]. Linkage of the rabbit [Hardison et al] and certain of the mouse [Edgell et al] embryonic and adult β-like globin genes of these species has also been established by analysis of overlapping cloned segments of DNA containing these genes. In every case where relevant data are available, the linked genes are all transcribed from the same DNA strand. The distance between linked genes varies from 3.0 to 14.0 kilobases (kb) although in certain species, eg goat [Lingrel et al] and sheep [Kretschmer et al] the available data suggest that the genes may be somewhat more widely spaced on the chromosome. The α and β genes of Xenopus laevis are linked and separated by only 7.7 kb of DNA [6].

Gene Order

In all the mammalian species examined to date, the genes in individual clusters are arranged 5′ to 3′ in the order of their expression during development. Examples include the human β-like genes (ϵ-${}^{G}\gamma$-${}^{A}\gamma$-δ-β), the human α-like genes, the rabbit β-like genes [β_4(embryonic)-β_3(embryonic)-β_2(pseudogene)-β_1(adult)], and the mouse β-like cluster. Naturally this arrangement has led to speculation that gene order is somehow related to sequential activation during development. However, an exception to this ordering has already become apparent. The chick β-like globin genes are arranged, embryonic-adult-adult-embryonic [Groudine et al].

CHARACTERISTICS OF INTERGENIC REGIONS

Pseudogenes

As noted above, these elements, found in all species so far examined, are clearly related to functional globin genes by virtue of sequence homology but cannot encode for a globin because of changes in promoter, splice, or coding sequences. Objecting to the diminutive term pseudogene, Edgell and colleagues argue that a more proper term might be β-homologous (or α-homologous) and point out that these elements might be important for regulation. The possibility that pseudogenes are scattered among functional genes in such a way that at least one is associated with a gene or genes expressed during each developmental period might suggest that the pseudogenes are involved in establishing the functional chromatin domains proposed by Groudine and Weintraub (see below). Alternatively, the pseudogenes might influence expression of functional genes by serving as alternative sites for binding of rate-limiting transcriptional components. These speculations or others, if found to be true, might save the pseudogenes from the evolutionary scrap heap. However, at this moment there is no evidence supporting a functional role for the pseudogenes. Answers to the following questions would be informative. Are the pseudogenes transcribed into detectable RNA products? If not, are RNA polymerase molecules associated with the pseudogenes? Is pseudogene chromatin in a transcriptionally active confirmation, eg sensitive to DNAse I and/or associated with HMG-14 and HMG-17?

Repetitive DNA Sequences

The presence of moderately repetitive DNA sequences interspersed throughout the genome of higher organisms has been known for several years. Thus the finding of such sequences scattered among the globin genes is no great surprise although the detailed knowledge about the globin gene genomic region provides the first opportunity to ask pointed questions about the function of these repetitive elements. The Alu I family described in this volume by Maniatis et al and Smithies et al has been proposed as functioning in DNA replication primarily because a part of its sequence is homologous to sequences found at the site of initiation of DNA synthesis of several papovaviruses. Other interesting features of the Alu I family are the ability of certain members to function as templates for RNA polymerase III in vitro and the presence of homologous sequences in heterogenous nuclear RNA. The proper attention accorded the Alu I family, however, should not obscure the presence of several other repetitive DNA families among the globin genes [Hardison et al]. Close to the human β-globin gene is a member of a long (6.4 kb) moderately repetitive DNA family [Adams et al], the function of which is totally unknown although the first guess would be that it is an incidental bystander to the nearby panorama of globin gene expression.

Homology of Flanking Sequences

Shortly after the first cloned DNA fragments containing globin genes were obtained, Leder and colleagues demonstrated that the mouse β^{maj} and β^{min} globin genes are embedded in non-homologous segments of DNA [7]. These results led to the suggestion that relatively little flanking sequence homology was required for coordinate expression of these genes. Homology in the flanking sequences for long distances around the two human γ and the two human α genes is thought to represent evidence of recent duplication and/or gene conversion events in the evolution of these genes [see Smithies et al and Maniatis et al]. The two members of each of these gene pairs are always expressed together. A contrasting example is the γ and β^{A} genes of sheep [Kretschmer et al]; these genes are very selectively expressed during the fetal and adult developmental periods respectively, but nonetheless are embedded within six kb segments of homologous DNA.

The presence of homology or lack thereof in flanking sequences of two genes may be most relevant to the evolutionary history of those genes. The functional significance of specific flanking sequences may not become apparent by simply comparing homology. This is suggested further when one recalls that DNA segments within cells are not simply linear arrays of nucleotides but rather are coiled into chromatin with extensive tertiary and quaternary structure. Thus small stretches of nucleotides separated by several hundred or even a few kilobases of DNA might be juxtaposed in chromatin to form a binding site for a regulatory protein. Such functional sites are unlikely to become obvious by focusing on the primary DNA sequence; other approaches will undoubtedly by required.

Regulatory Loci

That there are DNA sequences which are important in gene regulation and therefore which might be considered regulatory loci located in intergenic regions among the globin genes is a generally accepted hypothesis. As of yet, no experimental strategy has been devised which may be used to test this hypothesis. The ability to obtain 30–40 kb genomic fragments containing several globin genes by the cosmid cloning techniques may, as outlined by Flavell et al, allow a search for such loci by transfer of large DNA fragments into erythroid cells. Summarized in discussions at the conference by Maniatis were experiments in which DNA fragments containing human globin genes were introduced into mouse eyrthroleukemia (MEL) cells. These transformed cells contained only a low concentration of human globin mRNA sequences and there was no significant increase after the MEL cells were induced to undergo erythroid differentiation. If a higher level of expression were obtained with larger fragments and particularly if normal regulation occurred as exhibited by increased synthesis of human mRNA with induction, then one might infer that regulatory loci are present in the larger but not the smaller fragments already tested. Characterization of these loci by comparison of the function of several fragments of various size and composition might then be accomplished. The microinjection techniques outlined by Anderson et al whereby only one or a very few copies of a DNA fragment can be precisely inserted into an individual cell might facilitate such an analysis.

Salser et al described experiments in which genes which confer resistance to methotrexate were used to transform bone marrow stem cells in vitro. These stem cells were infused into irradiated mice and evidence was obtained suggesting that the transferred genes were expressed in vivo. This approach might, if generally successful, provide a convenient means to test the function of cloned globin gene fragments in normal bone marrow cells.

Until these or other strategies are developed, the only source of information about potential regulatory loci within the globin gene region comes from the analysis of deletion mutants of the human β-like gene complex. Several mutations which have eliminated part or all of the δ and β genes are accompanied by increased γ gene expression as reflected by augmented HbF production in heterozygoes and homozygoes for these mutations [Flavell et al and Bank et al]. The available data are compatible with the presence of a region about 2–3 kb upstream from the δ-globin gene which is involved in the postnatal suppression of γ gene expression. Another interesting deletion is one which begins in the middle of the $^{G}\gamma$ gene and includes the δ gene but leaves the β-globin gene intact. Nonetheless, the β-globin gene is not expressed. Reported by Ottolenghi was a deletion involving the intergenic region between the two γ genes. In the heterozygote, HbF production was increased.

Several criticisms may be raised against the interpretation of the deletion mutants with regard to the presence of regulatory loci. First, in all cases the deletions include several kb of DNA so that precise localization of regions affecting gene expression is impossible. Second, the deletion may not be critical in affecting γ gene expression but rather the juxtaposition of foreign DNA sequences adjacent to the γ-globin genes may be more relevant. Third, the possibility of two mutations, one a deletion removing the β-globin gene or other portions of the globin gene region and the second a point mutation leading to increased γ-gene expression, cannot be excluded. Thus while the deletion mutants are interesting, real definition of regulatory regions must await the development of more powerful experimental strategies.

CHROMATIN STRUCTURE

Through the efforts of Weintraub, Groudine, and their coworkers, evidence is accumulating which suggests that the globin gene region may be organized into domains of chromatin which extend up to as much as 100 kb. Defined by limited DNAse I digestion of intact nuclei are varying levels of sensitivity of sequences included within the globin chromatin domain. Expressed genes are highly sensitive, a property which depends on binding of the non-histone chromosomal proteins, HMG 14 and HMG 17, to the nucleosomes associated with these genes. Generally, flanking sequences and unexpressed globin genes within the β gene domain lack the HMG proteins and are only moderately sensitive to DNAse I compared to control DNA as represented by the gene for ovalbumin. Finally, certain short sequences within the domain appear to be hypersensitive to DNAse I; these might be exposed DNA sequences to which regulatory proteins might bind. Extension of these techniques to the study of other globin gene complexes, particularly those involved in the γ to β switch in human is awaited with great interest.

RNA METABOLISM

RNA Precursors

Proof for the existence and definition of the nature of precursors to globin mRNAs remained elusive until the canonical structure of the globin genes was defined. Substantial evidence now supports the hypothesis that transcription of β-like genes begins at the Cap site, continues through the three exons and two introns, and terminates at the polyadenylation site. The size of the "primary transcript" therefore is a function of the length of the introns. Direct evidence for such precursors for the mouse [Ross et al], rabbit [Flavell et al], and human β-globin mRNAs [Kantor et al] and also for the human γ mRNA [8] has been obtained.

Belief that transcription begins at the Cap site is based on several observations including the following. First, no RNA molecule of larger size has been definitively identified. Second, 50% of the mouse β-globin mRNA precursors are capped [9]. Because capping is thought to occur shortly after RNA synthesis begins, this supports the interpretation that the identified precursor is the primary transcript. Third, as discussed above, isolated gene fragments which include sequences at the 5′ end of globin genes function as promoters in cell-free extracts.

The techniques applied to date would not detect transcripts which are present as only a few copies per cell. Not yet eliminated completely is the possibility that transcripts might be processed during synthesis so that no complete molecule larger than the identified precursor exists. Also, most of the data relevant to precursor identification relates to β-like genes. Closely linked homologous genes which are expressed in parallel such as those for human γ and α-globin might be transcribed differently. Along these lines, it is worth noting that the nucleosomes associated with DNA sequences between the two chicken α genes bind HMG 14 and HMG 17 and are sensitive to DNAse I [Weintraub et al]. These properties are thought to be characteristic of transcriptionally active sequences. Possibly the two α genes may be transcribed to yield a polycistronic RNA molecule which is then processed to mature monocistronic α mRNA.

Processing Pathways

Summarized by Ross et al are pulse-chase studies which demonstrate step-wise or piecemeal removal of intervening sequence RNA from the mouse and human β-globin mRNA precursors. The existence of processing intermediates between the rabbit precursor and mature β-globin mRNA is established by the data presented by Flavell et al. RNA molecules intermediate in size between the initial transcript and mature mRNA are normally present in very low concentration.

Remaining to be resolved is the exact pathway of intervening sequence removal. Ross suggests that there may be at least two alternate pathways in the processing of the mouse β-globin mRNA precursor. Perhaps there are several possible splices which may be made. The frequency with which each is made might be related to the exact sequence of nucleotides at that splice site. Certain splices may be obligatory while others might simply be alternative or minor reactions. These considerations would suggest that merely demonstrating the existence of an intermediate does not establish its importance in the RNA processing pathway.

Processing Mutants: The β Thalassemias

The β thalassemias are human diseases characterized by deficient or absent production of β-globin. Summarized in this volume by Ross, Kantor, and Benz and their coworkers is functional evidence indicating that the quantitative deficiency of β-globin mRNA in these syndromes may result from impaired processing of the human β-globin mRNA precursor. The paper by Spritz contains the nucleotide sequence of a β-globin gene which is affected by a thalassemic mutation. The single nucleotide difference between it and the normal β-globin gene is located near the 3′ end of the small intron and creates an alternate splice site. Because the function of this gene has not yet been tested, the inference that this mutation causes a quantitative deficiency in β-globin mRNA due to impaired processing is only presumptive but seems highly likely.

Further characterization of the β thalassemic mutations may be a fruitful way to identify processing intermediates and to define their importance in the processing pathway. Because many such mutants exist, their study should be useful in clarifying this aspect of globin gene expression.

RNA Stability

In the reticulocyte, 95% of the total protein which is syntheiszed is globin. The mechanism by which globin mRNA accumulates to the exclusion of other mRNA species has been difficult to formulate because of data which suggested that globin mRNA, although initially stable during the early phases of maturation, was subsequently destabilized. Volloch and Housman have devised conditions whereby mouse erythroleukemia cells will survive in vitro allowing maturation to proceed to the reticulocyte stage. Measurements made under these conditions indicate that globin mRNA remains stable although non-globin mRNAs are destabilized as maturation nears completion. These two phenomena, namely the stability of globin mRNA and destabilization of non-globin mRNA species, appear to be responsible for the extreme specialization of the reticulocyte's protein synthetic apparatus.

REGULATION OF THE INDIVIDUAL GLOBIN GENES

Modulation of Processing or Transcriptional Regulation

The pattern of hemoglobin synthesis exhibited by erythroid cells is an accurate reflection of their mRNA content. For example, human fetal erythroid cells making HbF ($\alpha_2\gamma_2$) contain γ but not β mRNA while adult erythroid cells making HbA ($\alpha_2\beta_2$) contain β but not γ mRNA. Thus experimental attempts to define the molecular mechanism of regulation of the individual globin genes must be directed toward definition of the level at which control results in the selective accumulation of specific globin mRNAs.

Knowledge that there are physical precursors to the globin mRNAs which contain sequences not found in mature molecules has facilitated studies designed to determine whether selective processing of globin mRNA precursors is involved in the differential accumulation of particular globin mRNA species. Using a sensitive and highly specific technique, Curtis and coworkers were able to measure selectively the amounts of the precursors to mouse β^{maj} and β^{min} globin mRNA in mouse erythroleukemia cells. Both precursors appear after induction of erythroid maturation with Me_2SO in approximately the same ratio as the synthesis of the two globins. Furthermore, hemin induction, which leads to synthesis of β^{min} but not β^{maj} globin, results in the production of only the β^{min} mRNA precursor. These data provide strong evidence that modulation of expression of these two mouse β genes occurs at a stage prior to the accumulation of the RNA species which are thought to be the primary transcription products of these genes.

Drawing on evidence that small nuclear RNA molecules may have a role in splicing, Bank et al speculate that sequences transcribed from the human δ-globin gene intron might inhibit γ precursor processing while stimulating processing of the β-globin mRNA precursor. Data relevant to this hypothesis were reported by Kantor et al. The concentration of the precursors to human γ and β-globin mRNA were measured in erythroid cells from the fetal and adult developmental periods. In general the concentration of the individual precursors paralleled the concentration of the individual mRNA species. Thus fetal erythroid cells contained 130–170 molecules of γ precursor and approximately 28 molecules of β precursor; 6–12% of non-α ($\beta+\gamma$) mRNA in these cells was β-globin mRNA. However, in the adult bone marrow cells examined, γ mRNA accounted for only 1% of the non-α mRNA. Nonetheless, 30 molecules of γ mRNA precursor were found compared to 100 molecules of the β-globin mRNA precursor. Thus although the data support modulation prior to the stage of processing of the globin mRNA precursor, the results are by no means unequivocal.

Another approach utilized in an effort to learn whether there is transcriptional regulation of globin genes is the run-off assay employed by Groudine and co-workers. RNA was made radioactive by incubation of intact erythroid nuclei with [^{32}P]UTP thereby extending previously initiated RNA molecules. The RNA product was used to probe Southern blots of cloned genomic fragments containing the embryonic and adult β-like genes. A positive signal for a particular gene fragment indicates that it was being transcribed in cells from which the nuclei were derived. By this criteria, Groudine et al learned that the embryonic but not the adult β-like gene is transcribed in embryonic cells while the adult but not the embryonic β-like gene is transcribed in adult erythroid cells. Application of these techniques to the study of other globin gene families is awaited with considerable interest.

There are few direct studies that are relevant to transcriptional regulation of the individual globin genes. Study of isolated gene fragments in cell-free extracts has revealed equivalent transcription of fragments derived from several different genes provided certain consensus sequences are present (see above). Indeed, these heterologous systems would not be expected to provide insight into selective expression of particular genes in erythroid cells during development.

Comparison of the nucleotide sequences of the putative promoter region of genes which are selectively expressed is mainly interesting in a negative way. Human γ and β genes have diverged somewhat immediately upstream from Cap site [Maniatis et al and 10] but the γ and β^A genes of sheep [Kretschmer et al] and the β^A and β^C genes of goat [Lingrel et al] are virtually identical for the first 100–150 nucleotide pairs upstream. Thus significant divergence in this region is clearly not required for selective expression during development or during the hemoglobin switch induced by erythropoetic stress in goats. Whether the several nucleotide differences which are found in the initial portions of the coding blocks of the sheep and goat genes are important for transcriptional regulation remains to be defined.

Chromatin Domains

What is the relevance of the chromatin domain hypothesis to the expression of the individual globin genes? Reported by Groudine et al is the observation that both the embryonic and adult β-like genes are DNAse I sensitive in embryonic red cells while only the adult gene is sensitive in adult red cells. These data obtained with very limited DNAse I digestion are in contrast to results reported earlier [11] in which more extensive digestion of isolated nuclei with DNAse I suggested that the chicken adult β-like genes were relatively resistant in embryonic red cell nuclei. Analogous experiments with the sheep globin genes had suggested that the γ gene but not the β gene was DNAse I sensitive in nuclei from fetal erythroid cells while both the γ and β^{A} gene were sensitive in adult cells [12]. The different levels of DNAse I digestion may be detecting different properties of the globin chromatin domain. Clearly there is no simple correlation between DNAse I sensitivity and globin gene expression. Derived from the observations of Groudine et al is the concept that there may be a pre-activation state of a gene whereby it is included in a potentially active confirmation in chromatin. Other factors, including the HMG proteins, may come to be associated with the gene before it is actually transcribed.

FUTURE PROSPECTS

Suggesting the course of future scientific investigation is always hazardous but some predictions may be safer than others. Rapid accumulation of voluminous amounts of DNA sequence data seems a certainty based on the even more powerful techniques for determining sequences provided by the single stranded cloning vectors and the di-deoxy nucleotide method for obtaining a set of labeled DNA molecules. For example, within several years the sequence of the entire human β-like gene complex should be known. The returns for this still considerable investment of time and experimental ingenuity are less predictable. Possibilities include the further elucidation of the evolutionary relationships of the various globin genes, the nature of repetitive DNA sequences, and the identification of other potential genes near or among the globin genes.

A glaring deficiency in our knowledge of the transcriptional template in eukaryotic cells is the nature of the associated proteins. The role of the various histones in the nucleosome core particles has been clarified over the past several years and the association of HMG 14 and HMG 17 with actively transcribed genes is now established. Beyond this relatively little is known. The in vitro transcriptional systems should, as described by Luce et al, allow the purification of general transcriptional factors, and we may expect to learn where these bind to purified gene fragments. Identification and purification of gene specific transcriptional factors may be more difficult because the binding site for such factors may consist of complexes of chromatin and could include nucleotide sequences which are somewhat dispersed along the chromosome. Aside from a few attempts to identify DNA sequence specific binding proteins using cloned DNA fragments [13, 14], this is a largely unexplored area of investigation. To embark on such studies at present, one must be willing to abandon the relatively certain rewards of characterizing a selected portion of a particular genome by sequencing and mapping in exchange for the less certain prospects of defining transcriptional regulatory proteins.

How then are we to learn about gene function? Described within this volume are several potentially productive avenues of investigation. Definition of chromatin domains by DNAse I digestion is one approach. Clarification of the hypersensitive regions may be particularly informative. The ability to purify active nucleosomes by virtue of their binding to HMG 14 and HMG 17 may help. Obtaining a transcriptional map of cloned fragments should provide interesting information. Certainly one might hope to learn whether the various β-like genes in several species are modulated at the transcriptional level during development. Further study of the RNA processing pathways and their possible modulation during hemoglobin switching also seems desirable. Transfer of relatively large DNA fragments into erythroid cells could be informative in identifying DNA sequences which affect globin gene expression.

Our preoccupation with the globin genes at the Airlie House Meeting and in the manuscripts included within this volume should not cause one to lose sight of the fact that the globin genes are only a few of the many genes which are expressed during differentiation and maturation of erythroid cells. Identification of other expressed genes and their gene products by cloning, mapping, and DNA sequencing is a realistic goal. Perhaps such an effort would be rewarded by recognition of the common basis underlying the coordinated expression of the many genes required in the formation of the highly specialized red cell.

This volume is concerned with the structure and organization of globin genes and the molecular basis for their regulated expression during the terminal maturation of erythroid precursor cells. Substantial data support the concept that the crucial regulatory events which determine the pattern of expression of the individual globin genes do not occur in maturing precursors. Rather the regulatory events occur in stem cells, either in cells which are multipotent with respect to cell lineage (e.g., megakaryocytic and erythroid) or in committed erythroid progenitor cells. The companion volume *Hemoglobins in Development and Differentiation* contains information, presented and discussed at the Airlie House meeting, which relates to stem cell differentiation in vitro and in vivo. Considered are the factors, either intrinsic to the stem cells or within their microenvironment, which may be relevant to the regulated expression of the individual globin gene and, thereby, the switches in hemoglobin phenotype which occur during development.

REFERENCES

1. Stamatoyannopoulos G, Nienhuis AW: Cellular and molecular regulation of hemoglobin switching. New York, Grune and Stratton, 1979.
2. Leder P, Hansen JN, Konkel D, Leder A, Nishioka Y, Talkington C: Mouse globin system: A functional and evolutionary analysis. Science 20:1336–1342, 1980.
3. Konkel DA, Tilghman SN, Leder P: The sequence of the chromosomal mouse beta-globin major gene: Homologies in capping, splicing, and poly(A) sites. Cell 15:1125–1132, 1978.
4. Proudfoot NJ, Shander MHM, Manley JL, Geffer ML, Maniatis T: The structure and in vitro transcription of human globin genes. Science 209:1329–1336, 1980.
5. Hamer D, Kaehler M, Leder P: A mouse globin gene promoter is functional in SV40. Cell 21:697–708, 1980.
6. Jeffreys AJ, Wilson V, Wood D, Simons JP, Kay RM, Williams JG: Linkage of adult alpha-and beta-globin genes in X laevis and gene duplication by tetraploidization. Cell 21:555–564, 1980.

7. Tiemeier DC, Tilghman SM, Polsky FI, Seidman JG, Leder A, Edgell MH, Leder P: A comparison of two cloned mouse beta-globin genes and their surrounding and intervening sequences. Cell 14:237–245, 1978.
8. Courtney M, Williamson R: A nuclear precursor to human gamma-globin mRNA. Nucleic Acids Res 7:1121–1129, 1979.
9. Curtis PJ, Mantei N, Weissman C: Characterization and kinetics of synthesis of 15S beta-globin RNA, a putative precursor of beta globin mRNA. Cold Spring Harbor Symp Quant Biol 42:971–984, 1977.
10. Efstratiadis A, Posakony JW, Maniatis T, Lawn RM, O'Connell C, Spritz RA, DeRiel JK, Forget BG, Weissman SM, Slighton JL, Blechl AE, Smithies O, Baralle FE, Shoulders CC, Proudfoot NJ: The structure and evolution of the human beta–globin gene family. Cell 21:653–668, 1980.
11. Weintraub H, Groudine M: Chromosomal subunits in active genes have an altered confirmation. Science 93:848–853, 1976.
12. Young NS, Benz EJ, Kantor JA, Kretschmer P, Nienhuis AW: Hemoglobin switching in sheep: Only the gamma gene is in the active conformation in fetal liver but all the beta and gamma genes are in the active conformation in bone marrow. Proc Natl Acad Sci 75:5884–5888, 1978.
13. Weideli H, Gehring WJ: A new method for the purification of DNA-binding proteins with sequence specificity. Eur J Biochem 104:5–11, 1980.
14. Bowen B, Steinberg J, Laemmli UK, Weintraub H: Detection of DNA binding proteins by protein blotting. Nucleic Acids Res 8:1–20, 1980.

Organization of the Globin Genomic Regions

Organization and Expression of Globin Genes, pages 15–31

The Structure and Chromosomal Arrangement of Human Globin Genes

Tom Maniatis, Edward F. Fritsch, Joyce Lauer, Richard M. Lawn, Nicholas J. Proudfoot, Monica H. M. Shander, and C.-K. James Shen

Division of Biology, California Institute of Technology, Pasadena, California

INTRODUCTION

The application of gene mapping and molecular cloning procedures to the study of human globin genes has led to significant advances in the understanding of their structure and chromosomal organization (see [1, 2] for review). The known α-like and β-like globins are listed in Table I. The earliest embryonic hemoglobin tetramer, Gower 1, consists of ϵ (β-like) and ζ (α-like) polypeptide chains. Beginning at approximately eight weeks of gestation, the embryonic chains are gradually replaced by the adult α-globin chain and two different fetal β-like chains, designated $^{G}\gamma$ and $^{A}\gamma$. The γ chains differ only in the presence of glycine or alanine at position 136, respectively. During the transition period between embryonic and fetal development, Hb Gower 2 ($\alpha_2\epsilon_2$) and Hb Portland ($\zeta_2\gamma_2$) are detected. Eventually Hb F ($\alpha_2\gamma_2$) becomes the predominant Hb tetramer throughout the remainder of fetal life. Beginning just prior to birth, the γ-globin chains are gradually replaced by the adult β- and δ-globin polypeptides. At six months after birth, 97%–98% of the hemoglobin is Hb A ($\alpha_2\beta_2$), while Hb A_2 ($\alpha_2\delta_2$) accounts for approximately 2%. Small amounts of Hb F (1%) are also found in adult peripheral blood.

Recently, genes which encode each of the α-like and β-like polypeptides listed in Table I have been isolated from bacteriophage λ libraries of human DNA. Structural characterization of these genes and their flanking sequences have provided new information regarding the linkage arrangement and evolution of human globin genes. The purpose of this article is to review this information.

THE HUMAN β-LIKE GLOBIN GENE FAMILY

Gene Isolation

Human β-like globin genes were isolated by constructing and screening libraries of recombinant phage containing the products of a limit restriction endonuclease digestion of mammalian DNA [3, 4] or containing random, high molecular weight fragments of mammalian DNA [5]. The advantage of libraries of high molecular

TABLE I. Human Globin Polypeptides and Hemoglobin Tetramers

	Embryo	Fetus	Adult
α-Like Polypeptides	ζ, α	α	α
β-Like Polypeptides	ϵ	${}^{G}\gamma$, ${}^{A}\gamma$	$\delta\beta$
Hb Tetramers	$\zeta_2\epsilon_2$ (Gower 1) $\alpha_2\epsilon_2$ (Gower 2) $\zeta_2\gamma_2$ (Portland)	$\alpha_2\gamma_2$ (HbF)	$\alpha_2\delta_2$ (HbA_2) $\alpha_2\beta_2$ (HbA)

weight DNA is that the isolation of a set of overlapping clones, all of which contain a given gene, permits study of the sequences extending many kilobases (kb) from the gene. Moreover, even more distant regions along the chromosome can be obtained by rescreening a library using terminal fragments of the initially selected clones as hybridization probes. Thus, this approach to gene isolation makes it possible to study the organization of closely linked genes.

All five of the known human β-like globin genes have been obtained from recombinant clones of genomic DNA: β and δ [6] ${}^{G}\gamma$, ${}^{A}\gamma$, and ϵ [4, 7, 8, 9, 10, 11]. Identification of the ϵ gene was made possible by the recent determination of the partial amino acid sequence of ϵ-globin polypeptide [12].

Gene Linkage

Physical linkage between the δ and β genes was demonstrated by genomic blotting experiments [13] and by analysis of a recombinant bacteriophage containing both the δ and β genes [6]. Linkage of the ${}^{G}\gamma$ and ${}^{A}\gamma$ genes was determined by genomic blotting experiments [14] and was confirmed by analysis of clones containing both of these genes [8, 9, 10]. Linkage between the ${}^{A}\gamma$ and δ genes was established using a cloned hybridization probe from the ${}^{A}\gamma$-δ intergenic region and a well-defined deletion located in this region [8, 15], by mapping DNA fragments from the ${}^{A}\gamma$-δ intergenic region generated by nonlimit restriction enzyme digests of genomic DNA [16], and ultimately by cloning overlapping DNA fragments which span the ${}^{A}\gamma$-δ intergenic region [9]. Analysis of overlapping cloned DNA fragments also made it possible to demonstrate physical linkage between the ϵ gene, the isolation of which was initially reported by Proudfoot and Baralle [11], and the remainder of the gene cluster [9].

The linkage arrangement of the human β-like globin gene cluster is shown in Figure 1. Also shown is the set of overlapping phage recombinants which together contain the entire gene cluster [9]. As noted previously, all of the genes are transcribed from the same DNA strand and are arranged on the chromosome in the order of their expression during development: $5'$-ϵ-${}^{G}\gamma$-${}^{A}\gamma$-δ-β-$3'$. A similar pattern of gene organization has been found in the human α-like [17] and rabbit [18, 19] and mouse [20] β-like globin gene clusters.

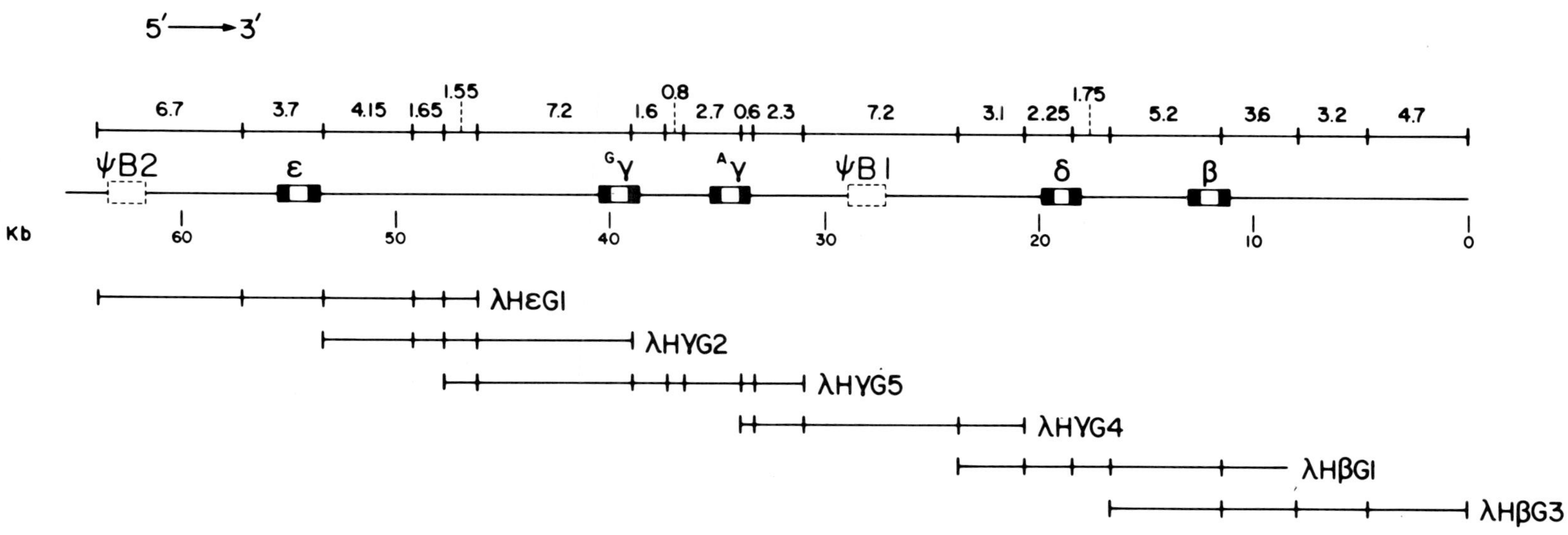

Fig. 1. A linkage map of the human β-like globin gene cluster. The map indicates Eco RI sites by vertical bars and the size of each fragment is given in kb. The locations of the ε-, $^{G}\gamma$-, $^{A}\gamma$-, δ-, and β-globin genes are indicated. Additional β-globin-like sequences (ψβ1 and ψβ2) are indicated by broken line boxes. The direction of transcription for all five genes is indicated by the arrow at the top. The human DNA inserts in the set of overlapping clones containing the linked genes are shown below the map. Reproduced from Cell by permission [9].

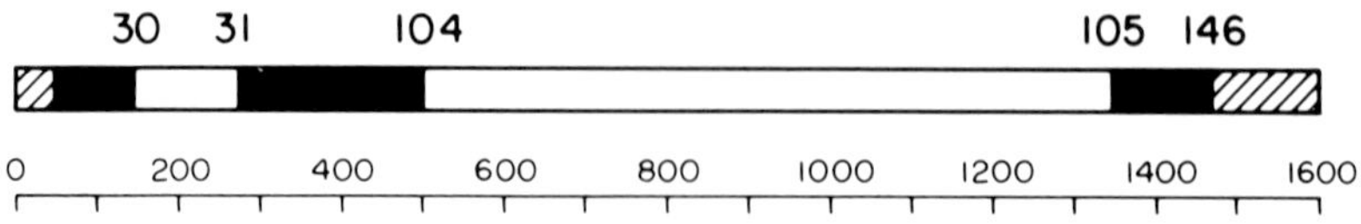

Fig. 2. Structure of human β-like globin genes. The canonical structure for the human β-like globin genes is drawn to approximate scale. Solid and open boxes represent coding (exon) and noncoding (intron) sequences, respectively. The β-like globin genes contain introns of approximately 122–130 and 800–900 bp, located between codons 30 and 31 and 104 and 105, respectively [31].

In addition to the five known globin genes, two β-like sequences (ψ β1 and ψ β2), which cannot be identified with known polypeptides, were identified in the gene cluster [9]. Although neither gene has been sequenced, it seems likely that one or both of them are pseudogenes, genes which cannot encode a functional globin polypeptide.

Gene Structure

The complete nucleotide sequences of the five human β-like globin genes and flanking regions have been determined [21, 22, 7, 23]. All five genes are interrupted by two introns at identical locations: the first, of 122–130 base pairs (bp) in length, is located between codons 30 and 31, and the second, of 700–900 bp, is between codons 104 and 105 (Fig. 2). As was shown for the mouse [24, 25, 26] and rabbit [27, 19] β-globin genes, the introns of the human β-like globin genes are transcribed as part of a nuclear mRNA precursor and are removed in steps by splicing to produce the mature globin mRNA [28, 29]. RNA-DNA hybridization experiments have shown the 5′ ends of the rabbit [27] and mouse [30] β-globin nuclear mRNA precursors are coterminal with their respective cytoplasmic mRNAs.

A detailed comparison of the human β-like globin gene sequences has been presented [31]. These comparisons made it possible to establish accurate evolutionary relationships among the human β-like globin genes and revealed interesting sequence homologies in regions which are potentially involved in globin gene transcription and splicing [31]. Alignment of the sequences 5′ to the human β-like globin genes provides an example of these homologies (Fig. 3). The most notable feature of this alignment is the presence of two blocks of sequence homology which are present in analogous positions adjacent to most eukaryotic genes. The first homology block (designated the ATA box) is found 29–30 bp 5′ to each gene and the second block (designated the CCAAT box) is located 70–78 bp 5′ to each gene. A possible role of these sequences in transcriptional initiation and/or RNA processing has been discussed [31].

In addition to the ATA and CCAAT box sequences which are found in many eukaryotic genes, other regions of homology which are shared among the β-like globin genes are scattered throughout the 5′ flanking sequence. The conservation of these sequences during the 200 million years since the time of ϵ,γ-δ,β divergence [31] suggests that they are functionally significant. The identification by sequence

Gene	Sequence	Ref.
Mouse β^{min}	CCTCACCC-TGCAAGGTAACACCCCTGGCATT-GGCCAATCTGCTC-AGAGAGGA-CAGAGTGGGCAGGAG-CCAG-CAT-TGGGTATATAAAGCTGAGCAGGGTCAGTTGCTTCTT	45
Mouse β^{maj}	CCTGATTC-CGTAGAGCCACACCC-TGGTAAG-GGCCAATCTGCTC-ACACAGGA-TAGAGAGGGCAGGAG-CCAG-GGC-AGAGCATATAAGGTGAGGTAGGATCAGTTGCTCCTC	45
Rabbit β	CCTCACCC-TGCAGAGCCACACCC-TGGTGTT-GGCCAATCTAC---ACACGGGG-TAGGGATTACATAGT-TCAG-GACTTGGGCATAAAAGGCAGAGCAGGG-CAGCTGCTGCTT	19
Goat β^{A}	CCTCACCC-TGTGGAACCACAACT-TGGCACG-AGCCAATCTGCTC-ACAGAAG--CAGGGAGGGCAGGAG-GCAG-GGC-TGGGCATAAAAGGAAGAGCCGGGCCAGCTGCTGCTT	51
Goat β^{C}	CCTCACCC-TGTGGAACCACAACT-TGGAATG-AGCCAATCTGCTC-ACAGAAG--CAGGGAGGGCAGGAG-GCAG-GGC-TGGGCATAAAAGGAAGAGCCGGGCCAGCTGCTGCTT	51
Human β	CCTCACCC-TGTGGAGCCACACCC-TAGGGTT-GGCCAATCTACTC---CCAGGAGCAGGGAGGGCAGGAG-CCAG-GGC-TGGGCATAAAAGTCAGGGCAGAGCCATCTATTGCTT	21
Human δ	(above: TAT) TCTCACAA-ACTAATG--AAACCC-TGCTCTTAAACCAACCTGCTC-ACT--GGAGCAGGGAGGACAGGA--CCA---------GCATAAAAGGCAGGGCAGAGTCGACTGTTGCTT	22
Human $^{A}\gamma$	(above: CAAACTT-GACCAATAGTCTT-AGAGTATC- ; TAGGATGAAGA) CTCCACCCATGGGTTGGC-CAGCC-TTGCCTT-GACCAATAGCCTTGACA-AGG--CAGTGAGGCCAGGGG-CCGGCGGCT--GGCATAAAAGGAAGCACCCTT-CAGCAGTTCCAC	7
Human $^{G}\gamma$	(above: CAAACTT-GACCAATAGTCTT-AGAGTATC- ; TAGGATGAAGA) CTCCACCCATGGGTTGGC-CAGCC-TTGCCTT-GACCAATAGCCTTGACA-AGG--CAGTGAGGCCAGGGG-CCGGCGGCT--GGCATAAAAGGAAGCACCCTT-CAGCAGTTCCAC	7
Human ε	(above: AGTAAAGA) CTCCACCCCTGAGGA--CACAGGT-CAGCCTT-GACCAATGACTTTTA-AGTAC--CATGGAGAACAGGGGGCCAG-AACTTCGGCATAAAAGGCCAGACAGAG-AGGCAGCAGCAC	23

Fig. 3. Alignment of the 5′ flanking sequences of mammalian β-like globin genes. An alignment of the sequences preceding the mRNA capping site of ten different mammalian β-like globin genes is shown. Dashes indicate gaps introduced to achieve the most consistent alignment (see Table II). Similarly, the nucleotides above a given sequence were displaced to give the best alignment of the sequences on either side. Regions of sequence homology which are common to all globin genes sequences to date (the ATA and CCAAT boxes) are indicated by shading. Numbers in parentheses give the positions of the indicated bases (arrows) in nucleotides from the mRNA capping site (where known) (see [31] for details). The numbers to the right of the sequences are the references from which the sequences were taken.

comparisons of regions of possible functional importance may be useful in the search for structural differences between normal and mutant globin genes.

Comparison of the complete nucleotide sequences of the $^{G}\gamma$ and $^{A}\gamma$ genes has revealed that the coding, intervening, and flanking sequences of these two genes are virtually identical [7]. The distribution of the infrequent sequence differences has led to formulation of a model for gene correction [7].

Repetitive Sequences

The universal occurrence of repetitive sequences in eukaryotic DNA and their interspersion with single-copy sequences have led to the proposal that repetitive sequences play a role in the regulation of gene expression. Previous studies of repetitive sequence elements have described their genomic distribution and their differential transcription during development [32]. The isolation of globin gene clusters by molecular cloning provides an opportunity to study repeat sequences in relation to well-defined sets of developmentally regulated genes.

Cross hybridization experiments between the intragenic sequences of the human β [9] and α [C.-K.J. Shen, unpublished] gene clusters revealed a nonglobin repeat sequence which is interspersed within the globin gene clusters and also repeated many times within the human genome. As shown in Figure 4, this repeat sequence is located on the 5′ side of the ϵ-globin gene and flanks both the $^{G}\gamma$-$^{A}\gamma$ and γ-β-globin gene pairs. Nucleotide sequence analysis of these repetitive sequences [33; F.E. Baralle and N.J. Proudfoot, unpublished; O. Smithies, unpublished] indicates that they are members of a particular repeat sequence family (the Alu family) which is reiterated approximately 300,000 times in the human genome [34]. In addition, the repeats are transcribed in vitro by RNA polymerase III [35]. They show sequence homology with an abundant class of small nuclear RNAs [33, 36] and with double-stranded heterogeneous nuclear RNA [33]. Finally, the repeats contain a sequence which is homologous to a sequence found near the replication origin of SV40, polyoma, and BK DNA tumor viruses [33]. A similar set of repetitive sequences has been identified with a cluster of rabbit β-like globin genes [37]. At present, there is little information regarding the expression or function of these interesting repetitive elements in vivo (see [38] for discussion).

THE HUMAN α-LIKE GLOBIN GENE FAMILY

Gene Structure and Linkage

The existence of two α-globin genes per haploid genome was established on the basis of genetic studies [1]. The structure and chromosomal arrangement of the human α-like globin genes have been examined by genomic blotting experiments [39, 40] and by analysis of cloned DNA segments [17]. The two adult α-globin genes are referred to as $\alpha 1$ and $\alpha 2$. Restriction endonuclease analysis demonstrated that the $\alpha 1$ and $\alpha 2$ genes are each interrupted by two introns of approximately 95 and 125 bp (Fig. 5) [17]. Determination of the complete nucleotide sequence of an independently isolated $\alpha 2$ gene indicated that the introns are located between the codons for amino acids 31 and 32 and 99 and 100 [41]. The locations of introns in the human [41] and mouse [42] α-globin genes are identical, and are analogous to the positions of introns in β-like globin genes.

Genomic blotting experiments demonstrated that the linked adult α-globin genes are transcribed from the same DNA strand and are 3.7 kb apart [39, 40]. The α-globin gene linkage was confirmed by isolation of recombinant bacteriophage clones containing the two adult α-globin genes, the embryonic ζ1 gene, and an α-like sequence which could not be identified with any known globin polypeptide (ψα1) [17]. Nucleotide sequencing studies indicated that ψα1 is a globin pseudogene [43, see below].

A second ζ-globin gene, ζ2, was identified by genomic blotting experiments using a labeled restriction fragment containing part of the ζ1 gene as a hybridization probe [17]. Physical linkage of the ζ2 gene to the other α-like globin genes was demonstrated by genomic blotting using a cloned restriction fragment located 5′ to the ζ1 gene as a hybridization probe. The ζ2 gene is located approximately 12 kb 5′ to the ζ1 gene (Fig. 6). The ζ1-ζ2 gene linkage was confirmed by analysis of a bacteriophage recombinant (λHζG1) bearing a 10.9 kb Bgl II fragment which overlaps λHαG1 and contains the ζ2 gene [44].

Both ζ-globin genes appear to be functional. Detection of ζ-globin protein in an infant with a homozygous deletion which removes the ζ1 gene but spares the ζ2 gene indicates that ζ2 is a functional gene [45]. The nucleotide sequence of the ζ1 gene [C. O'Connell, N. Proudfoot, and T. Maniatis, unpublished results] agrees with the amino acid sequence of ζ-globin polypeptide [J. Clegg, personal communication]; the ζ1 gene may therefore also be functional. Genomic blotting-melting experiments suggest that the two genes are very homologous [17]. Since only one ζ-globin polypeptide sequence has been identified, the ζ1 and ζ2 genes may encode identical polypeptide chains. These questions will be resolved when the two genes are sequenced in their entirety.

ψα1, An α-Globin Psuedogene

As mentioned above, ψα1 was identified by hybridization to the adult α-globin cDNA probe [17]. Limited nucleotide sequence analysis of ψα1 indicated that this gene does not encode any of the known α-like globin polypeptides [17]. Subsequently, the complete nucleotide sequence of ψα1 was determined [43]. This sequence is shown in Figure 7 along with the sequence of the normal α2 globin gene [41] for comparison. The arrangement of coding and intervening sequences within the two genes is very similar, and the genes are approximately 73% homologous. Numerous single base changes are distributed throughout the entire gene. Similarly, deletions and additions of sequence occur in all regions, although at lower frequency. The largest stretch of nonhomology between the ψα1 and α2 coding sequences is a 20 bp deletion which removes the codons for amino acids 38–45. This deletion alters the translational reading frame at codon 45. In this reading frame, UGA termination codons are in phase at codons 75, 79, and 83. Thus, ψα1 cannot encode a functional α-globin polypeptide chain.

Although ψα1 contains two introns in exactly the same position as in the α2 gene, it is unlikely that a ψα1 RNA transcript would be normally spliced. Comparison of a large number of eukaryotic genes has led to the formulation of a consensus sequence for intron-exon junctions [46, 47]. Table II compares the junction sequences of the two introns in the human α and ψα1 globin gene with the consensus sequence. Although a number of nucleotides in ψα1 are compatible with the con-

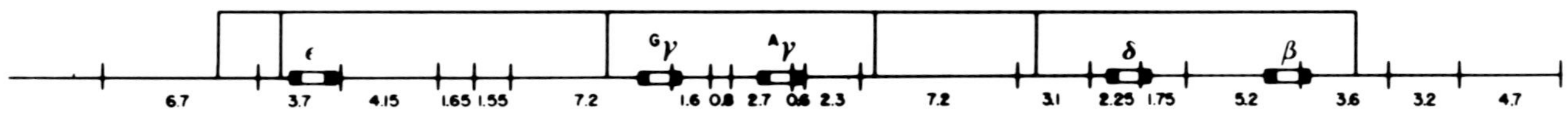

Fig. 4. Intergenic repeat sequences in the human β-like globin gene cluster. The locations of the cross-hybridizing sequences are shown on this map of Eco RI sites in the 65 kb segment of DNA. Lines are drawn between hybridizing fragments. The approximate location of the repeat sequence within each fragment is indicated by the position of the vertical line.

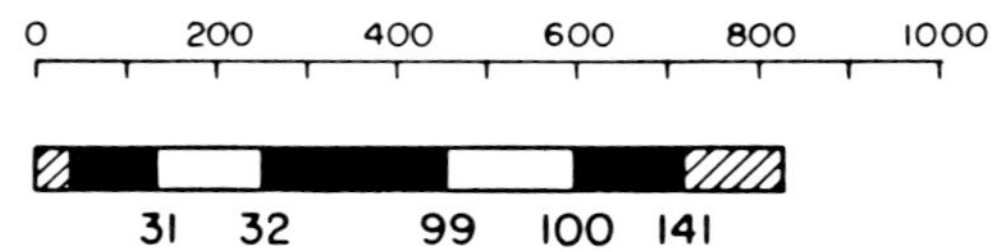

Fig. 5. Structure of human α-globin genes. The canonical structure for the human α-like globin gene is drawn to approximate scale. Solid and open boxes represent coding (exon) and noncoding (intron) sequences, respectively. The α-like globin genes contain introns of approximately 95 and 125 bp, located between codons 31 and 32 and 99 and 100, respectively.

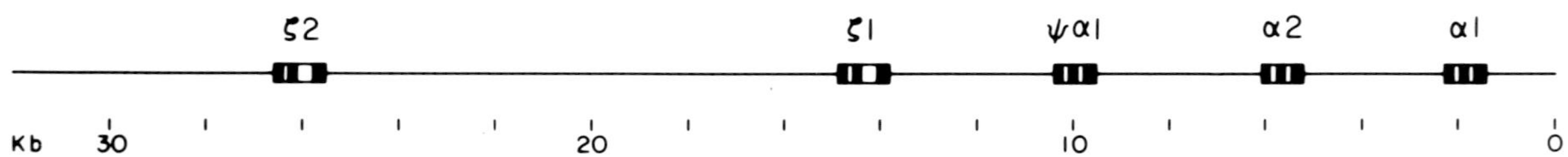

Fig. 6. Linkage map of the human α-like globin gene cluster. The location of the two adult ($\alpha 1$ and $\alpha 2$), two embryonic ($\zeta 1$ and $\zeta 2$), and (one) pseudo ($\psi\alpha 1$) globin genes are indicated by black (exon sequence) and white (intron sequences) boxes. All five genes have the same relative transcriptional orientations ($5' \rightarrow 3$, left to right) and the genes are arranged in the order of their expression during development.

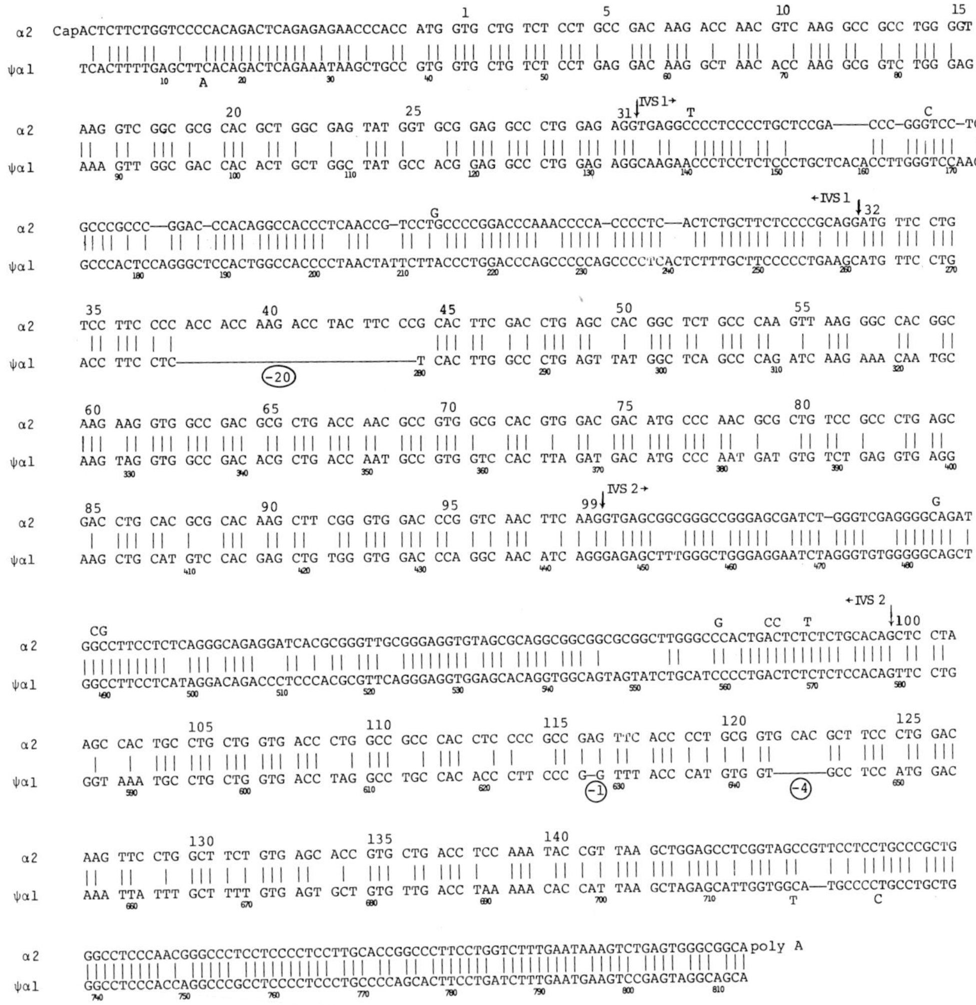

Fig. 7. Sequence comparison of $\psi\alpha1$ and $\alpha2$. The top rows are the $\alpha2$ [41] sequence and the bottom rows are $\psi\alpha1$. Vertical lines indicate sequence homology. The limits of the intervening sequences are denoted by arrows. The nucleotides written above the line of sequence denote the addition of the indicated nucleotides at that position, while horizontal line denotes a deletion. These additions and deletions were necessary to align the two sequences. The three deletions in the coding sequence of $\psi\alpha1$ that alter the translational reading frame are denoted by (5′) –20, –2, –4 (3′). Large numbers indicate amino acid position in human α-globin. Small numbers indicate the distance of the nucleotide from the putative capping site in $\psi\alpha1$. Reprinted from Cell by permission [43].

TABLE II. Comparison of the Intron/Exon Junctions of Human α and $\psi\alpha1$ Globin Genes

	5′ Exon	Intron	Exon 3′
Consensus	A A G►G T A A G T	....	T Py T Py Py Py T C A G►G
$\alpha2$ IVS 1	G A G G T G A G G	....	C T C C C C G C A G G
$\psi\alpha1$ IVS 1	G A G G C A A G A	....	C C C C C T G A A G C
$\alpha2$ IVS 2	A A G G T G A G C	....	C T C T G C A C A G T
$\psi\alpha1$ IVS 2	A G G G A G A G C	....	C T C T C C A C A G T

sensus splicing sequence, neither intron contains the 5′ GT sequence characteristic of nearly all intron/exon junctions. Thus, even though the 3′ junctions of both $\psi\alpha1$ gene introns appear to have the necessary features for splicing, it is possible that neither intron would be excised from a primary transcript.

Comparison of the 5′ flanking sequences of $\psi\alpha1$ and $\alpha2$ reveals that $\psi\alpha1$ has both of the sequences which are highly conserved among all globin genes (CCAAT box and ATA box [31]), but the distance between these two sequences in the $\psi\alpha1$ gene (13 bp) differs significantly from that in most other globin genes (50 ± 15 bp). Interestingly, $\psi\alpha1$ is specifically transcribed in vitro but at a significantly reduced level compared to $\alpha2$ [48]. The possibility that $\psi\alpha1$ is transcribed in vivo has not been tested.

Globin pseudogenes have also been detected in the rabbit [19, 49], mouse [20], and goat [50] β-like globin gene clusters. It is noteworthy that, in every mammalian β-like gene cluster a pseudogene is found between the embryonic (or fetal) genes and the adult genes. This conservation of location suggests that pseudogenes have some as yet unidentified function. Alternatively, pseudogenes may be the products of gene duplication and subsequent divergence [see references 43 and 49 for discussion].

A novel mouse α-globin pseudogene has recently been described [52, 53]. As in the case of the pseudogenes described above, the mouse α-globin pseudogene has frameshift mutations that would result in premature translational termination. However, unlike the other pseudogenes, the mouse gene is missing both introns. The mechanism by which this pseudogene arose, and its location with respect to the normal α-globin gene cluster, are unknown.

GENE DUPLICATION AND CORRECTION

Globin Gene Duplication

A common feature of globin gene clusters is the occurrence of two immediately adjacent genes which are coordinately expressed during a given developmental stage. Examples of this are the human δ-β, ${}^{G}\gamma$-${}^{A}\gamma$, $\alpha1$-$\alpha2$, and $\zeta1$-$\zeta2$ globin gene pairs. The δ and β genes are highly homologous in the coding region, but the noncoding sequences within and surrounding the two genes have diverged considerably [6, 31]. Extensive divergence of noncoding regions has also been observed in some other closely linked, coordinately expressed globin gene pairs [19, 54]. In contrast, the two members of the ${}^{G}\gamma$-${}^{A}\gamma$ gene pair are virtually identical to one another throughout their coding, intervening, and flanking sequences [7]. Although the nucleotide sequences of linked human α-globin genes have not yet been determined, restriction mapping and heteroduplex analysis of the $\alpha1$ and $\alpha2$ genes indicate that the sequences within and flanking these two genes are virtually identical. Each α-globin gene is located within an approximately 4 kb region of homology interrupted by two small regions of nonhomology (Fig. 8A [17, 43]). The leftward endpoints of this duplication were initially mapped by examination of heteroduplexes [17] and subsequently determined by nucleotide sequence analysis [43]. As shown in Figure 8B, the sequence on the 3′ side of the poly(A) addition site of $\psi\alpha1$ and $\alpha2$ differs in only one out of the 124 nucleotides sequenced. In contrast, the sequences on the 5′ side of this site have diverged considerably. This abrupt change in sequence homology suggests that the α-globin gene duplication unit ends

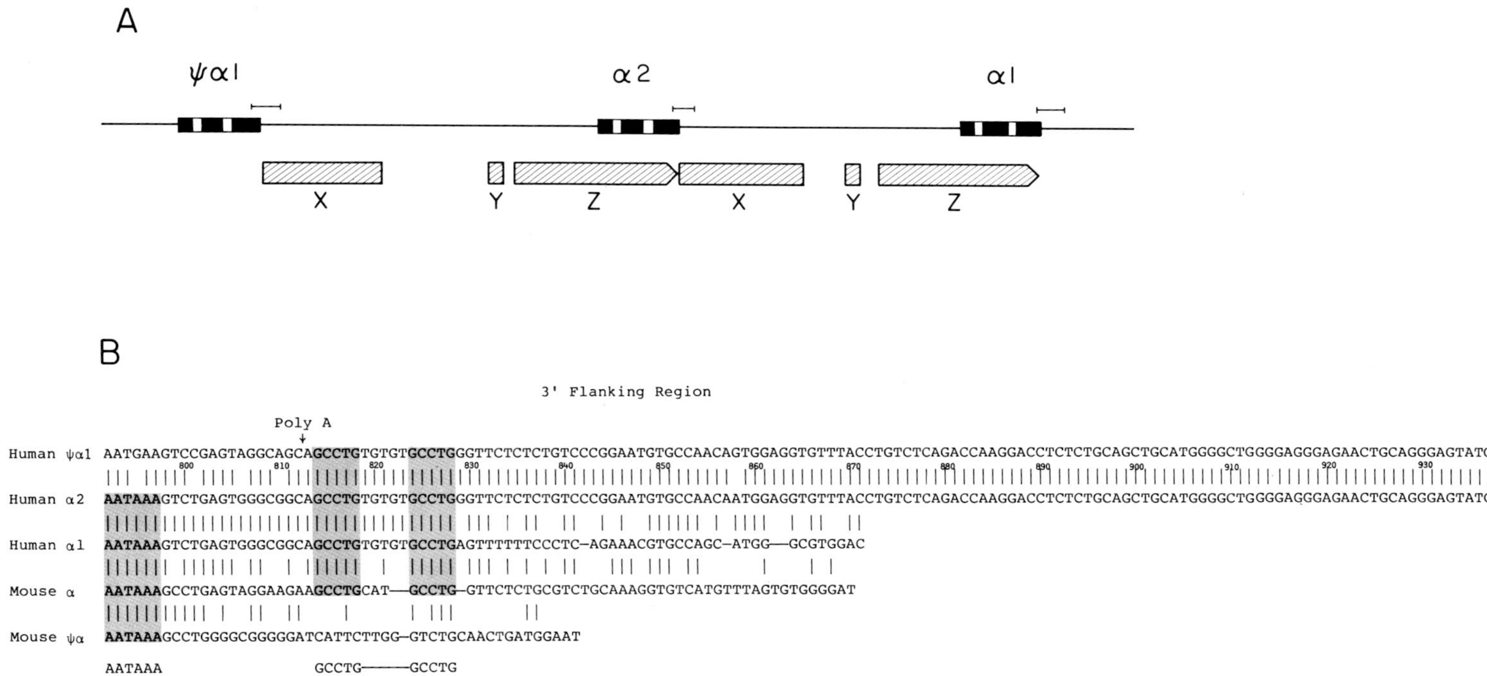

Fig. 8. Shows α-globin gene duplication. (A) The linkage arrangement of human $\psi\alpha1$-, $\alpha2$-, and $\alpha1$-globin genes. Rectangular boxes denote regions of homology between each duplication unit. The point at the end of the Z box denotes the 3′ end of the duplication unit. Horizontal brackets denote the three regions that have been sequenced. (B) The sequence adjacent to the poly(A) addition site in human $\psi\alpha1$-, $\alpha2$-, and $\alpha1$-globin genes and mouse α- and $\psi\alpha$-globin genes. Reprinted from Cell by permission [43].

precisely at the poly(A) addition site of the $\psi\alpha 1$ gene. This conclusion is supported by comparison of the 3′ flanking sequences of $\alpha 2$ and $\alpha 1$. In this case, the regions between the AATAAA sequence and the poly(A) addition sites are identical. On the 3′ side of the poly(A) addition site, the sequence identity continues for 15 bp and then diverges abruptly (60% homology in the first 43 bp). It is interesting to note that the 15 bp sequence 3′ to the poly(A) addition sites of $\psi\alpha 1$, $\alpha 2$, and $\alpha 1$ consists of a repeated pentanucleotide (GCCTG) separated by the sequence TGTGT. The occurrence of this sequence in all three genes and its location with respect to the endpoints of the α-globin gene duplication suggests that this sequence might be associated with the mechanism by which the genes were duplicated or corrected (see [43] for discussion).

The virtual identity of sequences within the $^{G}\gamma$-$^{A}\gamma$ and $\alpha 1$-$\alpha 2$ gene pairs appears to be the product of a mechanism for gene matching during evolution. Based on the nearly identical distribution of restriction sites surrounding the α-globin genes in a number of primate species, it has been suggested that the α-globin gene duplication occurred prior to the time of primate divergence [55]. Differences between the α-globin amino acid sequences of various primate species are consistent with sequence drift following primate divergence [48]. However, intraspecies comparisons show much less divergence, indicating that the α-globin genes within a species have been corrected against one another. Gene conversion and expansion/contraction of gene number by homologous but unequal crossing-over have been proposed as mechanisms for gene correction [57].

Analysis of the complete nucleotide sequences of cloned $^{G}\gamma$ and $^{A}\gamma$ genes has led to formulation of a specific intrachromosomal gene conversion model to explain sequence matching between linked genes [7]. The $^{G}\gamma$ and $^{A}\gamma$ genes on one chromosome are identical in the region 5′ to the center of the large intron, yet show greater divergence 3′ to that position. Examination of the boundary between the conserved and divergent regions revealed a block of "simple sequence" DNA (TG_n) [7]. Slightom et al [7] have proposed that this simple sequence is a "hot spot" for initiation of recombination events which lead to unidirectional gene conversion.

As described above, an active mechanism for gene matching must also be invoked to explain the observed homology between the human α-globin genes. Although α-globin gene matching could occur by gene conversion, sequence data which would address this possibility are not available. Evidence that α-globin gene sequence matching could occur by expansion and contraction of gene number by unequal crossing-over is provided by the frequent occurrence of one-gene [39, 40] and three-gene [58, 59] chromosomes in some human populations [see 17, 55 for discussion].

DISCUSSION

Molecular cloning studies have provided information which is essential to the understanding of globin gene regulation at the molecular level. Although comparison of the nucleotide sequences of globin genes and their intervening and flanking sequences suggests that certain sequences are required for transcription, splicing, and polyadenylation of globin mRNAs, the actual identification of such sequences requires the use of in vivo or in vitro assays for gene expression.

Two in vivo assays for globin gene expression have been developed. The first assay involves the stable introduction of globin genes into mouse fibroblasts using the herpes virus thymidine kinase gene as a selectable marker [60–62]. The rabbit β-globin gene was introduced into mouse L-cells using this procedure. Although the rabbit gene is transcribed in the mouse cells and the RNA transcript is accurately spliced, the amount of rabbit β-globin mRNA synthesized is very low. In addition, some cell lines produce globin mRNAs which are missing the 5′ terminal 45 nucleotides [62]. It is not known whether this defect is due to abnormal transcriptional initiation of mRNA processing.

A second assay for in vivo globin gene expression involves the use of the tumor virus SV40 as a mammalian cloning vector. Transient introduction of globin genes into mammalian cells can be achieved by replacing the genes for SV40 viral capsid proteins with globin genes and growing the recombinants in the presence of an SV40 helper virus [63–65]. The recombinant virus is amplified many thousand fold, facilitating the study of globin gene expression. Mature globin mRNA and β-globin mRNA and β-globin polypeptides are made in large amounts in monkey fibroblasts infected by the SV40-globin recombinants. It seems likely that the sequences necessary for globin gene transcription and RNA splicing will be identified by using the SV40 vector system in conjunction with procedures for site-directed mutagenesis [66, 67].

The recent development of cell-free extracts for RNA polymerase II-dependent transcription of cloned eukaryotic genes provides an in vitro approach to the study of globin gene expression [68, 69]. Recently, one of these in vitro transcription systems [69] has been used to study the in vitro transcription of human globin genes [48]. This study revealed that all of the human globin genes can be specifically transcribed in vitro. Therefore, all of the human globin genes appear to comprise individual transcription units. In addition, with the exception of the δ-globin gene which appears to be transcribed somewhat inefficiently, the embryonic, fetal, and adult genes are transcribed with roughly equal efficiencies in an extract prepared from HeLa cells, which do not ordinarily express globin genes. Thus, the interaction of different globin promoters with RNA polymerase II in vitro is approximately the same, and the mechanisms that mediate tissue-specific transcription do not operate in vitro.

The fidelity of the in vitro transcription reaction is demonstrated by the fact that the 5′ ends of the in vitro transcripts indicate that they are identical to the in vivo transcripts. This was shown by RNA mapping experiments and direct analysis of the 5′ ends of in vitro transcripts derived from the human β-globin gene [48]. The β-globin RNA synthesized in vitro contains the same 5′ cap structure as authentic β-globin mRNA.

Experiments to identify the nucleotide sequence required for accurate in vitro transcription of the human β-globin gene are in progress [70]. Our preliminary results indicate that the CCAAT box is not necessary for the in vitro transcription of the normal β-globin gene, while a region of sequence including the ATA box appears to be essential.

In summary, the isolation and structural characterization of normal and abnormal globin genes, and the application of site-directed mutagenesis procedures to normal globin genes, provide the means of identifying sequences involved in human globin gene expression. In addition, these studies have provided detailed physical maps of the α- and β-like globin gene clusters and hybridization probes containing specific regions of the gene clusters. These hybridization probes have been used to map deletions that are associated with certain types of α- or β-thalassemias [see 1, for references]. The most provocative conclusion derived from these studies is that deletions which alter the normal pattern of globin gene expression during development map many kilobases away from the genes which are affected. Since these deletions apparently act in cis to the affected genes, important regulatory sequences may be removed by the deletions or, alternatively, the deletions may alter some structural feature of the chromosome which is necessary for normal gene activation or repression [1, 8, 15, 71, 72, 73]. The latter interpretation would be consistent with experiments in the chicken globin gene system which suggest that globin gene activation is associated with a structural alteration of a large region of chromosomal DNA [74].

SUMMARY

We review the current status of our understanding of the structure and chromosomal arrangement of human globin genes. All of the known α-like and β-like globin genes have been isolated from bacteriophage λ libraries of human fetal liver DNA. Each of the β-like globin genes has the same transcriptional orientation and each is arranged in a closely linked cluster in the order $5'$-$\psi\beta2$-ϵ-$^{G}\gamma$-$^{A}\gamma$-$\psi\beta1$-δ-β-$3'$. Similarly, the α-like globin genes are arranged in the order $5'$-$\zeta1$-$\zeta2$-$\psi\alpha1$-$\alpha2$-$\alpha1$-$3'$ and all of the genes have the same transcriptional orientation. Thus, in both gene clusters and $5'$ and $3'$ organization of the genes reflects the order of their expression during development. A number of globin sequences were detected which cannot be identified with any known globin polypeptide chain ($\psi\beta1$, $\psi\beta2$, and $\psi\alpha1$). The nucleotide sequence of one of these genes ($\psi\alpha1$) indicates that it is a globin pseudogene, a gene which has acquired mutations which prevent its expression. Restriction mapping and heteroplex analyses have revealed extensive sequence homology within and flanking the $\alpha1$ and $\alpha2$ genes. Similarly, direct nucleotide sequence analysis indicates that the $^{G}\gamma$ and $^{A}\gamma$ genes are virtually identical. In both cases the extensive sequence homology appears to be the result of a mechanism for sequence matching during evolution.

ACKNOWLEDGMENTS

This work was supported by a grant from the National Institutes of Health.

REFERENCES

1. Maniatis T, Fritsch EF, Lauer J, Lawn RM: The molecular genetics of human hemoglobins. Ann Rev Genet 14:145–178, 1980.
2. Maniatis T: Recombinant DNA procedures in the study of eukaryotic genes. In L Goldstein, Prescott, DM (ed): "Cell Biology: A Comprehensive Treatise," New York: Academic Press, 1980, pp 564–608.
3. Blattner FR, Blechl AE, Denniston-Thompson K, Faber HE, Richards JE, Slightom JL, Tucker PW, Smithies O: Cloning human fetal γ-globin and mouse α-type globin DNA: Preparation and screening of shotgun collections. Science 202:1279–1283, 1978.

4. Smithies O, Blechl A, Denniston-Thompson K, Newell N, Richards J, Slightom J, Tucker D, Blattner F: Cloning human fetal γ-globin and mouse α-type globin DNA: Characterization and partial sequencing. Science 202:1284–1289, 1978.
5. Maniatis T, Hardison RC, Lacy E, Lauer J, O Connell C, Quon D, Sim GK, Efstratiadis A: The isolation of structural genes from libraries of eucaryotic DNA. Cell 15:687–701, 1978.
6. Lawn RM, Fritsch EF, Parker RC, Blake G, Maniatis T: The isolation and characterization of linked δ- and β-globin genes from a cloned library of human DNA. Cell 15:1157–1174, 1978.
7. Slightom J, Blechl A, Smithies O: Human fetal $^{G}\gamma$ and $^{A}\gamma$ globin genes: Complete nucleotide sequences suggest that DNA can be exchanged between these duplicated genes. Cell 21:627–638, 1980.
8. Fritsch EF, Lawn RM, Maniatis T: Characterization of deletions which affect the expression of fetal globin genes in man. Nature 279:598–603, 1979.
9. Fritsch EF, Lawn RM, Maniatis T: Molecular cloning and characterization of the human β-like globin gene cluster. Cell 19:959–972, 1980.
10. Ramirez F, Burns AL, Mears JG, Spence S, Starkman D, Bank A: Isolation and characterization of cloned human fetal globin genes. Nucleic Acids Res 7:1147–1162, 1979.
11. Proudfoot N, Baralle F: Molecular cloning of the human ϵ-globin gene. Proc Natl Acad Sci USA 76: 5435–5439, 1979.
12. Gale R, Clegg J, Huehns E: Human embryonic hemoglobins Gower 1 and Gower 2. Nature 280: 162–164, 1979.
13. Flavell RA, Kooter JM, De Boer E, Little PFR, Williamson R: Analysis of the human β-δ-globin gene loci in normal and Hb Lepore DNA: Direct determination of gene linkage and intergene distance. Cell 15:25–41, 1978.
14. Little PFR, Flavell RA, Kooter JM, Annison G, Williamson R: Structure of the human fetal globin gene locus. Nature 278:227–231, 1979.
15. Tuan D, Biro PA, de Riel JK, Lazarus H, Forget BG: Restriction endonuclease mapping of the human γ globin gene loci. Nucleic Acids Res 6:2519–2544, 1979.
16. Bernards R, Little PFR, Annison G, Williamson R, Flavell RA: Structure of the human $^{G}\gamma$-$^{A}\gamma$-δ-β-globin gene locus. Proc Natl Acad Sci USA 76:4827–4831, 1979.
17. Lauer J, Shen C-KJ, Maniatis T: The chromosomal arrangement of human α-like globin genes: Sequence homology and α-globin gene deletions Cell 20:119–130, 1980.
18. Lacy E, Hardison RC, Quon D, Maniatis T: The linkage arrangement of four rabbit β-like globin genes. Cell 18:1273–1283, 1979.
19. Hardison RC, Butler ET, Lacy E, Maniatis T, Rosenthal N, Efstratiadis A: The structure and transcription of four linked rabbit β-like globin genes. Cell 18:1285–1297, 1979.
20. Jahn CL, Hutchinson CA III, Phillips SJ, Weaver S, Haigwood NL, Voliva CF, Edgell MH: DNA sequence organization of the β-globin complex in the BALB/c mouse. Cell 21:159–168, 1980.
21. Lawn RM, Efstratiadis A, O'Connell C, Maniatis T: The nucleotide sequence of the human β-globin gene. Cell 21:647–651, 1980.
22. Spritz R, DeRiel J, Forget B, Weissman S: Nucleotide sequence of the human δ globin gene. Cell 21:639–646, 1980.
23. Barelle FE, Shoulders C, Proudfoot NJ: The primary structure of the human epsilon gene. Cell 21:621–626, 1980.
24. Tilghman SM, Curtis PJ, Tiemeier DC, Leder P, Weissman C: The intervening sequence of a mouse β globin gene is transcribed within the 15 S β-globin mRNA precursor. Proc Natl. Sci USA 75:1309–1313, 1978.
25. Kinniburgh AJ, Mertz JE, Ross J: The precursor of mouse β-globin messenger RNA contains two intervening RNA sequences. Cell 14:681–693, 1978.
26. Kinniburgh AJ, Ross J: Processing of the mouse β-globin mRNA precursor: At least two cleavage-ligation reactions are necessary to excise the large intervening sequence. Cell 17:915–921, 1979.
27. Flavell RA, Bernards R, Grosveld GC, Hooijmakers-VanDommeien HAM, Kooter JM, De Boer E: The structure and expression of globin genes in rabbit and man. In Axel R, Maniatis T, Fox CF (eds): "Eukaryotic Gene Regulation, ICN-UCLA Symposium on Molecular and Cellular Biology, XIV," New York: Academic Press, 1979, pp 335–354.
28. Maquat LE, Kinniburgh AJ, Beach LR, Honig GR, Lazerson J, Ershier WB, Ross J: Processing of the human β-globin mRNA precursor to mRNA is defective in three patients with β^{+} thalassemia. Proc Natl Acad Sci USA, 77:4287–4291, 1980.

29. Kantor JA, Turner PH, Nienhuis AW: Beta+ thalassemia: Mutations which affect processing of the β-globin mRNA precursor. Cell 21:149–157, 1980.
30. Weaver RF, Weissman C: Mapping of RNA by a modification of the Berk-Sharp procedure: The 5′ termini of 15 S β-globin mRNA precursor and mature 10 S β-globin mRNA have identical map coordinates. Nucleic Acids Res 7:1175–1193, 1979.
31. Efstratiadis A, Posakony JW, Maniatis T, Lawn RM, O'Connell C, Spritz RA, DeRiel JK, Forget B, Weissman SM, Slightom JL, Blechl AE, Smithies O, Baralle RE, Shoulders CC, Proudfoot NJ: The structure and evolution of the human β-globin gene family. Cell 21:653–668, 1980.
32. Davidson E, Britten R: Regulation of gene expression: Possible role of repetitive sequences. Science 204:1052–1059, 1979.
33. Jelinek WR, Toomey TP, Leinwand L, Duncan CH, Biro PA, Choudary PN, Weissman SM, Rubin CM, Houck CM, Deininger PL, Schmid CW: Ubiquitous, interspersed repeated sequences in mammalian genomes. Proc Natl Acad Sci USA 77:1398–1402, 1980.
34. Houck CM, Rinehart FF, Schmid CW: An ubiquitous family of repeated DNA sequences in the human genome. J Mol Biol 132:289–306, 1979.
35. Duncan C, Biro PA, Chowdary PV, Elder JT, Wang RRC, Forget BG, De Riel JK, Weissman SM: RNA polymerase III transcriptional units are interspersed among human non-α-globin genes. Proc Natl Acad Sci USA 76:5095–5099, 1980.
36. Jelinek WG, Leinwand L: Low molecular weight RNAs hydrogen-bonded to nuclear and cytoplasmic poly(A)-terminated RNA from cultured Chinese hamster ovary cells. Cell 15:205–214, 1978.
37. Shen, C-K, Maniatis T: The organization of repetitive sequences in a cluster of rabbit β-like globin genes. Cell 19:379–391, 1980.
38. Fritsch EF, Shen C-KJ, Lawn RM, Maniatis T: Intergenic repeat sequences in mammalian globin gene clusters. Cold Spring Harbor Symp Quant Biol, in press, 1980.
39. Orkin SH: The duplicated human α-globin genes lie close together in cellular DNA. Proc Natl Acad Sci USA 75:5950–5954, 1978.
40. Embury S, Lebo R, Dozy A, Kan YW: Organization of the α-globin genes in the Chinese α-thalassemia syndromes. J. Clin Invest 63:1307–1310, 1979.
41. Liebhaber S, Goosens M, Kan YW: Cloning and complete nucleotide sequence of the human 5′-alpha globin. Proc Natl Acad Sci USA, in press, 1980.
42. Nishioka Y, Leder P: The complete sequence of a chromosomal mouse α-globin gene reveals elements conserved throughout vertebrate evolution. Cell 18:875–882, 1979.
43. Proudfoot NJ, Maniatis T: The structure of a human α-globin pseudogene and its relationship to α-globin gene duplication. Cell 21:537–544, 1980.
44. Lauer J. PhD Thesis, California Institute of Technology, 1980.
45. Pressley L, Higgs D, Clegg J, Weatherall D: Gene deletions in α-thalassaemia prove that the 5′ ζ locus is functional. Proc Natl Acad Sci USA 77:3586–3589, 1980.
46. Breathhach R, Benoist C, O'Hare, K, Gannon F, Chambon P: Ovalbumin gene: Evidence for a leader sequence in mRNA and DNA sequences at the exon-intron boundaries. Proc Natl Acad Sci USA 75:4853–4857, 1978.
47. Lerner M, Boyle J, Mount S, Wolin S, Steitz J: Are snRNPs involved in splicing? Nature 283: 220–224, 1980.
48. Proudfoot NJ, Shander MHM, Manley JL, Geffer ML, Maniatis T: The structure and in vitro transcription of human globin genes. Science 209:1329–1336, 1980.
49. Lacy E, Maniatis T: The nucleotide sequence of an α-globin pseudogene. Cell 21:545–553, 1980.
50. Lingrel et al. This Symposium.
51. Haynes JR, Rosteck P, Lingrel JB: Unusual sequence homology at the 5′ ends of the developmentally regulated β^A, β^C and γ globin genes in the goat. Proc Natl Acad Sci USA, in press, 1980.
52. Nishioka Y, Leder A, Leder P: An unusual α-globin-like gene that has cleanly lost both globin intervening sequences. Proc Natl Acad Sci USA 77:2806–2809, 1980.
53. Vanin EF, Goldberg GI, Tucker PW, Smithies O: A mouse α-globin related pseudogene (ψα30.5) lacking intervening sequences. Nature 286:222–226, 1980.
54. Konkel DA, Maizel JV, Leder P: The evolution and sequence comparison of two recently diverged mouse chromosomal β-globin genes. Cell 18:865–873, 1979.
55. Zimmer EA, Martin SL, Beverley SM, Kan YW, Wilson AC: Rapid duplication and loss of genes coding for the α chains of hemoglobin. Proc Natl Acad Sci USA 77:2158–2162, 1980.

56. Dayhoff MO: "Atlas of Protein Sequence and Structure," Washington, D.C.: National Biomedical Research Foundation, 1972.
57. Hood L, Campbell JH, Elgin SCR: The organization, expression and evolution of antibody genes and other multigene families. Ann Rev Genet 9:305–353, 1975.
58. Goosens M, Dozy A, Embury S, Zacharides A, Hadjiminas M, Stamatoyannopoulos G, Kan YW: Triplicated α-globin loci in humans. Proc Natl Acad Sci USA 77:518–521, 1980.
59. Higgs DR, Old JM, Pressley L, Clegg JB, Weatherall DJ: A novel α-globin gene arrangement in man. Nature 284:632–635, 1980.
60. Wigler M, Sweet R, Sim GK, Wold B, Pellicer A, Lacy E, Maniatis T, Silverstein S, Axel R: Transformation of mammalian cells with genes from procaryotes and eucaryotes. Cell 16:777–785, 1979.
61. Wold B, Wigler M, Lacy E, Maniatis T, Silverstein S, Axel R: Expression of an adult rabbit β-globin gene stably inserted into the genome of mouse L-cells. Proc Natl Acad Sci USA 76: 5684–5688, 1979.
62. Mantei N, Boll W, Weissman C: Rabbit β-globin mRNA production in mouse L-cells transformed with cloned rabbit β-globin chromosomal DNA. Nature 281:40–46, 1979.
63. Mulligan RC, Howard BH, Berg P: Synthesis of rabbit β-globin in cultured monkey kidney cells following infection with a SV40 β-globin recombinant clone. Nature 277:108–114, 1979.
64. Hamer DH, Leder P: Expression of the chromosomal mouse β^{maj} globin gene cloned in SV40. Nature 281:35–40, 1979.
65. Hamer DH, Smith KD, Boyer SH, Leder P: SV40 recombinants carrying rabbit β-globin gene coding sequences. Cell 17:725–735, 1979.
66. Shortle D, Nathans D: Local mutagenesis, a method for generating viral mutants with base substitutions in preselected regions of the viral genome. Proc Natl Acad Sci USA 75:2170–2174, 1978.
67. Weissman C, Nagota S, Taniguchi T, Weber H, Meyer F: The use of site directed mutagenesis in reversed genetics. In Seltow J, Hollaender A (eds): "Genetic Engineering," Vol 1. New York: Plenum, 1979, pp 13–150.
68. Weil PA, Luse DS, Segall J, Roeder RG: Selective and accurate initiation of transcription at the Ad2 Major Late Promoter in a soluble system dependent on purified RNA polymerase II and DNA. Cell 18:469–484, 1979.
69. Manley JL, Fire A, Cano A, Sharp PA, Gefter ML: DNA dependent transcription of adenovirus genes in a soluble whole-cell extract. Proc Natl Acad Sci USA 77:3855–3858, 1980.
70. Parker V, Proudfoot NJ, Shander MHM, Maniatis T: Unpublished observations.
71. Bernards R, Kooter JM, Flavell RA: Physical mapping of the globin gene deletion in $\delta\beta$ thalassemia. Gene 6:265–280, 1979.
72. Van der Ploeg LHT, Konings A, Oort M, Roos D, Bernini L, Flavell R: δ-β thalassemia studies showing that deletion of the γ and δ genes influences β-globin gene expression in man. Nature 283: 637–642, 1980.
73. Tuan D, Murnone MJ, De Riel JK, Forget BG: Heterogeneity in the molecular basis of hereditary persistence of fetal hemoglobin. Nature 285:335–337, 1980.
74. Stalder J, Larsen A, Engel JD, Dolan M, Groudine M, Weintraub H: Tissue specific DNA cleavages in the globin chromatin domain introduced by DNAse 1. Cell 20:451–460, 1980.

Organization and Expression of Globin Genes, pages 33–43

Cloning and Characterization of DNA Sequences Surrounding the Human β-Like Globin Genes: Identification of an Unusually Long Moderately Repetitive DNA Sequence

Jesse W. Adams, Russel E. Kaufman, Peter J. Kretschmer, Helen C. Coon, W. French Anderson, and Arthur W. Nienhuis

Clinical Hematology Branch (J.W.A., R.E.K., A.W.N.) and Laboratory of Molecular Hematology (P.J.K., H.C.C., W.F.A.), National Heart, Lung, and Blood Institute, Bethesda

INTRODUCTION

The human embryonic, fetal, and adult hemoglobins are produced during ontogenic development in a sequential and regulated fashion [1–4]. Two complexes of genes encode for these hemoglobins, the α-like genes on chromosome 16 [5] and the non-α cluster on chromosome 11 [6]. The general map and organization of these genomic regions has been discussed in detail in the paper by T. Maniatis and colleagues in this volume. The non-α cluster includes genes for the following globins: ε found in embryonic hemoglobin, $^{G}\gamma$ and $^{A}\gamma$ found in fetal hemoglobin (Hb F), δ globin found in Hb A_2, and β globin found in the normal adult hemoglobin, Hb A. By both restriction endonuclease mapping of genomic DNA and molecular cloning of the human globin genes, the ε, two γ, the δ, and the β globin genes have been shown to be closely linked on a 46 kilobase (kb) segment of DNA [7–15]. Each globin gene is approximately 1,600 base pairs (bp) in length, for in addition to the DNA regions coding for a globin, each gene is interrupted by two introns of approximately 100 and 800–900 bp. Thus, the five genes together occupy approximately 17% of the 46 kb segment of chromosome 11 in which they are found. The role of the remaining DNA between the genes and the DNA flanking the entire complex remains unexplained, although those segments between and around structural genes have been postulated to contain DNA sequences which may be important for gene regulation.

GENERAL EXPERIMENTAL STRATEGY

Our general approach to this problem has been to obtain purified segments of the genomic region by molecular cloning recombinant DNA techniques [16]. To this end a human genomic DNA fragment library was constructed with DNA from an

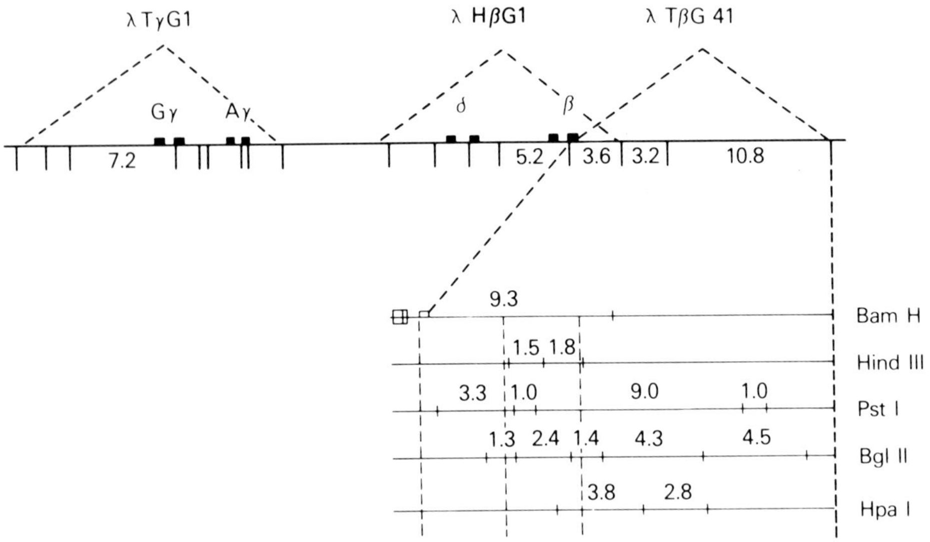

Fig. 1. Restriction endonuclease maps of recombinant bacteriophage. On the top line the DNA which contains the human $^{G}\gamma$, $^{A}\gamma$, δ, and β globin genes is diagrammed with the Eco RI sites marked. The sizes in kb of some of the Eco RI fragments are given. The fragments which are cloned in λTγG1, λHβG1, and λTβG41 are shown, and a more detailed restriction endonuclease map of λTβG41 is given below.

individual with homozygous β-thalassemia. DNA fragments generated by partial Eco RI digestion were cloned into Escherichia coli using the bacteriophage vector Charon 4A, and then the library was screened with a probe specific for β-like globin gene sequences. Recombinants containing the ϵ, the two γ, or the 3′ end of the β gene along with 17 kb of flanking DNA have been isolated and characterized (Fig. 1) [15]. These clones, along with the recombinant containing the δ and β globin genes (λHβG1) isolated by Lawn et al [9], have been used to examine the nature of the nonglobin gene DNA using the general techniques of Southern blotting and heteroduplex analysis with the electron microscope.

COMPARISON OF THE γ and δβ GENE REGIONS REVEALS A LACK OF SEQUENCE HOMOLOGY

The extent of sequence homology in the DNA flanking the γ globin genes has been compared to the DNA flanking the δ and β globin genes by performing heteroduplex analysis using the electron microscope [17]. The recombinant we have isolated which contains the two γ genes (λTγG1) has been compared to λHβG1 [15]. The orientation of the inserted DNAs in these two recombinants is such that the right arm of Charon 4A is 5′ with respect to the two globin genes and the relative position of the genes in each recombinant is approximately the same (Fig. 1), making them highly suitable for heteroduplex formation and analysis. Shown in Figure 2 is a typical DNA heteroduplex. Aside from the arms of Charon 4A, only two regions of duplex formation were noted. The location of these within the inserted DNA fragments in the two recombinants was defined by measurement of eight heteroduplexes using an electronic graphic calculator. One area where duplex formation occurred was localized to the position of the $^{A}\gamma$ and β globin

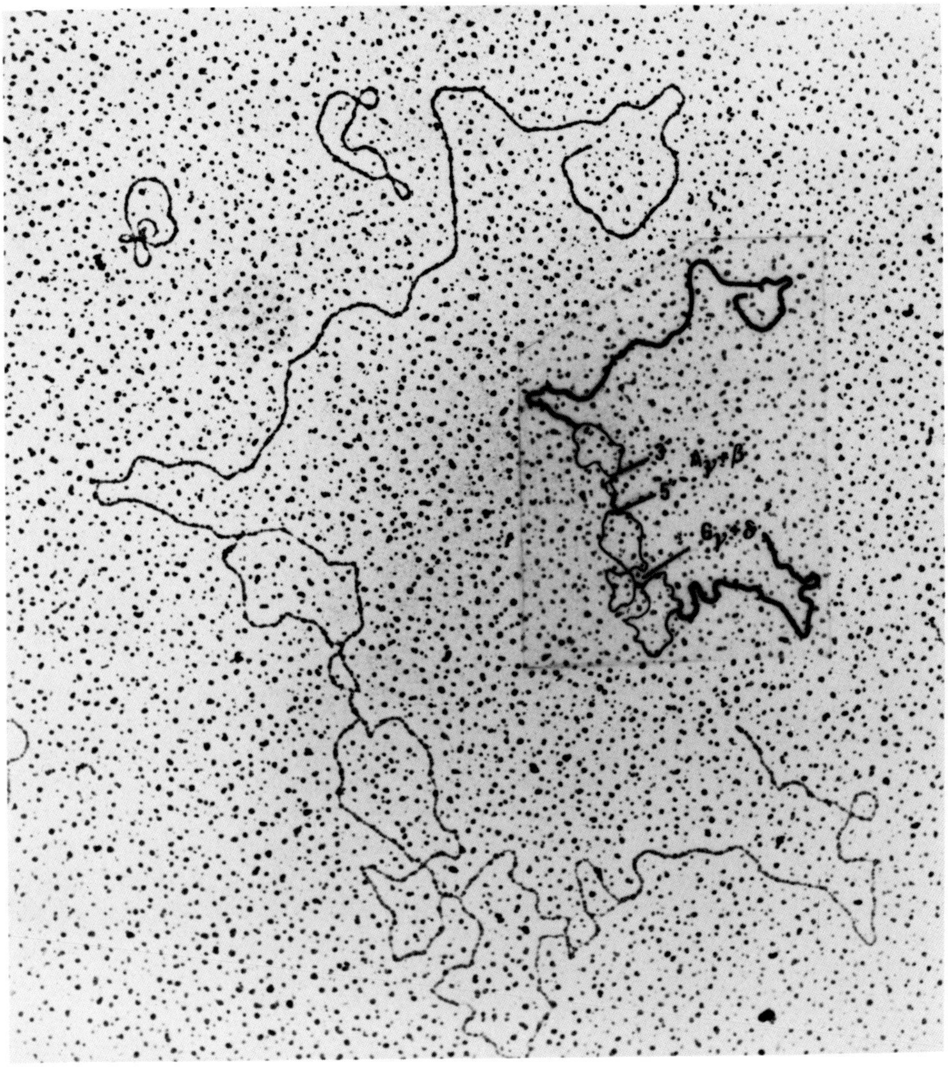

Fig. 2. Heteroduplex formed between λTγG1 and λHβG1. Heavy lines on the drawing (inset) represent regions of hybridization. Large loops on either side of short hybridized structural gene regions demonstrate lack of homology in these flanking sequences between the $^{G}\gamma$ and δ genes and $^{A}\gamma$ and β genes. × 26,400.

genes. A nonannealing region of 800–1,000 bp represented the large introns of these two genes which are surrounded by two short annealing regions ranging from 300 to 600 by 5′ to the intron and 150 to 280 by 3′ to the intron. Seven of the eight heteroduplexes analyzed exhibited this pattern while no annealing occurred in this region in the other heteroduplex. Heteroduplex formation, which on measurement corresponded to the 5′ coding sequences in the $^{G}\gamma$ and δ globin genes, ranged in length from 200 to 450 bp and was observed in four of the eight heteroduplexes examined. The large intron sequences of these two genes did not form heteroduplex nor did the coding sequences 3′ to the intron.

Formation of the heteroduplexes examined in this study occurred in 47.6% formamide at 25°C. That these are relatively nonstringent conditions is illustrated by the fact that the coding sequences of the $^{A}\gamma$ and β genes annealed despite 20%–25% sequence divergence on direct comparison of the nucleotide sequence of these genes. Thus, failure of the heteroduplex to extend into the flanking regions of these genes must be taken to indicate that there is very extensive sequence divergence in these regions. That homology may exist is, of course, possible, particularly if deletions and insertions have shifted the position of homologous segments in the flanking regions relative to one another, thereby reducing the probability of heteroduplex formation.

Leder and colleagues initially argued that lack of homology surrounding the mouse β^{maj} and β^{min} genes had at least two implications [18]. First, the regulatory sequences necessary for expression of the globin genes at a high level in erythroid cells must be close to the coding portions of these genes since extensive divergence in the flanking sequences did not appear to alter the level of expression. Second, non-homology of flanking and intron sequences was suggested as being protective of closely linked, highly homologous genes in preventing cross-over which might lead to deletion. Indeed, in the case of the human α globin genes, extensive sequence homology between the regions surrounding these genes and also between a third closely linked α-like gene is reflected by the occurrence of deletions in man in the form of the α-thalassemia syndromes and also in vitro in E coli during the replication of recombinant bacteriophages containing these genes [19]. That flanking sequence divergence is not obligatory for highly selective expression of two globin genes is illustrated by our results obtained in the analysis of the sheep γ and β^{A} globin genes [20]. These genes are included in eight kb segments of DNA which form linear, almost perfect, heteroduplexes under the exact conditions in which we formed the heteroduplexes between the human γ and $\delta\beta$ gene regions. Thus, there are already examples of paired globin genes which exhibit either divergence or conservation of surrounding DNA sequences, making firm conclusions about the significance of homology or lack thereof rather hazardous.

MAP OF RESTRICTION ENDONUCLEASE SITES DOWNSTREAM FROM THE β GLOBIN GENE

Isolation of the cloned DNA fragment in λTβG41 which includes 17 kb of DNA downstream from the β globin gene allowed us to extend the previously published restriction endonuclease map of the β-like gene cluster [7–15]. This region of the genome is of particular interest because of restriction endonuclease site polymorphisms found here which have been of use in prenatal diagnosis of sickle cell anemia [21] and β-thalassemia [22] and also because the deletions which result in $\delta\beta$ thalassemia [23] and pancellular hereditary persistence of fetal hemoglobin (HPFH) [11, 12] are known to terminate in this region. To facilitate the mapping of λTβG41, several DNA fragments were subcloned into E coli using the plasmid vector, pBR322 [15]. The resulting recombinant plasmids contained the indicated fragments oriented 5′ to 3′ in λTβG41 (Fig. 1): pRK12-3.6 Kb Eco RI fragment; pRK11-3.2 kb Eco RI fragment; pRK19-1.5 kb Eco RI-Bam HI fragment; and pRK20-9.3 kb Bam HI–Eco RI fragment. Single and double digestion of these plasmids with a variety of enzymes yielded the map of restriction endonuclease sites shown in Figure 1.

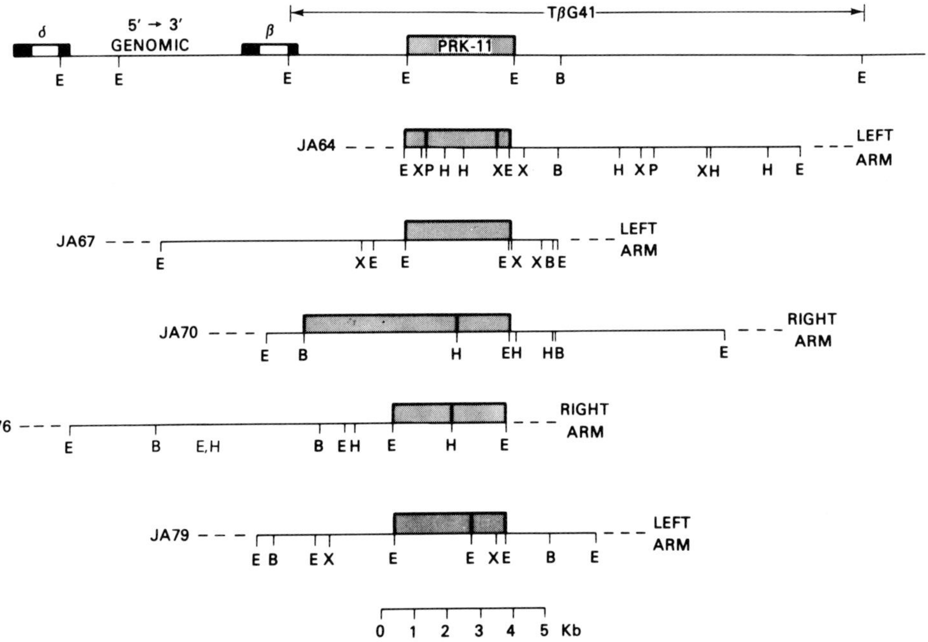

Fig. 3. Restriction endonuclease maps of several recombinant bacteriophages which contain copies of the long repeated DNA sequence. At the top is a map of certain sites in and around the human δ and β globin genes and 3′ flanking DNA. The regions which are contained in λTβG41 and pRK11 are indicated. Below are shown the restriction endonuclease maps for the clones λJA64, 67, 70, 76, and 79. The fragments which hybridized to the 3.2 kb β-3′ Eco RI probe from pRK11 are indicated by the shaded boxes and arrayed vertically immediately below pRK11. The shaded area in JA70 is larger due to a lack of one Eco RI restriction site: this does not imply a greater length of homology to the probe. The 5′ to 3′ orientation of the five cloned DNAs was assigned tentatively based on the Bam HI site 1.4 kb 3′ to the 3.2 kb Eco RI fragment, and the cloned DNAs are drawn accordingly. These orientations were confirmed by electron microscopy. The restriction sites are labeled as follows: E is Eco RI, X is Xba I, P is Pst I, H is Hind III, B is Bam HI. The Eco RI and Hind III sites in the 5′ end of λJA76 could not be unambiguously located and so are not precisely marked.

In DNA from most individuals, the β globin gene is found in a 7.6 kb Hpa I fragment. However, Kan et al [21] first described a polymorphism which yields a 13.0 kb Hpa I fragment. This polymorphism affects the Hpa I site 3′ to the β globin gene and is found linked to 60%–87% [21, 24] of genes affected by the mutation which causes sickle cell anemia, but is rarely linked to the normal β globin gene. The second Hpa I site 3′ to the β globin gene is 3.8 kb from the first (Fig. 3). Thus, a point mutation causing loss of the first Hpa I site should yield a fragment of 11.4 kb. The difference between 13.0 kb and 11.4 is probably not significant considering the inaccuracies of sizing the large DNA fragments on Southern blots of genomic DNA. Thus this polymorphism could be the result of a point mutation. This is also true of a second polymorphism found in this region. Thirty-three percent of non thalassemic Sardinian individuals have a 22.0 kb Bam HI fragment rather than the usual 9.3 kb Bam HI fragment which contains the 3′ end of the β globin gene [22]. Again, the site affected is the Bam HI site 3′ to the β globin gene and contained in our recombinant λTβG41. We did not find another Bam HI site downstream within this clone; this result is compatible with the Bam HI polymorphism being a point mutation.

Several deletion mutations which extend into the region of DNA contained in λTβG41 have been partially characterized. The deletion in one form of δβ thalassemia extends from the middle of the δ globin gene to approximately 1.0 kb 3′ to the β globin gene [23]. The restriction endonuclease sites mapped 3′ to that deletion matched those found in λTβG41. However, the 3′ extent of the larger deletions found in individuals with pancellular HPFH have not been mapped [11, 12]. In one case, a number of restriction endonuclease sites were mapped 3′ to the deletion [12]. We attempted to match these to sites found in λTβG41 without success. Because of our interest in the deletional syndromes, we decided to use fragments from λTβG41 as probes in Southern blot analysis of genomic DNA from normal individuals and from an individual homozygous for pancellular HPFH. By this strategy we hope to map the 3′ end of the deletion.

DISCOVERY OF A LONG MODERATELY REPETITIVE DNA SEQUENCE 3′ TO THE β GLOBIN GENE

The first probe utilized to analyze genomic DNA was a 2.9 kb Pst I: Eco RI fragment, which begins just beyond the end of the β globin gene. (Fig. 1). This fragment was found to anneal diffusely to restricted human DNA, indicating that it contains sequences repeated many times within the human genome [15]. Furthermore, analysis of clones containing human β-like genes localized this sequence to several other positions within the cluster in keeping with the results of Duncan et al [25], who have shown that this fragment, as well as others within the cluster, contains a copy of the Alu I family [26] which may be transcribed in vitro with RNA polymerase III. (See also Forget et al, this volume.)

When the next fragment downstream, namely the 3.2 kb Eco RI fragment (Fig. 1) was used as a probe against human genomic DNA, its content of moderately repetitive sequences was again apparent. Thus this fragment annealed diffusely to all size classes of restricted genomic DNA, and discrete bands also were visible on the radioautograph, including a particularly intense band at 3.2 kb. That the entire fragment might be part of a long repeated DNA sequence was suggested by these results and subsequently proved to be the case. We were prompted to further characterize this repeated DNA sequence and the family to which it belongs because of its location adjacent to the β globin gene and also because of its apparent long length. Although moderately repetitive DNA constitutes 15%–20% of the human genome, most families average from 300 to 500 bp in length [27]; thus this long repetitive sequence appeared distinctly unusual and, therefore, potentially interesting.

The strategy employed to study this repeated DNA family involved the isolation of several other members from our human DNA fragment library [28]. This was accomplished by using the 3.2 kb Eco RI fragment from λTβG41 as a probe. In a plaque hybridization screening about 3% of the plaques in the library hybridized to this probe. Five positive plaques were selected and these recombinant bacteriophage were mapped as to the position of several restriction endonuclease sites. The map of these clones is compared to that of λTβG41 in Figure 3. Pertinent is the fact that four of the five clones contained Eco RI sites 3.2 kb apart which defined fragments to which the 3.2 kb Eco RI fragment from λTβG41 annealed and, furthermore, all relevant clones contained a Bam HI site 1.5 kb from the most 3′ Eco

RI site. These results supported our interpretation that we were examining a family of long repeated DNA sequences and suggested that the Bam HI site was included within the repeated DNA.

Length of Individual Members of This Repeated DNA Family

Heteroduplexes formed between several pairs of the recombinants which contain a member of the long repeat family were examined with the electron microscope. One example is shown in Figure 4. This heteroduplex formed between λJA67 and λTβG41 shows the right arm of the vector, then a division of the two strands which rejoin to form a 5.04 ± 0.12 (95% confidence limit) duplex, then a loop of single-stranded DNA (from λTβG41), and finally the left arm of the bacteriophage vector. This molecule demonstrates the 5′ end of these copies of the repeated DNA sequence. By examining heteroduplexes formed between different pairs of recombinants, we reached a conclusion that the member of this moderately repetitive DNA family located 3′ to the β globin gene is 6.4 ± 0.2 kb long. Other members of the family might be somewhat longer or shorter although the five members selected at random appear to be of a similar length. For convenience, we refer subsequently to this family of repeated DNA sequences as the 6.4 kb family.

Number and Homology of Copies in the 6.4 kb Family

Two different types of experiments were performed to estimate the number of members of the 6.4 kb family [28]. In both, the 3.2 kb Eco RI fragment from λTβG41 was used as a probe after it was rendered radioactive by a nick-translation reaction. With this probe as tracer, its complementary sequences were found to anneal in a solution hybridization reaction at a $Cot_{1/2}$ of 0.25–0.4 under conditions in which unique sequence DNA anneals with a $Cot_{1/2}$ of 1,200. These results suggest that there are 3,000–4,800 copies of the sequence found in the 3.2 kb Eco RI fragment from λTβG41 in the human genome. This result was substantiated by a second experimental strategy. Serial dilutions of human genomic DNA and of DNA from pRK11 (the plasmid containing the 3.2 Eco RI fragment from λTβG41) were spotted onto a nitrocellulose filter and then incubated with the radioactive 3.2 kb fragment. The signal from genomic DNA was as intense as a 100-fold dilution of the control plasmid, indicating that the sequences in the 3.2 kb Eco RI fragment constitute as much as 1% of human genomic DNA. Duplexes formed in solution between the 3.2 kb Eco RI fragment from λTβG41 (tracer) and genomic DNA (driver) melted with a slightly flattened curve, giving a Tm 3°C below that of perfectly matched duplexes of similar length, suggesting a moderate sequence mismatch between different members of the 6.4 kb family.

Internal Organization of the 6.4 kb Family

Long moderately repetitive DNA sequences in other species have been shown to be composed of shorter moderately repetitive sequences, arranged in tandem in the case of sea urchin [29] or scrambled in the case of Drosophila [30]. That such an organization does not exist in the 6.4 kb family in the human genome is suggested by two lines of evidence. First, the linear perfect heteroduplexes observed with the electron microscope do not suggest a scrambled array of shorter locally repetitive sequences. Second, when parts of the member of the 6.4 kb family found 3′ to the β

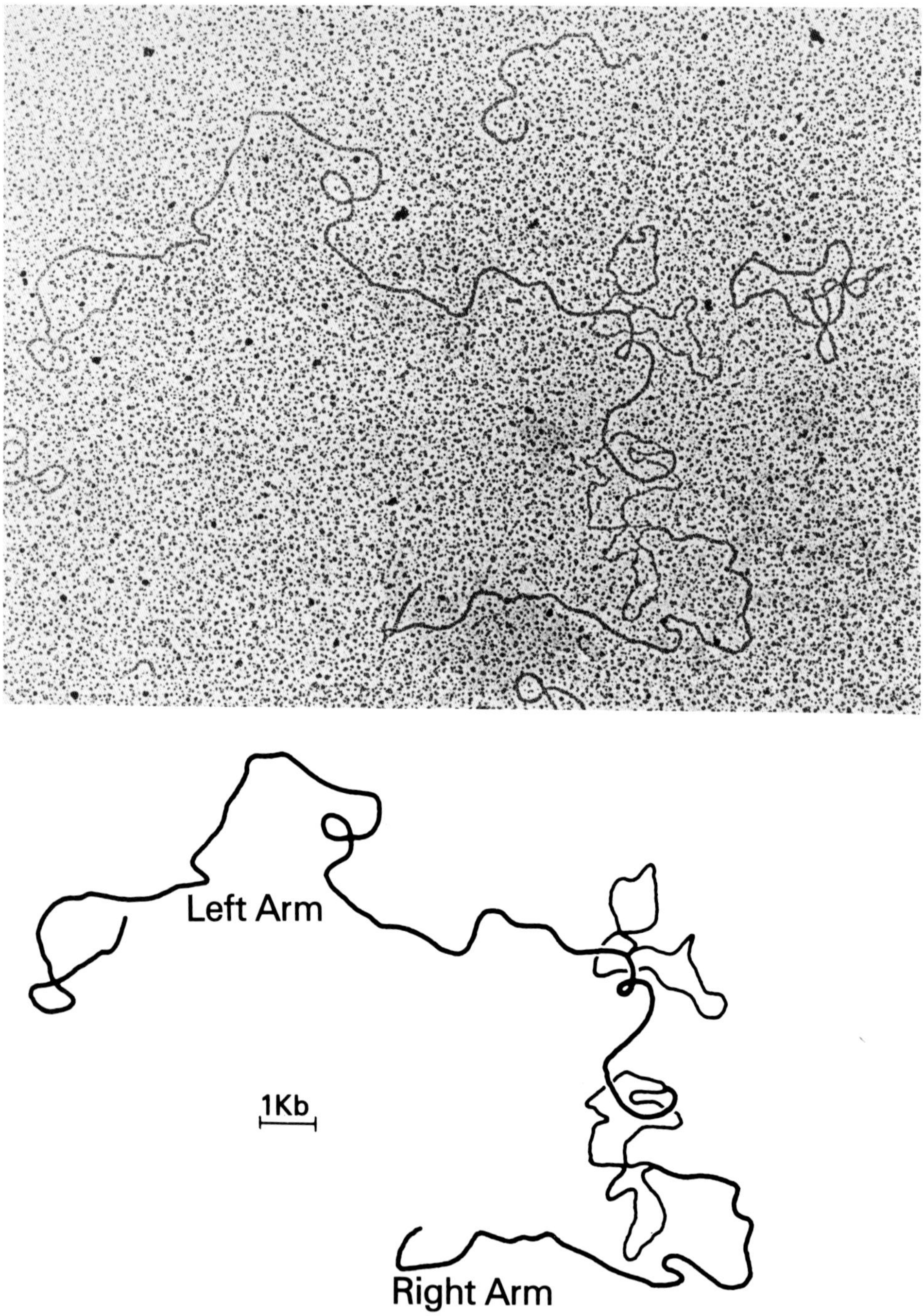

Fig. 4. Heteroduplex between λJA67 and λTβG41. This heteroduplex shows the right arm of the vector, then a division of the two strands which rejoin to form a 5.0 kb duplex followed by a single-stranded loop-out (from λTβG41) which marks the take off of the left arm. This molecule demonstrates the 5′ end of these copies of the repeated DNA sequence. × 30,000.

globin gene were used as a probe against restricted DNA from a recombinant bacteriophage containing an entire copy in a Southern blotting experiment, only those fragments which corresponded to the position in the repetitive sequence from which the individual probe were derived gave a positive signal. For example, the 3.2 kb Eco RI fragment from λTβG41 annealed only to itself in λTβG41, and to the corresponding fragments in the other clones. A similar result was obtained when the 1.4 kb Bam HI-Eco RI fragment was used as a probe.

Relevance of the 6.4 kb Repeat Family

The use of recombinant DNA-molecular cloning technology allows the rapid accrual of information about the structure of the human genome. Thus the anatomical substratum on which gene expression and regulation occurs is being defined, although the function of particular regions of the genome is not always readily apparent. This is certainly true of the 6.4 kb family.

Speculations about the role of moderately repetitive DNA are far ranging. A theory developed by Davidson and Britten [31] is based on their observations that certain moderately repetitive DNA families in the sea urchin genome are expressed in a highly selected and tissue-specific fashion; their transcripts are found predominantly in nuclear RNA. The hypothesis suggests that these transcripts modulate gene expression by influencing transcriptional rate or processing of the transcripts of structural genes. Such a role for the 6.4 kb repeat family seems highly unlikely based on its extremely long length and also based on preliminary experiments which suggest that this family of sequences is transcribed either not at all or at a very low level [28].

Other speculations about moderately repetitive DNA sequences are that they may play a role in chromosome structure or perhaps in folding of chromatin to define certain chromatin domains. Alternatively, such sequences have been postulated to have no important function, but rather to represent regions of the genome which are unimportant and drifting evolutionarily [32, 33]. Alternatively, the dispersion of long, moderately repetitive DNA sequence throughout the genome might provide catalytic sites for rearrangements or transpositions of large blocks of DNA.

Other studies which might provide some insight into the function of the 6.4 kb family include the following. We are analyzing clones containing members of this family to determine whether they contain structural genes expressed in human erythroid cells. If this proves to be the case, we will expand our analysis of this family to determine whether there is any pattern in the linkage relationship of individual members of the family to expressed genes in particular tissues. Ultimately, one might wish to construct genetic elements which contain all or part of a member of the repetitive family linked to a defined structural gene and test such constructs in vivo. The ability to introduce DNA into eukaryotic cells by a variety of techniques is already feasible [34] although relatively little information is yet available which bears on the question of the level of expression of such transformed genes.

SUMMARY

These studies add to the rapidly developing knowledge regarding the nature of the DNA sequences surrounding the β-like globin genes of man. Divergence of the flanking sequences surrounding the human γ compared to the human δ and β globin genes is extensive, as revealed by electron microscopic analysis of heteroduplexes formed between recombinants containing these genes. Downstream, or in the 3′ direction from the β globin gene, are moderately repetitive DNA sequences. One such sequence which begins 3.0 kb from the β globin gene is a member of a family of very long moderately repetitive units; this particular copy is 6.4 kb in length. This family has approximately 3,000–4,800 members and as such constitutes 0.5%–1.0% of the human genome. Its individual members are only moderately divergent from one another as revealed by thermal stability studies. No evidence was found for internal repetition within the long repeat unit.

REFERENCES

1. Nienhuis AW, Benz EJ Jr: Regulation of hemoglobin synthesis during the development of the red cell. N Engl J Med 297:1318–1328, 1371–1381, 1430–1436, 1977.
2. Weatherall DJ, Clegg JB: Recent developments in the molecular genetics of human hemoglobin. Cell 16:467–479, 1979.
3. Forget BG: Molecular genetics of human hemoglobin synthesis. Ann Intern Med 91:605–616, 1979.
4. Bank A, Mears JG, Ramirez F: Disorders of human hemoglobin. Science 207:486–493, 1980.
5. Deisseroth A, Nienhuis A, Turner P, Velez R, Anderson WF, Ruddle F, Lawrences J, Creagan R, Kucherlapati R: Localization of the human α-globin structural gene to chromosome 16 in somatic cell hybrids by molecular hybridization assay. Cell 12(1):205–218, 1977.
6. Deisseroth A, Nienhuis A, Lawrence J, Giles R, Turner P, Ruddle FH: Chromosomal localization of β-globin gene on human chromosome 11 in somatic cell hybrids. Proc Natl Acad Sci USA 74: 1456–1460, 1978.
7. Flavell RA, Kooter JM, Deboer E, Little PFR, Williamson R: Analysis of the β- δ-globin gene loci in normal and Hb Lepore DNA: Direct determination of gene linkage and intergene distance. Cell 15:25–41, 1978.
8. Mears JG, Ramirez F, Leibowitz D, Bank A: Organization of human δ- and β-globin genes in cellular DNA and the presence of intragenic inserts. Cell 15:15–23, 1978.
9. Lawn RM, Fritsch EF, Parker RC, Blake G, Maniatis T: The isolation and characterization of linked δ- and β-globin genes from a cloned library of human DNA. Cell 15:1157–1174, 1978.
10. Little PFR, Flavell RA, Kooter JM, Annison G, Williamson R: Structure of the human fetal globin gene locus. Nature 278:227–231, 1979.
11. Fritsch EF, Lawn RM, Maniatis T: Characterization of deletions which affect the expression of fetal globin genes in man. Nature 279:598–603, 1979.
12. Tuan D, Biro PA, DeRiel JK, Lazarus H, Forget BG: Restriction endonuclease mapping of the human γ-globin gene loci. Nucleic Acids Res 6:2519–2544, 1979.
13. Blattner FR, Blechl AE, Denniston-Thompson K, Faber HE, Richards JE, Slightom JL, Tucker PW, Smithies O: Cloning human fetal γ-globin and mouse α-type globin DNA: Preparation and screening of shotgun collections. Science 202:1279–1284, 1978.
14. Ramirez F, Burns AL, Mears JG, Spense S, Starkman D, Bank A: Isolation and characterization of cloned human fetal globin genes. Nucleic Acids Res 7:1147–1162, 1979.
15. Kaufman RE, Kretschmer PJ, Adams JW, Coon HC, Anderson WF, Nienhuis AW: Cloning and characterization of DNA sequences surrounding the human γ- δ- and β-globin genes. Proc Natl Acad Sci USA 77:4229–4223, 1980.
16. Maniatis T, Hardison RC, Lacey E, Lauer J, O'Connell C, Quon D, Sim GK, Efstratiadis A: The isolation of structural genes from libraries of eukaryotic DNA. Cell 15:687–701, 1978.
17. Davis RW, Simon M, Davidson N: Electron microscope heteroduplex methods for mapping regions of base sequence homology in nucleic acids. Methods Enzymol 21:413–428, 1971.
18. Tiemeier DC, Tilghman SM, Polski FI, Seidman JG, Leder A, Edgell MH, Leder P: A comparison of two cloned mouse β-globin genes and their surrounding and intervening sequences. Cell 14: 237–245, 1978.
19. Lawn J, Shen CJ, Maniatis T: The chromosomal arrangement of human α-like globin genes: Sequence homology and α-globin gene deletions. Cell 20:119–130, 1980.

20. Kretschmer PJ, Coon HC, Davis A, Harrison M, Nienhuis AW: Hemoglobin switching in sheep: Isolation of the fetal γ-globin gene and demonstration that the fetal γ and adult β^{A}-globin genes lie within 8 kilobase segments of homologous DNA. J Biol Chem, in press, 1981.
21. Kan YW, Dozy AM: Polymorphism of DNA sequence adjacent to human β-globin structural gene: Relationship to sickle mutation. Proc Natl Acad Sci USA 75:5631–5635, 1978.
22. Kan YW, Lee KY, Furbetta M, Angius A, Cao A: Polymorphism of DNA sequence in the β-globin gene region. N Engl J Med 302:185–188, 1980.
23. Bernards R, Kooter JM, Flavell RA: Physical mapping of the globin gene deletion in $(\delta - \beta)^{\circ}$-thalessemia. Gene 6:265–280, 1979.
24. Feldenzer J, Mears JG, Burns AL, Natta C, Bank A: heterogeneity of DNA fragments associated with the sickle globin gene. J Clin Invest 64:751–755, 1979.
25. Duncan C, Biro PA, Choudary PV, Elder JT, Wang NRC, Forget BG, DeRiel JK, Weissman SM: RNA polymerase III transcriptional units are interspersed among human non-α-globin genes. Proc Natl Acad Sci USA 76:5095–5099, 1979.
26. Houk CM, Rinehart FP, Schmid CW: A ubiquitous family of repeated DNA sequences in the human genome. J Mol Biol 132:289–306, 1979.
27. Schmid CW, Deininger PL: Sequence organization of the human genome. Cell 6:345–358, 1979.
28. Adams JA, Kaufman R, Krestchmer PJ, Harrison P, Nienhuis AW: A family of long moderately repetitive DNA sequences, one copy of which is next to the human β-globin gene. Nucleic Acids Res, in press, 1981.
29. Scheller RH, Thomas TL, Lee AS, Klein WH, Niles WD, Britten RJ, Davidson EH: Clones of individual repetitive sequences from sea urchin DNA constructed with synthetic ECO R1 sites. Science 196:197–200, 1977.
30. Wensink PC, Tabata S, Pachi C: The clustered and scrambled arrangement of moderately repetitive elements in Drosphila DNA. Cell 18:1231–1248, 1977.
31. Davidson EH, Britten RJ: Regulation of gene expression: Possible role of repetitive sequences. Science 204:1052–1059, 1979.
32. Doolittle WF, Sapienza C: Selfish genes, the phenotype paradigm and genome evolution. Nature 284:601–603, 1980.
33. Orgel LE, Crick FHC: Selfish DNA: The ultimate parasite. Nature 284:605–607, 1980.
34. Wigler M, Pellicer A, Silverstein S, Axel R: Biochemical transfer of single copy eukaryotic genes using total cellular DNA as donor. Cell 14:725–731, 1978.

Organization and Expression of Globin Genes, pages 45–57

Structural Organization of the Normal and Defective Globin Genes of the Goat

Jerry B. Lingrel, Joel R. Haynes, Eric A. Schon, Michael L. Cleary, and Patricia M. Gallagher

Department of Biological Chemistry, University of Cincinnati, College of Medicine, Cincinnati

INTRODUCTION

Understanding the mechanism of globin gene switching during human embryonic, fetal, and adult life may be helpful in developing new therapeutic approaches for the treatment of hemoglobinopathies such as sickle cell anemia and β-thalassemia. Such an understanding will require, among other things, a detailed description of the structural organization of the various globin genes and the construction of cell-free systems which exhibit the differential expression seen in vivo. The latter endeavor would be greatly aided by an animal model where sufficient amounts of tissue can be obtained. The goat, whose hemoglobins go through a series of changes during development, provides such a system [1–14]. Early in embryonic life Hb E, $\alpha_2\epsilon_2$ is synthesized. At about day 21 following conception, this hemoglobin is replaced by Hb F, $\alpha_2\gamma_2$ which, in turn, is replaced just before birth by Hb C, $\alpha_2\beta^C$. During the first year of life, Hb C is replaced by Hb A, $\alpha_2\beta^A_2$. These switches are summarized in Figure 1 along with those occurring in man. Thus, the goat switches its globins during development in a manner similar to that of humans. Also, as this animal is relatively large, adequate amounts of erythropoietic tissue can be obtained from each developmental stage.

Our approach to understanding the mechanism of globin gene expression during development is to isolate the globin genes of goat using recombinant DNA technology and to study their sequence organization. From such investigations it is hoped that clues concerning regulatory mechanisms can be obtained. With this information in hand, we plan to develop cell-free systems which mimic in vivo regulation.

ISOLATION OF THE GOAT GLOBIN GENES

As an initial approach for isolating the globin genes of the goat, a library [15] of recombinants containing nearly random regions of the goat genome was constructed [16]. A partial Eco R1 digest of the total goat genome was prepared and cloned, using the vector λCharon 4A. Charon 4A contains two internal Eco R1 fragments which are not required for lytic infection and these can be replaced by

	HUMAN		GOATS	
Embryo:	Hb Gower I	$\zeta_2\epsilon_2$	Hb E	$\alpha_2\epsilon_2^A$
	Hb Gower 2	$\alpha_2\epsilon_2$		$\alpha_2\epsilon_2^B$
	Hb Portland	$\zeta_2\gamma_2$		
Fetus:	Hb F	$\alpha_2\gamma_2$	Hb F	$\alpha_2\gamma_2$
Preadult:			Hb C	$\alpha_2\beta_2^C$
Adult:	Hb A	$\alpha_2\beta_2$	Hb A	$\alpha_2\beta_2^A$
	Hb A_2	$\alpha_2\delta_2$		

Fig. 1. Summary of hemoglobin switches in man and goat.

DNA fragments of 12–22 kilobases (kb) [17]. The goat DNA recombinants were incorporated into infectious particles using in vitro packaging extracts [18] and the resulting phage amplified. Phage which contained globin gene inserts were identified by screening the collection of recombinants with cDNA prepared to a mixture of α, β^A, and β^C globin mRNA. Approximately 15 unique recombinants were isolated and these contained either different globin loci or represented overlapping regions of the same gene. Restriction endonuclease maps were determined for each of the clones and the type of globin gene contained in each recombinant was determined by sequencing a portion of the globin coding region [19].

CHARACTERIZATION OF CLONES CONTAINING THE GOAT ϵ, γ, β^A, AND β^C GLOBIN GENES

The primary structure of the goat, α, γ, β^A, and β^C globins are known [20] and a comparison of the amino acid sequences coded for by each of the recombinants to the published sequences allowed us to identify clones containing all the known β-like globin genes. In addition to the known genes, four additional globin coding sequences were observed. Two of these were tentatively identified as goat embryonic genes, while the other two appear to be defective globin genes.

The restriction maps of four overlapping recombinants containing the β^A globin gene are shown in Figure 2. The entire β^A globin gene is present and is located at the 3′ end of the recombinants. This gene is very similar in structural organization to globin genes of other species and is interrupted by two intervening sequences, one of approximately 125 nucleotides and the other of approximately 900 nucleotides. A total length of approximately 18.4 kb is represented by the overlapping clones and an additional coding sequence termed β^Z appears about 12 kb upstream from the β^A gene.

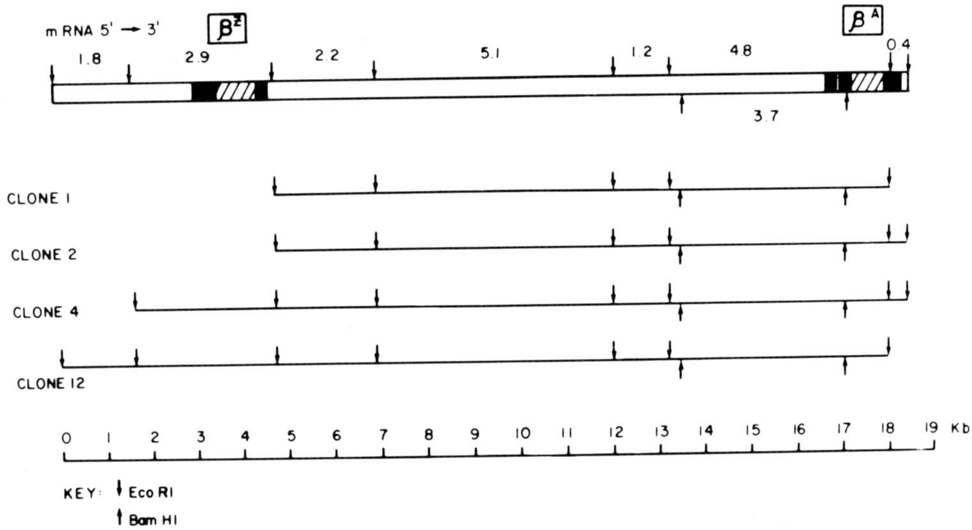

Fig. 2. Overlapping clones containing the β^{Z} and β^{A} globin genes and their restriction endonuclease maps. Darker areas represent sequences found in mature mRNA and hatched regions indicate intervening sequences.

The β^{C} globin gene is present in two overlapping recombinants and a third clone overlaps one of these (Fig. 3). Together these clones cover a total of 35 kb. Two β-like globin genes, β^{Y} and β^{X}, are located upstream from the β^{C} gene. Again, the structural organization of the β^{C} gene is very similar to that of other characterized globin genes.

Two overlapping clones were obtained which contain the γ globin gene (Fig. 4) and together they cover a total length of 24 kb. Another recombinant (Fig. 5) was obtained and found to contain all but the 3′ end of a globin gene. The nucleotide sequence of a portion of the coding region was determined and the amino acids coded for by this sequence were compared to those of the known goat globins. The sequence did not correspond to any of the characterized globin genes. However, a striking homology, shown in Table I (80% in the region sequenced) between the amino acid sequence coded for by this clone and that of human ϵ globin was observed and for this reason we have tentatively identified it as an embryonic globin gene.

A region 5′ to the β^{C} globin gene was found to hybridize to globin cDNA and this region is designated β^{Y} (Fig. 3). The entire β^{Y} globin gene was sequenced and the amino acids coded for by these nucleotide sequences were compared to those of the known goat globins. Again, the primary sequence of the globin did not correspond to any of the known goat globins. However, similar to the ϵ globin gene, a marked homology (72%) between the human ϵ globin and the sequences coded for by the β^{Y} gene was observed (Table I). Thus, it is likely that β^{Y} represents a second embryonic gene. It should be pointed out that the structural features of this gene, including size and position of intervening sequences and the presence of a Hogness-Goldberg box preceding the transcription initiation site (Fig. 6), are similar to the known globin genes. Of course at the present time it is unknown if either presumptive embryonic globin gene is expressed in goats, but it is known that two embryonic β-like globin genes do occur in this animal [Kitchen, personal communication].

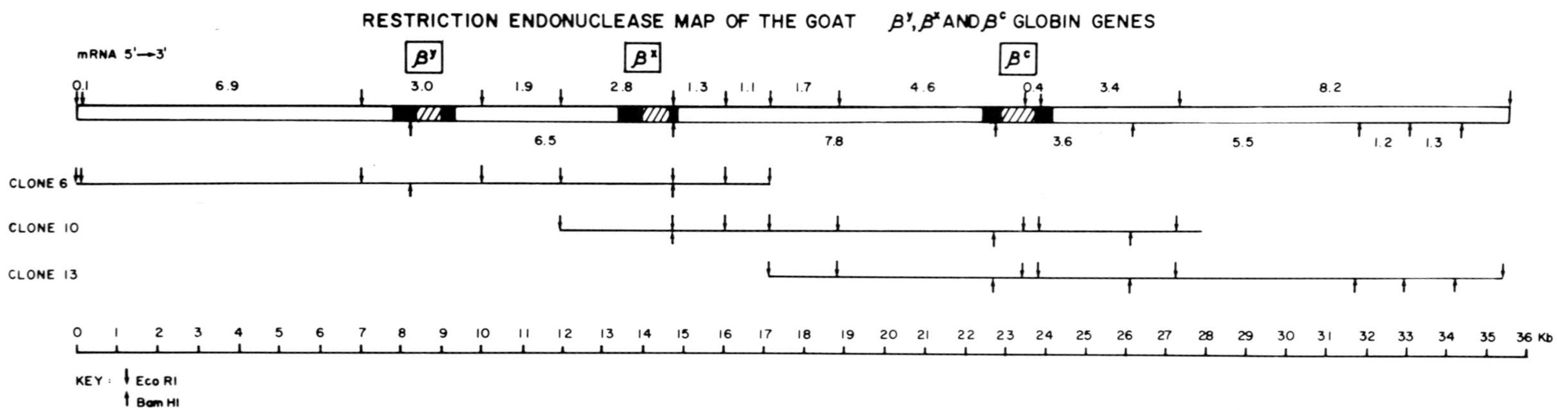

Fig. 3. Overlapping clones containing the β^{Y}, β^{X}, and β^{C} globin genes and their restriction endonuclease maps. Darker areas represent sequences found in mature mRNA and hatched regions indicate intervening sequences.

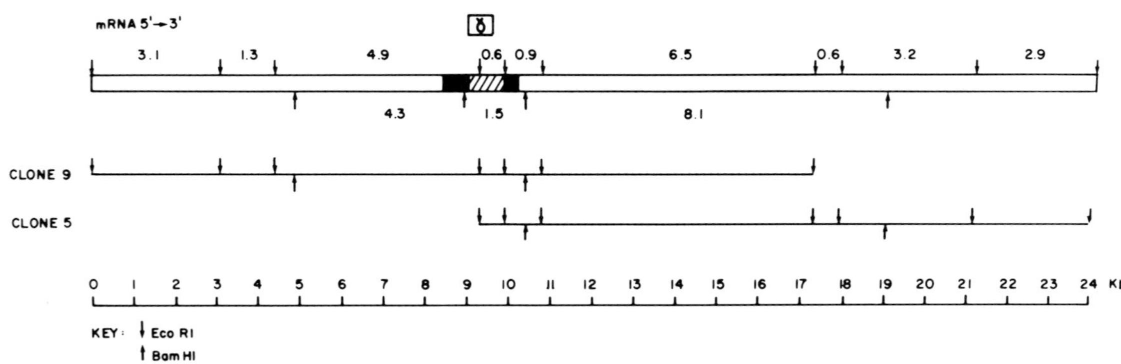

Fig. 4. Overlapping clones containing the γ globin genes and their restriction endonuclease maps. Darker areas represent sequences found in mature mRNA and hatched region indicates intervening sequence.

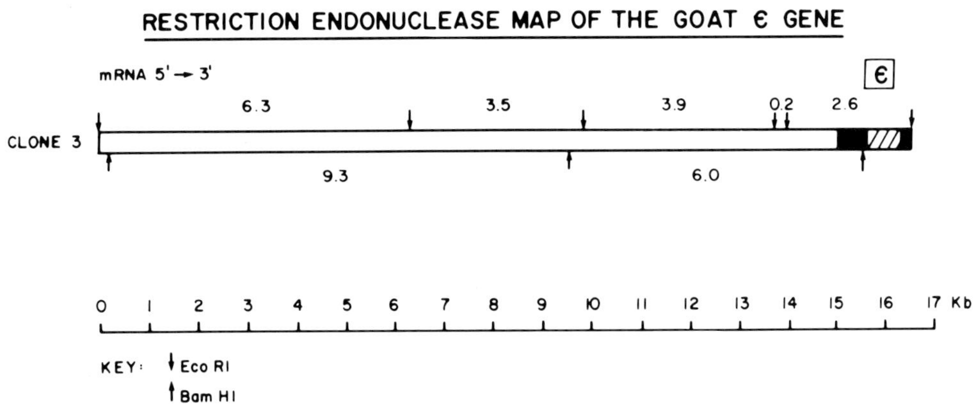

Fig. 5. Restriction endonuclease map of a clone containing the putative ε globin gene. Darker areas represent sequences found in mature mRNA and hatched region indicates intervening sequence.

PRESENCE OF DEFECTIVE GLOBIN GENES

In addition to the five genes of the β globin family which are known or thought to occur in goats (ϵ, β^{Y}, γ, β^{C}, and β^{A}), two other coding sequences were observed. Extensive nucleotide sequencing of these genes was performed and based on this information, it is concluded that neither codes for a functional globin [19; Cleary, Haynes, and Lingrel, unpublished; Schon, Cleary, and Lingrel, unpublished].

One of these genes, designated β^{X}, is located approximately 7 kb upstream from the β^{C} gene (Fig. 2). The complete nucleotide sequence of the gene, excluding the second intervening sequence, has been determined. The insertion of a T residue in the first coding block has been observed which would result in a frame shift in translation. Thus a functional globin could not be synthesized by this gene. If one does not consider the frame shift mutation then the two histidine residues which are required for binding the heme group are not coded for by the β^{X} gene. Rather, arginine and aspartic acid occur at these positions. Because the histidines are conserved in all globins studied to date, it is unlikely that such a globin could function in oxygen transport. In addition to the histidines, two other highly conserved amino acids, glycine at position 63 and phenylalanine at position 102, are replaced by aspartic acid and isoleucine, respectively. Besides these changes, the expected termination codon is altered so that a polypeptide seven amino acids longer than the normal globin would be synthesized. Whether this alteration in structure would be sufficient to affect the function of a globin, if it were made, is unknown.

TABLE I. Amino Acid Homologies (%)

	Goat ϵ	Goat β^{Y}
Human ϵ	80	72
Goat β^{A}	69	68
Goat β^{C}	76	67
Goat γ	71	69
Goat β^{X}	53	55
Goat β^{Z}	—	28

	-40 Hogness-Goldberg Box -1+1
β^A	GGCTGGGCATAAAAGGAAGAGCCGGGCCAGCTGCTGCTTACA
β^C	GGCTGGGCATAAAAGGAAGAGCCGGGCCAGCTGCTGCTTACA
γ	GGCTAAGCATAAAAGGAAGAGCCGGGCCAGCTGCTGCTTACA
β^Y	GATGAGGAATAAAAGGCCATGCAGTGAAGCAGCGGCACAGAC
β^X	CGTGGGGCTCACAAGGAAGACAAGGGCCCCCTGCTGCTTACA

Fig. 6. Nucleotide sequence of the 5′ termini of the genes of the β globin family. The nucleotide designated 1 is the suspected site of transcription initiation.

Based on other information, it is considered unlikely that this gene is even transcribed. The nucleotide sequence at the 5′ end of the β^X gene is shown in Figure 6 along with a similar region of the β^A, β^C, β^Y, and γ globin genes. The Hogness-Goldberg box [21], a nucleotide sequence found at the 5′ end of other eukaryotic genes coding for mRNA and thought to be involved in the initiation of transcription [22–33], is present in the β^A, β^C, β^Y, and γ globin genes but is markedly altered in the β^X gene. Both an A and T are replaced by Cs, eliminating the AT-rich character of the sequence. If this region is required for transcription then these nucleotide differences may affect the expression of this gene. While it has not yet been determined if an RNA is synthesized from this gene in vivo, it is not transcribed in a cell-free RNA polymerase II [34]. The goat γ, β^C, β^A, and β^Y globin genes are transcribed by such a system.

The β^Z gene is located 12 kb upstream from the β^A gene (Fig. 2). This gene has been found to contain several stop codons within the coding region. These occur at amino acid positions 6, 36, and 38. Thus, no functional globin could be synthesized from this gene as well.

It is interesting to note that of the seven non-α globin genes identified in the goat, two are defective. Defective or pseudoglobin genes have also been identified in humans [35], mice [36, 37], and rabbits [33]. If the number of normal and defective β-globin genes of the goat is extended to other gene families, then it is possible that approximately one-third of the mammalian genome is composed of defective or nonfunctional DNA. It should, of course, be pointed out that while these genes are defective in globin synthesis, they could serve some other function, possibly a regulatory role. Even if these genes are not expressed they provide a pool of sequences for continued evolution.

SEARCH FOR REGULATORY SEQUENCES

Based on the regulatory mechanisms operating in bacteria, one would expect that nucleotide sequences involved in controlling globin gene expression would be located near the transcription initiation site. Thus, if differences in sequences are found in this region in two or more developmentally regulated genes, then such sequences might be responsible for the differential expression of these genes. For this reason, the region surrounding the transcription initiation site of the β^A and β^C

globin genes was examined. The nucleotide sequences are shown in Figure 7. A portion of the coding sequence is also shown and differences are observed in this region consistent with the 14-amino-acid difference between the β^A and β^C genes. However, the nucleotide sequences proximal to the coding region are identical over a large area. The location of the transcription initiation site is based on the finding by Weaver and Weissman [38] that the 5′ end of the 15S globin mRNA precursor maps at the A designated by nucleotide #1. Mature β globin mRNA contains the same 5′ sequence, and because the A is capped, this site is generally referred to as the cap site. Identification of the transcription initiation site, of course, is based on the conclusion that no longer precursor to globin mRNA exists. However, as no RNA larger than 15S has been found to contain globin mRNA sequences in erythroid cells [39], it is likely that the 15S RNA is the initial transcription product. The presence of the Hogness-Goldberg box 24 nucleotides proximal to the cap site of the β^A and β^C globin genes also supports the assignment of the transcription initiation site as this sequence is located at the same relative position in front of the known initiation site of adenovirus late mRNA [22, 23]. In addition, Luse and Roeder [40] have found that transcription initiation occurs at the cap site when the mouse β^{maj} globin gene is transcribed by an in vitro RNA polymerase II system. Because of the close homology between the mouse β^{maj} globin gene and those of the goat, we assume that this same site is the transcription initiation point of the goat globin genes as well. Based on this identification, the nucleotide sequences of the β^A and β^C globin genes are identical for 80 nucleotides in front of the transcription initiation point. Differences do exist at nucleotide positions -80 and -82 and it is possible that these differences are involved in the differential expression of these genes. Nevertheless, it is clear that the regulatory signals are not present in the first 80 nucleotides preceding the initiation of transcription.

It should be noted that the mouse β^{maj} and β^{min} globin genes, two genes which are expressed at the same developmental stage, do not have identical sequences surrounding their 5′ ends. Instead, significant differences exist, including insertions and deletions as well as single nucleotide changes [24].

It is of course possible that sequences required for transcription are present at the 5′ end even though regulatory sequences may not be. Along this line it is interesting to note that the Hogness-Goldberg box which precedes the transcription initiation site in many eukaryotic genes is present in the goat globin genes. This conservation suggests it may perform a specific function. In initial studies of the in vitro transcription of the histone H2A gene, a gene transcribed by polymerase II, deletion of the Hogness-Goldberg box reduced but did not abolish transcription [44]. As new 5′ termini were generated by this deletion, the Hogness-Goldberg box may be involved in specifying the transcription initiation site. Furthermore, deletion of the cap site along with a few nucleotides in front of it and most of the 5′ leader sequences also reduced but did not eliminate transcription. Thus, it is likely that several regions may be required for accurate transcription.

It should be noted that a sequence of about 30 nucleotides within the Xenopus 5S ribosomal gene is sufficient for its transcription [41, 42] and that sequences in both the 5′ and 3′ halves of a Xenopus tRNA gene are required for transcription initiation [43]. These genes are transcribed by RNA polymerase III; polymerase II genes, however, may not be transcribed in a similar manner.

While these observations may identify sequences required for transcription, they may not define sequences involved in the regulation of gene expression. In this

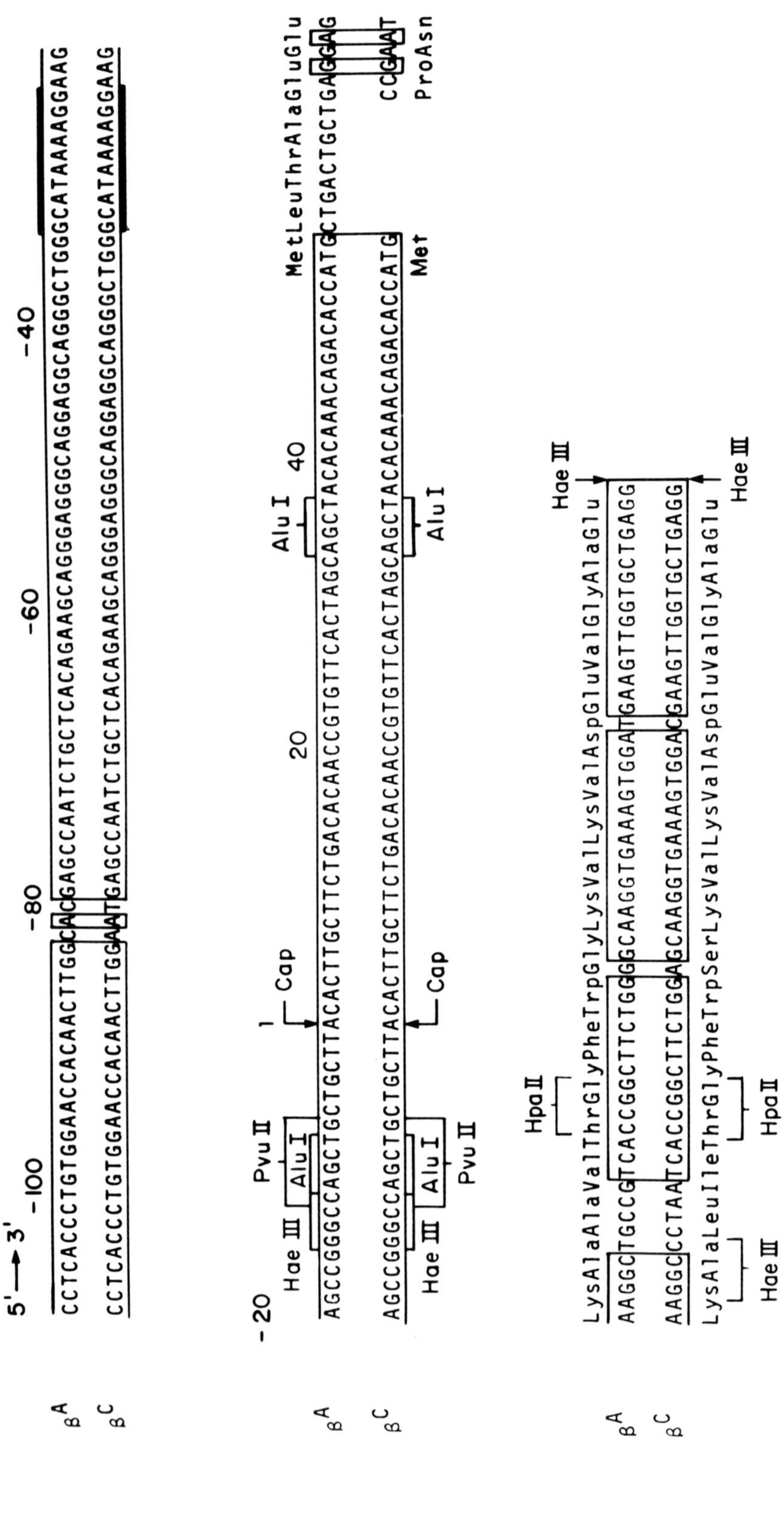

Fig. 7. Nucleotide sequence of the 5′ region of the goat β^{A} and β^{C} globins.

respect, it is interesting to note that a cis-acting element, located 3′ to the human γ globin gene, has been implicated in its regulation [45]. Two deletion mutants, which lack sequences downstream from the γ genes, have been studied. One is hereditary persistence of fetal hemoglobin, where γ chain synthesis continues during adult life, and the other $\delta\beta$-thalassemia, where it does not. As the deletion in hereditary persistence of fetal hemoglobin extends further upstream toward the γ globin genes, it is postulated that this sequence is involved in switching the γ genes off during development. Thus, those sequences which are involved in the differential expression of genes may not necessarily be located near the transcription initiation site.

LINKAGE OF THE NON-α GLOBIN GENES

As pointed out earlier, linkage of some of the non-α globin genes of the goat have been established. These are summarized in Figure 8. The β^A and β^Z genes are linked with the β^Z gene located 11 kb in the 5′ direction of the β^A gene. The β^Y, β^X, and β^C genes are also linked. The β^X gene is located 7 kb upstream from the β^C gene and the β^Y globin gene is located 4 kb further in the 5′ direction. Unfortunately, no other linkages have been found. In humans [35, 45–49] and rabbits [33] the non-α globin genes are arranged in the order of their expression during development. As this may also be true for the globin genes of goat, they have been arranged in the order shown in Figure 7. The gene arrangement of the non-α globin genes of humans and rabbits is shown for comparison. While the rabbit and human genes are contained in a region of 44 kb and 65 kb respectively, the recombinants containing globin genes of the goat already cover a distance of 96 kb. Because clones linking all the non-α globin genes have not yet been found, this region must be even larger. The reason for this larger size could be the greater number of non-α globin genes present in the goat, compared to humans and rabbits, or to increased distances between the genes.

SUMMARY

Recombinants containing the γ, β^C, and β^A globin genes of the goat have been isolated from a recombinant library prepared from genomic DNA. The recombinants were selected by screening phage with goat globin cDNA and DNA from a plasmid carrying the mouse β globin cDNA sequence. The type of globin gene present in each clone was established by comparing the amino acids coded for by the DNA sequence of the recombinants to those of published amino acid sequences. In addition to clones containing the γ, β^C, and β^A globin genes, several other β-like globin regions were identified which did not code for any of the known goat globins. As the amino acid sequence coded for by two of these genes showed close homology to human ϵ globin, it is concluded that they are embryonic globin genes. The other two genes, designated β^X and β^Z, appear defective based on one or more of the following: the 5′ Hogness-Goldberg box is altered, invariant amino acids are replaced by other amino acids, termination codons are altered or stop codons appear within the coding sequences. It is unknown at the present time whether these genes are transcribed in the goat, but if they are, their translation would result in proteins unable to transport oxygen. The finding of a number of defective globin genes may indicate that much of the DNA carried in eukaryotic cells is nonfunctional. It is, of course, possible that some other function besides the synthesis of globin is carried out by these genes.

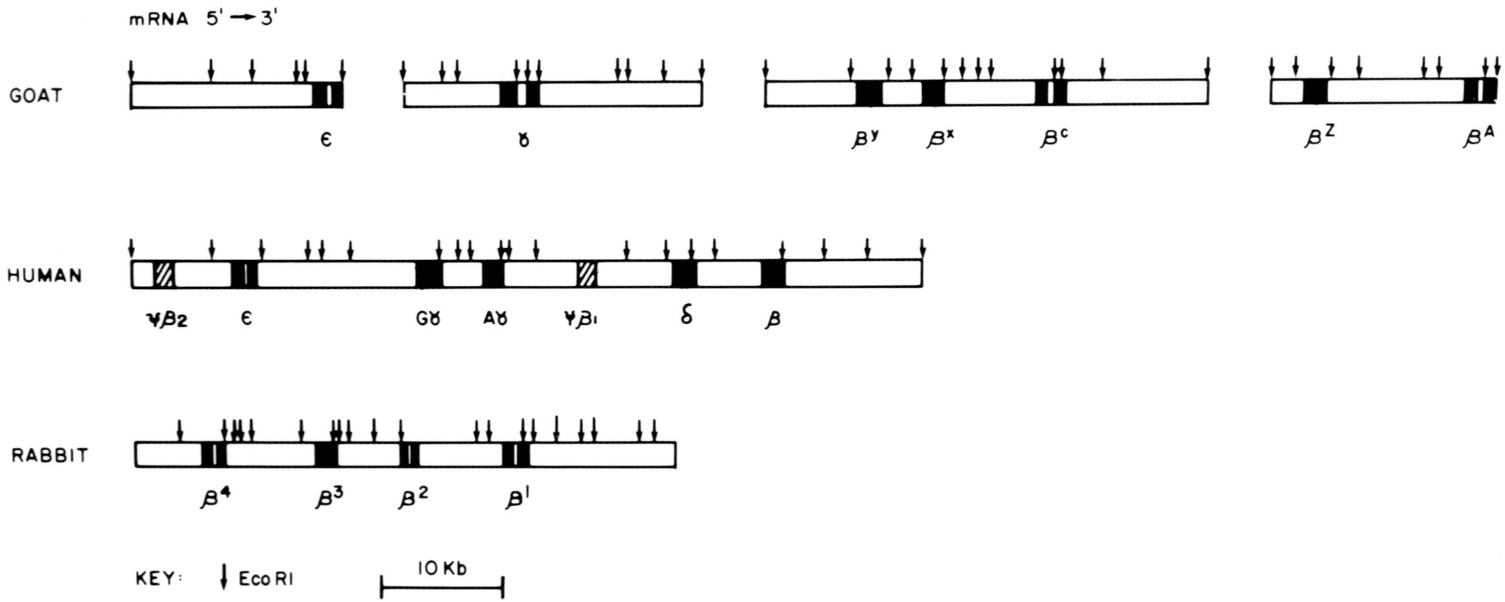

Fig. 8. Comparison of the linkage arrangements of the non-α globin genes of the goat, human, and rabbit. The human and rabbit globin gene arrangements are taken from references 35, 48, 49, and 33, respectively. Darker areas represent sequences found in mature mRNA and hatched region indicates intervening sequence.

Another interesting feature of the goat non-α globin genes is that they occupy a rather large region of the chromosome when compared to those of the rabbit and human. The maximum size of this locus has not yet been established because recombinants overlapping the various genes have not yet been identified.

In an attempt to identify sequences which are involved in the differential expression of these genes during development, we have sequenced the 5′ end of the β^C and β^A globin genes. The nucleotide sequence is identical for 131 nucleotides preceding the translation initiation site and this region includes the transcription initiation point and the Hogness-Goldberg box region. Because these genes are active at different times during development, it can be argued that sequences involved in their differential expression are not present in the immediate 5′ region.

ACKNOWLEDGMENTS

We wish to thank Donald VanLeeuwen and Gail Sears for their excellent technical assistance. This work was supported by grants AM 20119 and GM 10999.

REFERENCES

1. Kitchen H, Brett I: Embryonic and fetal hemoglobin in animals. Ann NY Acad Sci 241:653–671, 1974.
2. Kleihauer E, Stoffler G: Embryonic hemoglobins of different animal species. Mol Gen Genet 101: 59–69, 1968.
3. Huisman THJ, Adams HR, Dimmock MO, Edwards FB, Wilson JB: The structure of goat hemoglobins. I. Structure studies of the β chains of the hemoglobins of normal and anemic goats. J Biol Chem 242:2534–2541, 1967.
4. Garrick MD, Charlton JP: Inheritance of structural divergence for the alpha-chain locus. Biochem Genet 3:393–402, 1969.
5. Adams HR, Boyd EM, Wilson JB, Miller A, Huisman THJ: The structure of goat hemoglobins. III. Hemoglobin D, a β chain vairant with one apparent amino acid substitution (21 Asp → His). Arch Biochem Biophys 127:398–405, 1968.
6. Wrightstone RN, Wilson JB, Miller A, Huisman THJ: The structure of goat hemoglobins. IV. A third β chain variant (β^E) with three apparent amino acid substitutions. Arch Biochem Biophys 138:451–456, 1970.
7. Van Vliet G, Huisman THJ: Changes in the haemoglobin types of sheep as a response to anaemia. Biochem J 93:401–409, 1964.
8. Blunt MH: Changes in types of hemoglobin during experimental hemorrhagic anemias in sheep. Am J Physiol 209:986–990, 1965.
9. Huisman THJ, Lewis JP, Blunt MR, Adams HR, Miller A, Dozy AM, Boyd EM: Hemoglobin C in newborn sheep and goats: A possible explanation for its function and biosynthesis. Pediatr Res 3: 189–198, 1969.
10. Thurman RF, Boyer SH, Crosby EF, Shepard MR, Noyes AN, Stahlman F: Hemoglobin switching in nonanemic sheep. II. Evidence for presumptive identity between the A C factor and erythropoietin. Blood 36:598–606, 1970.
11. Gabuzda TG, Schuman MA, Silver RK, Lewis HB: Erythropoietic kinetics of sheep studied by means of induced changes in hemoglobin phenotype. J Clin Invest 47:1895–1904, 1968.
12. Nienhuis AW, Anderson WF: Hemoglobin switching in sheep and goats: Changes in functional globin messenger RNA in reticulocytes and bone marrow cells. Proc Natl Acad Sci USA 69: 2184–2188, 1972.
13. Wood WG, Pearce K, Clegg JB, Weatherall DJ, Robinson JS, Thorburn GD, Dawes GS: Switch from fetal to adult haemoglobin synthesis in normal and hypophysectomised sheep. Nature 264: 799–801, 1976.
14. Huisman THJ: The in vivo production of hemoglobin C in ruminants. Ann NY Acad Sci 241: 549–555, 1974.
15. Maniatis T, Hardison RC, Lacy E, Lauer J, O'Connell C, Quon D, Sim GK, Efstratiadis A: The isolation of structural genes from libraries of eucaryotic DNA. Cell 15:687–701, 1978.

16. Robbins J, Rosteck P, Haynes JR, Freyer G, Cleary ML, Kalter HD, Smith K, Lingrel JB: The isolation and partial characterization of recombinant DNA containing genomic globin sequences from the goat. J Biol Chem 254:6187–6195, 1979.
17. Blattner FR, Williams BG, Blechl AE, Thompson KD, Faber HR, Furlong LA, Grunwald DJ, Kiefer DO, Moore DD, Schumm JW, Sheldon EL, Smithies D: Charon phages: Safer derivatives of bacteriophage lambda for DNA cloning. Science 196:161–169, 1977.
18. Sternberg N, Tiemeier D, Enquist L: In vitro packaging of a D*am* vector containing Eco R1 DNA fragments of Escherichia coli and phage P1. Gene 1:255–280, 1977.
19. Haynes JR, Rosteck P, Schon EA, Gallagher PM, Burks DJ, Smith K, Lingrel JB: The isolation of the β^A, β^C and γ globin genes and a presumptive embryonic globin gene from a goat DNA recombinant library. J Biol Chem 255:6355–6367, 1980.
20. Dayhoff MO, Eck RV: "Atlas of Protein Sequence and Structure," Vol. 5, Third Supplement. Silver Spring, Maryland: The Biomedical Research Foundation, 1978, pp 154–158.
21. Goldberg M: PhD Thesis. Sequence analysis of Drosophila histone genes. Stanford University, 1979.
22. Ziff EB, Evans RM: Coincidence of the promoter and capped 5′ terminus of RNA from the adenovirus 2 major late transcription unit. Cell 15:1463–1475, 1978.
23. Manly JL, Sharp PA, Gefter ML: RNA synthesis in isolated nuclei: In vitro initiation of adenovirus 2 major late mRNA precursor. Proc Natl Acad Sci USA 76:160–164, 1979.
24. Konkel DA, Maizel JV, Leder P: The evolution and sequence comparison of two recently diverged mouse chromosomal β-globin genes. Cell 18:865–873, 1979.
25. Gannon R, O'Hare K, Perrin F, LePennec JP, Bennoist C, Cochet M, Breathnach R, Royal A, Garapin A, Cami B, Chambon P: Organization and sequences at the 5′ end of a cloned complete ovalbumin gene. Nature 278:428–434, 1979.
26. Lomedico P, Rosenthal N, Efstratiadis A, Gilbert W, Kolodner R, Tizard R: The structure and evolution of the two nonallelic rat preproinsulin genes. Cell 18:545–558, 1979.
27. Tsujimoto Y, Suzuki Y: Structural analysis of the fibroin gene at the 5′ end and its surrounding regions. Cell 16:425–436, 1979.
28. Nishioka Y, Leder P: The complete sequence of a chromosomal mouse α-globin gene reveals elements conserved throughout vertebrate evolution. Cell 18:875–882, 1979.
29. Lai EC, Stein JP, Catterall JF, Woo SLC, Mace ML, Means AR, O'Malley BW: Molecular structure and flanking nucleotide sequences of the natural chicken ovomucoid gene. Cell 18: 829–842, 1979.
30. Bernard O, Hozuni N, Tonegawa S: Sequences of mouse immunoglobulin light chain genes before and after somatic changes. Cell 15:1133–1144, 1978.
31. Sure I, Lowry J, Kedes LH: The DNA sequence of sea urchin (S purpuratus) H2A, H2B and H3 histone coding regions. Cell 15:1033–1044, 1978.
32. Schaffner W, Kanz G, Daetwyler H, Telford J, Smith HO, Birnstiel ML: Genes and spaces of cloned sea urchin histone DNA analyzed by sequencing. Cell 14:655–671, 1978.
33. Hardison RC, Butler ET, Lacy E, Maniatis T, Rosenthal N, Efstratiadis A: The structure and transcription of four linked rabbit β-like globin genes. Cell 18:1285–1297, 1979.
34. Luse DS et al. This volume.
35. Fritsch EF, Lawn RM, Maniatis T: Molecular cloning and characterization of the human β-like globin gene cluster. Cell 19:959–972, 1980.
36. Nishioka Y, Leder A, Leder P: An unusual alpha globin-like gene that has cleanly lost both globin intervening sequences. Proc Natl Acad Sci USA 77:2806–2809, 1980.
37. Hutchinson C et al. This volume.
38. Weaver RF, Weissmann C: Mapping of RNA by a modification of the Berk Sharp procedure: The 5′ termini of 15S β-globin mRNA precursor and mature 10S β-globin mRNA have identical map coordinates. Nucleic Acids Res 7:1175–1193, 1979.
39. Haynes JR, Kalb VF, Rosteck P, Lingrel JB: The absence of a precursor larger than 16S to globin mRNA. FEBS Lett 91:173–177, 1978.
40. Luse DS, Roeder RC: Accurate transcription initiation on a purified mouse β-globin DNA fragment in a cell-free system. Cell 20:691–699, 1980.
41. Sakonju S, Bogenhagen DF, Brown DD: A control region in the center of the 5S RNA gene directs specific initiation of transcription: I. The 5′ border of the region. Cell 19:13–25, 1980.
42. Bogenhagen DF, Sakonju S, Brown DD: A control region in the center of the 5S RNA gene directs specific initiation of transcription: II. The 3′ border of the region. Cell 19:27–35, 1980.

43. Kressmann A, Hofstetter H, Di Capua E, Grosschedl R, Birnstiel ML: A tRNA gene of Xenopus laevis contains at least two sites promoting transcription. Nucleic Acids Res 7:1749–1763, 1979.
44. Grosschedl R, Birnstiel ML: Identification of regulatory sequences in the prelude sequences of an H2A histone gene by the study of specific deletion mutants in vivo. Proc Natl Acad Sci USA 77:1432–1436, 1980.
45. Fritsch EF, Lawn RM, Maniatis T: Characterization of deletions which affect the expression of fetal globin genes in man. Nature 279:598–603, 1979.
46. Lawn RM, Fritsch EF, Parker RC, Blake G, Maniatis T: The isolation and characterization of linked δ- and β-globin genes from a cloned library of human DNA. Cell 15:1157–1174, 1980.
47. Little PFR, Flavell RA, Kooter JM, Annison G, Williamson R: Structure of the human fetal globin gene locus. Nature 278:227–231, 1979.
48. Bernards R, Little PFR, Annison G, Williamson R, Flavell RA: Structure of the human Gγ-Aγ-δ-β-globin gene locus. Proc Natl Acad Sci 76:4827–4831, 1979.
49. Tuan D, Biro PA, De Riel JK, Lazarus H, Forget BG: Restriction endonuclease mapping of the human γ globin gene loci. Nucleic Acids Res 6:2519–2544, 1979.

Organization and Expression of Globin Genes, pages 59–68

The Structure and Organization of the Sheep Globin Genes

Peter J. Kretschmer, Helen C. Coon, Austine Davis, Maria Harrison, and Arthur W. Nienhuis

Laboratory of Molecular Hematology (P.J.K., H.C.C.) and Clinical Hematology Branch (A.D., M.H., A.W.N.), National Heart, Lung, and Blood Institute, Bethesda, Maryland

INTRODUCTION

The sheep offers several advantages as a model for the analysis of the mechanism and regulation of hemoglobin switching [1, 2]. The relatively large size and long gestational period of this species provides an opportunity to employ in vivo manipulations in an effort to influence the switching phenomena and also an opportunity to obtain substantial amounts of erythroid tissue for molecular analysis. Furthermore, this species exhibits an embryonic to fetal hemoglobin switch early in gestation, and then later, a fetal to adult hemoglobin switch during the perinatal period, which are quite analogous to those seen in man. A further switch in hemoglobin phenotype is observed in response to erythropoietic stress in those animals which have an allele for the adult β^A globin; Hb A ($\alpha_2\beta_2^A$) is replaced with Hb C ($\alpha_2\beta_2^C$) transiently until the erythropoietic stress is removed [3, 4]. This Hb A to Hb C switch, because it can be manipulated by defined stimuli, has provided a model to relate the commitment of erythroid stem cell to the ultimate hemoglobin phenotype of their progeny red cells. [3–5].

Selectively produced in the sheep during various developmental periods or during stress are the embryonic, fetal (γ), adult (β^A), or stress (β^C) globins [1, 2]. Thus the sheep genome must include at least four β-like globin genes, those for the γ, β^A, and β^C globins as well as at least one for an ϵ-like globin. In contrast to the situation in humans where a general concept of β-like globin gene cluster, derived from genetic analysis, proved highly useful in developing a map of this region using the Southern blotting and molecular cloning techniques [6–9], nothing is known about the number, linkage distances, or order of the sheep globin genes on the chromosome. Nonetheless, this information seemed essential for the further development of this model for the study of hemoglobin switching. Therefore, we decided to use the recombinant DNA molecular cloning technology to isolate and characterize the β-like globin genes of sheep, thereby hoping to define their general organization and also to obtain DNA fragments which might be useful for in vitro analysis of transcription and chromatin structure. This report contains a description of the progress we have achieved to date along with the conclusions which may already be formulated about the β-like globin genes of sheep.

GENERAL EXPERIMENTAL STRATEGY

The sheep genome was cloned as 15–20 kilobase (kb) Eco RI fragment, obtained by partial Eco RI digestion of fetal liver DNA, into Escherichia coli using the bacteriophage vector, Charon 4A [10, 11]. Clones containing β-like globin genes were identified in a plaque hybridization assay in which either globin cDNA or appropriate restriction endonuclease fragments derived from recombinant plasmids containing the cloned γ, β^B, or β^C globin genes [12] were used as probes. The clones were characterized by digestion of isolated DNA with restriction endonucleases, followed by agarose gel electrophoresis and then Southern blot hybridization [13] using globin coding sequence-specific probes. By this means a general map of restriction endonuclease sites was derived and the location and orientation of the globin gene sequences established. Fragments of further interest were subcloned into E coli using the plasmid vector, pBR 322 [10, 11]. Further restriction endonuclease mapping provided fragments to be used for DNA sequence analysis using the chemical degradation method of Maxim and Gilbert [14, 15] as modified by Konkel et al [16, 17]. Analysis of heteroduplexes formed between DNA isolated from two recombinant bacteriophages was by electron microscopy [11].

CHARACTERIZATION AND COMPARISON OF THE CHROMOSOMAL REGIONS CONTAINING THE SHEEP γ AND β^A GLOBIN GENES

Isolation and Mapping of Recombinant Bacteriophage Containing the γ and β^A Genes

Among the several recombinants containing globin gene sequences was one, designated $\lambda S\beta^A G21$ [10], containing an inserted DNA fragment of 18.8 kb in length. The included globin gene was first identified as encoding for β^A globin by analysis of the thermal stability of duplexes formed between it and various globin cDNAs [10], but this identification has been subsequently verified by DNA sequence analysis [11]. Another recombinant containing 12.4 kb of the sheep genome is referred to as $\lambda S\gamma G31$ as it has been shown to contain a γ globin gene by DNA sequence analysis [11].

Analysis of Heteroduplexes Formed Between DNA Isolated From $\lambda S\gamma G31$ and $\lambda S\beta^A G21$

The orientation and relative position of the globin genes with respect to the right arm of the vector, Charon 4A, are quite similar for these two recombinants. This proved advantageous, for these clones are, therefore, highly suitable for heteroduplex formation and analysis. A representative heteroduplex selected from among the 15 examined in detail is shown in Figure 1. Of note is that the coding sequences of these genes have been shown to be highly homologous in hybridization reactions and, furthermore, the large introns annealed to one another in

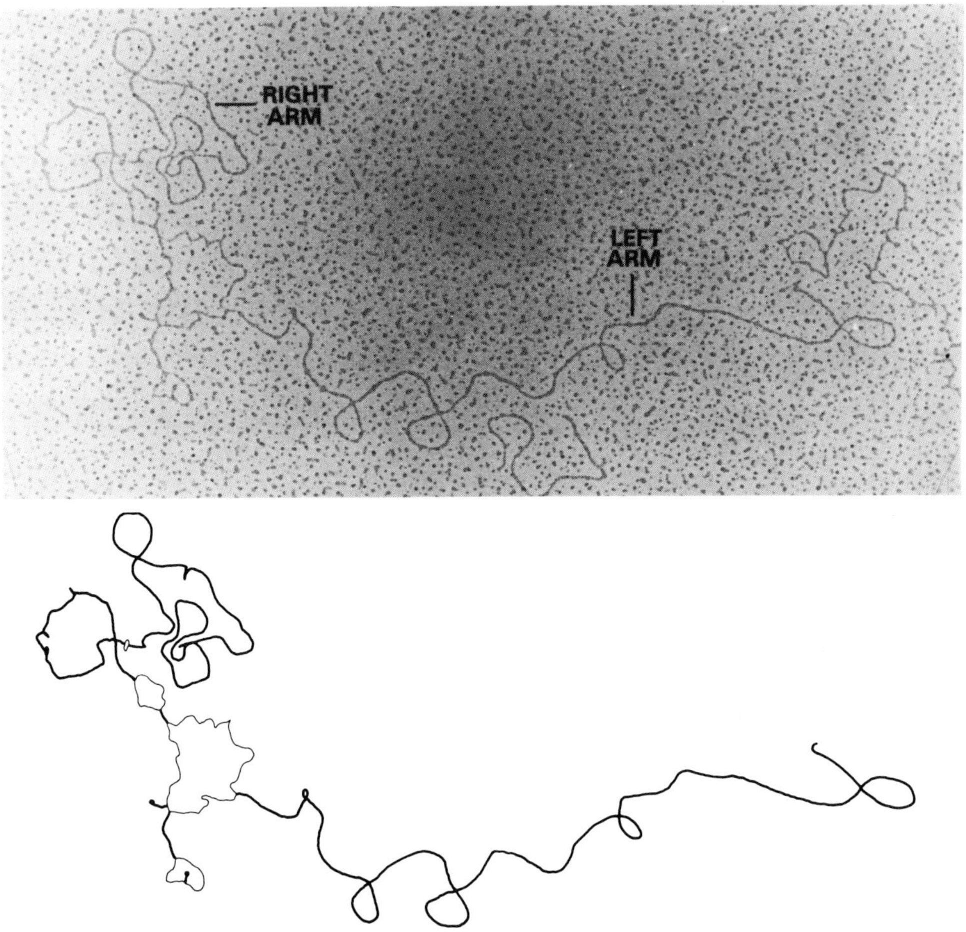

Fig. 1. A heteroduplex formed between λSβ^AG21 and λSγG31. The position of the right and left arms of the vector is indicated in the actual electron micrograph above. Below is shown a sketch of this molecule in which the double-stranded regions of homology are more darkly outlined compared to the single-stranded regions of nonhomology.

Southern blotting experiments (data not shown). Therefore, we expected as a minimum to observe heteroduplex formation between the globin gene sequences in these two recombinants. However, heteroduplex formation was far more extensive, beginning at the point of the inserted DNA fragments adjacent to the right arm of the vector and extending 8 kb, including the entire globin genes and approximately 1 kb beyond in the 3′ direction (Figs. 1 and 2). A number of small irregularities were noted within this region in a majority of the heteroduplexes examined; these were located by careful measurements and their positions are shown in Figure 2. Three stem and loop structures, one large and the other two small, were noted 3′ to the globin genes, indicating the presence of inverted repeated DNA sequences. All of these were shown to be in λSβ^AG21 by study of heteroduplexes formed between its DNA and that of the vector, Charon 4A.

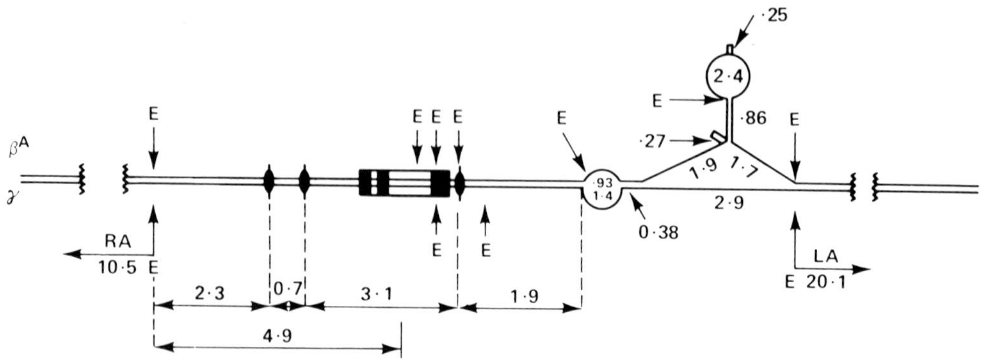

Fig. 2. A summary of heteroduplexes formed between λSβ^A^G21 (above) and λSγG31 (below). The letter E denotes the position of Eco RI restriction endonuclease sites in the respective recombinant bacteriophages. The position of the globin genes is shown by the box; the dark areas indicate the exons whereas the open areas indicate the introns. RA and LA are the right and left arms of Charon 4A, respectively. The dark oval symbols indicate three areas of dishomology within the 8 kilobase segments of homologous DNA.

The duplication which produced the γ and β^A genes of sheep is likely to have occurred some 50 million years ago based on analogy with other species and the 26 amino acid sequence differences exhibited by these two globins. In contrast, the structurally and functionally more closely related mouse β^{maj} and β^{min} genes are likely to have arisen by duplication some 10 million years ago as only six differences in amino acid sequence are found between these two globins. Nonetheless, sequence divergence in the flanking and intron DNA of the two mouse globin genes is so extensive that heteroduplex formation does not occur under conditions quite analogous to those used in our studies [18]. Leder and colleagues [18, 19] have argued that dishomology in flanking and intron sequences protects otherwise closely homologous genes from deletion due to unequal cross-over. The closely linked human γ, δ, and β genes show a similar high level of divergence within the flanking DNA as revealed by heteroduplex analysis [20]. Perhaps because the sheep globin genes are more widely spaced on the chromosome (see below), this homology in surrounding DNA sequence does not promote rearrangements. Extensive homology is certainly compatible with the highly selective and noncoordinate expression exhibited by the sheep γ and β^A globin genes during development. We speculate that the small areas of nonhomology revealed by careful study of heteroduplexes formed between recombinants containing these two genes (Figs. 1 and 2) might be useful in attempting to localize DNA sequences involved in their regulated expression.

DNA Sequence Analysis of the γ and β^A Globin Genes

Sequencing of these genes was undertaken to unequivocally establish their identity and to attempt to define transcription signals comparable to those found with other globin genes. Eighty percent of the coding portions of these two genes has been sequenced. Indeed, the nucleotide sequences predict the previously reported amino acid sequence of the two globins exactly with the exception of one residue in the γ chain (number 119 is aspartic acid rather than glycine. Initiation and termination codons are present in the proper reading frame and in the expected location. The location of the two introns, predicted by restriction endonuclease mapping, was defined precisely as corresponding to the position of amino acids 29-30 and 103-104. The putative polyadenylation signal, AATAAA [21], was found 100 nucleotides beyond the terminator codon for both genes. Shown in Figure 3 is the nucleotide sequence of the 5′ exon and flanking portion of these two genes. Fifty base pairs (bp) upstream from the initiation codon is the cap site (ACA) which marks the beginning of a 10 bp box of capping homology noted in this position relative to other globin genes [16, 17]. Some 30 bp further upstream from the cap site is the Hogness-Goldberg box which includes the sequence ATAAAA; RNA polymerase binding is thought to occur at this site [16, 17, 22]. Finally, a recently defined area of sequence homology noted flanking several globin genes was found some 45 bp further upstream from both the sheep β^A and γ genes (Maniatis et al, this volume).

Thus the DNA sequence data confirm the identity of these two genes as being those for the β^A and γ globins and provide strong support that these genes are functional since the expected transcriptional signals were identified. DNA sequences immediately 5′ to the genes were highly homologous as only seven nucleotide differences were identified in the 150 bp upstream from the initiation codon. This observation implies that sequences important for differential expression of these genes during development are not found in this region, although one, of course, does not know whether one or two nucleotide differences would be sufficient to alter function. Furthermore, extensive sequence differences are found just downstream from the initiator codon in the coding region; these conceivably could be important for functional interaction with proteins which modulate transcription.

THE SHEEP β-LIKE GENES ARE WIDELY SEPARATED ON THE CHROMOSOME

Two sets of observations indicate that the sheep globin genes are not as closely linked to one another as are the human and rabbit β-like genes. First, we have identified ten recombinants containing part or all of a different globin gene. The inserted DNA fragment in these ranges from 11–19 kb; only in one case were two separate globin genes included in a single recombinant. Second, we have obtained four recombinants containing all or part of the γ globin gene found in $\lambda S\gamma G31$ (Fig. 4). Together they cover a 28 kb region of the sheep genome with the γ globin gene in the approximate middle. No linked globin genes were found. Thus, we can be fairly confident that the γ gene is at least 12–14 kb away from another globin gene in either direction. This contrasts with the human genome in which the individual genes are 3.5–14 kb apart [9], and the rabbit genome in which the average distance between β-like genes is approximately 5 kb [23]. Considering the many β-like genes already found in the sheep genome (see below), only a few of which have been matched to a globin product, a large and complex organization of the sheep β-like globin gene cluster seems quite likely.

IDENTIFICATION AND PARTIAL CHARACTERIZATION OF OTHER β-LIKE GLOBIN GENES ISOLATED FROM THE SHEEP GENOME

In the course of screening our sheep genomic DNA fragment library, we have identified five recombinants which contain β-like globin genes, but the restriction endonuclease maps of these suggest that they do not encode for the γ, β^A, or β^C

β^A GTATGGCTATCACCATTCAAGCCTCACCCTGTGGAACCACAACTTGGCACGAGCCAATCTGCTCACAGAAGCAGGGAGGGCAGGAGGCAGG
γ GTATGACTGTCATCATTCAAGCCTCACCCTGTGGAACCACACCTTGGCCTGAGCCAATCTGCTCACAGAAGCAGGGAGGGCAGGAGGCAGG

β^A GCTGGGCATAAAAGGAAGAGCCGGGCCAGCTGCTGCCTACACTTGCTTCTGACACAACTGTGTTCACTAGCAGCTACACAAACAGACACC
γ GCTAAGCATAAAAGGAAGAGCCGGGCCAGCTGCTGCTTACACTTGCTTCTGACACAACTGTGTTCACTAGCAGCTACACAAACAGACACC

10 20 29
Met Leu Thr Ala Glu Glu Lys Ala Ala Val Thr Gly Phe Trp Gly Lys Val Lys Val Asp Glu Val Gly Ala Glu Ala Leu Gly Arg
β^A ATGCTGACTGCTGAGGAGAAGGCTGCCGTCACCGGCTTCTGGGGCAAGGTGAAAGTGGATGAAGTTGGTGCTGAAGCCTTGGGCAGGTAGG
γ ATGCTGACTGCTGAGGAGAAGGCTTCTGTCATCTCCCTATTTGCCAAGGTGAACGTGGAAGAAGTCGGTGGTGAGGCCCTGGGCAGGTAGG
Met Leu Thr Ala Glu Glu Lys Ala Ser Val Ile Ser Leu Phe Ala Lys Val Asn Val Glu Glu Val Gly Gly Glu Ala Leu Gly Arg

Fig. 3. Nucleotide sequence of the 5′ flanking and 5′ exon of the β^A and γ globin genes. The amino acid sequence of the globins beginning with the methionine residues specified by the initiation codon is indicated above (β^A) or below (γ) the nucleotide sequences. The under-or-over-scored amino acids are residues which differ between the two globins. The boxes demarcate sequences observed proximal to the coding portions of other globin genes which are thought to be important for initiation of transcription (see text). After amino acid residue 29, the nucleotide sequence deviates from that necessary to encode for the globins; therefore this position marks the beginning of the small intron present in both genes.

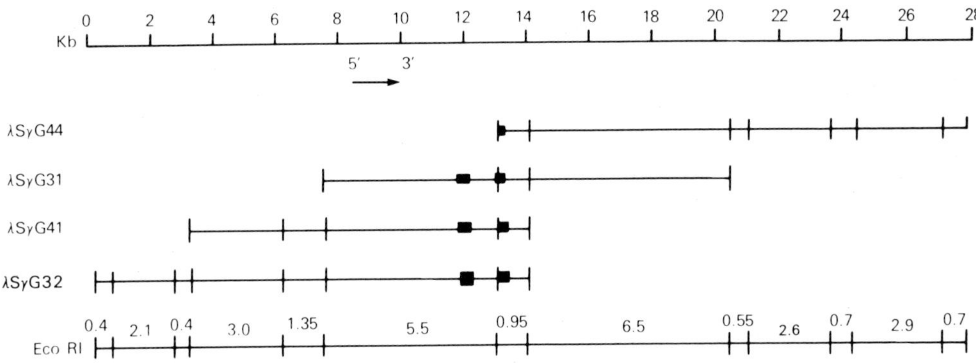

Fig. 4. The Eco RI restriction endonuclease map of four recombinant bacteriophages, the inserted DNA of which overlaps and contains a gene which encodes for sheep γ globin. The scale above is in kilobases (kb) while the numbers above the line in the lower part of the figure indicate the length in kb of the individual Eco RI restriction endonuclease fragments.

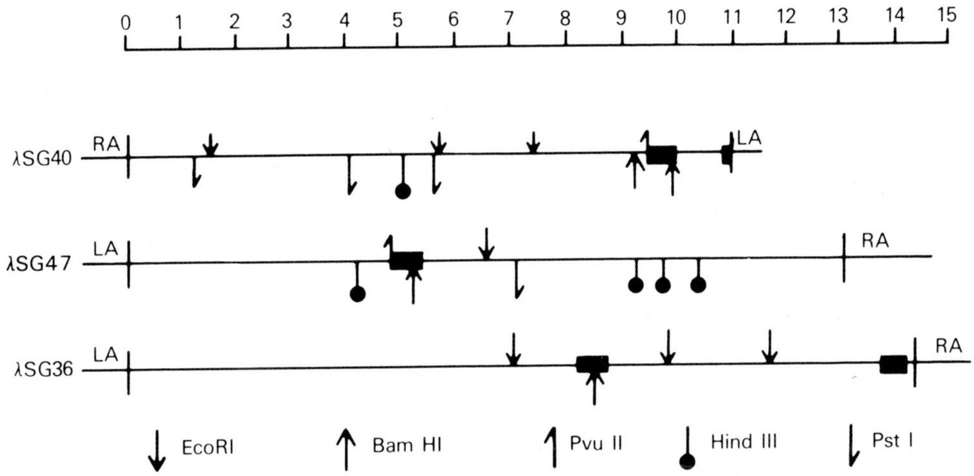

Fig. 5. Restriction endonuclease maps of three recombinant bacteriophages isolated from the sheep genome, which contains β-like globin genes. The shaded areas indicate the position of the globin coding sequences. These were localized by restriction endonuclease digestion and Southern blotting of DNA from the individual bacteriophage. Portions of the globin gene-like sequences in λSG40 and λSG47 have been determined allowing precise positioning of these sequences within the recombinant. The position of the globin gene-like sequences in λSG36 is by analogy with characterized comparable recombinants isolated from the goat genome [see Lingrel, et al, this volume].

globins. One of these, λSεG4, is thought to contain a gene for an embryonic β-like globin, based on the annealing of its coding sequences to β gene sequence probes in Southern blotting experiments and the high thermal stability of duplexes formed in solution, between its coding sequences and one component of a mixed globin cDNA made with polyA$^+$ RNA from embryonic red blood cells [10].

The restriction endonuclease maps of three other recombinants are shown in Figure 5. Two linked genes, separated by approximately 3 kb of DNA, are contained with λSG36. These genes from the sheep genome have not been characterized further, but a highly similar pair of genes isolated from the goat genome have been sequenced extensively [see paper by Lingrel et al in this volume]. The most 5′ of the pair encodes for a globin which, by analogy with the human ϵ globin, probably is produced during the embryonic period of erythropoiesis. The second gene from this goat genome has several defects which suggest that it does not encode for a functional globin. We suspect that the two genes from the sheep genome found in λSG36 will also have these properties.

Portions of the genes contained in λSG40 and λSG47 have been sequenced. The data, beginning from the Pvu II site near the putative cap signal and extending into the first coding block, are shown for the gene in λSG40 in Figure 6 and are compared to the similar region of the γ and β^A globin genes. Noteworthy, is a codon which specifies histidine in the position corresponding to the second or third amino acid depending on whether the methionine residue encoded by the initiator codon is removed. This histidine residue, not found in the sheep γ, β^A, or β^C globins [24], is an important residue for interaction with diphosphoglycerate (DPG) [24]. Sheep red cells do not contain this substance nor do sheep Hb F, HB A, or Hb C interact with DPG in solution [26], in keeping with the fact that these globins lack this critical histidine residue. The gene contained in λSG40 clearly could encode for a globin which had the ability to bind to DPG. Characterization of the embryonic globins of sheep and further sequencing of the gene in λSG40 will be necessary to determine whether this gene encodes for functional embryonic globin or whether it represents a pseudogene which marks an evolutionary period in which a globin which interacted with DPG was produced by sheep or an antecedent species.

SUMMARY

Several features of the sheep β-like globin genes have become apparent. The cluster must be highly complex, for five genes have already been identified for which a specified function cannot yet be assigned. Furthermore, the fairly long distances between individual genes suggest that the cluster will extend perhaps as much as a few hundred kilobases along the chromosome. Nonetheless, two genes quite definitively identified as those which encode for the β^A and γ globins are embedded within an 8 kb segment of DNA with sufficient homology to form heteroduplexes visible with the electron microscope. Unresolved at this time is whether there is only one or possibly more genes encoding for γ and β^A globin, and the number and localization of the β^C gene or genes within the cluster remain to be established. As additional clones are analyzed, the general organization and linkage of the β-like cluster of sheep may become apparent. These results seem a necessary prerequisite for further development of this model as a useful means to approach the problem of hemoglobin switching.

REFERENCES

1. Young NS, Nienhuis AW: Hemoglobin switching in sheep and man. In Gordon A, Silber R, LoBue J (eds): "Year in Hematology," Vol 2, New York: Plenum, 1978, pp 103–141.
2. Hammerberg G, Brett I, Kitchen H: Ontogeny of hemoglobin in sheep. Ann NY Acad Sci 241: 672–682, 1974.

```
λSβAG21  CAACTGTGTTCACTAGCA GCTACACAAACAGACACCATGCTG          ACTGCTGAGGAGAAGGCTGCC
                                              MetLeu          ThrAlaGluGluLysAlaAla
                                              MetValHisPheThrAlaGluGluLysAlaAla
λSG40          CAATCACTAGCAAGCTCTCAGGCTTGGCATCATGGTGCATTTTACTGCTGAGGAGAAGGCCGCC
λSγG31   CAACTGTGTTCACTAGCA GCTACACAAACAGACACCATGCTG          ACTGCTGAGGAGAAGGCTTCT
                                              MetLeu          ThrAlaGluGluLysAlaSer

λSβAG21  GTCACCGGCTTCTGG GGCAAGGTGAAAGTGGATGAAGTTGGTGCTGAAGCCTTGGGCAGGTAGGTATCC
         ValThrGluPheTrp GlyLysValLysValAspGluValGlyAlaGluAlaLeuGlyArg
         ValThrSerProTrpSer LysMetAsnValGluGluAlaGlyGlyGluAlaLeuGlyArg
λSG40    GTCACCAGCCCGTGGAGC AAGATGAATGTGGAAGAGGCCGGAGGTGAAGCCTTGGGCCGGTAAGCCTTG
λSγG31   GTCATCTCCCTATTT GCCAAGGTGAACGTGGAAGAAGTCGGTGGTGAGGCCCTGGGCAGGTAGGTATCC
         ValIleSerLeuPhe AlaLysValAsnValGluGluValGlyGlyGluAlaLeuGlyArg

λSβAG21  CACTTACAAGACAGGTTTAAGGAGAGTGAATG
λSG40    GTCTCAATGCATGGGATGAGGGTGATATACCG
λSγG31   TGCTTACTAGACAGGTTTAAGG
```

Fig. 6. Comparison of the nucleotide sequence of the 5′ portion of the "globin gene" in λSG40 to the genes for sheep γ and β^A globins. The boxes demarcate areas in which 70% or more of the nucleotides are identical. The amino acid sequence for the γ and β^A globins is given below the nucleotide sequence of those genes while that for the globin potentially encoded by the gene in λSG40 is shown above the nucleotide sequence in that recombinant. In the area where the β^A and γ genes differ most strikingly, the gene in λSG40 is most like the β^A. A small intron beginning with the di-nucleotide GT is found in λSG40 in an analogous position to the introns in the β^A and γ genes. Noteworthy is the fact that the nucleotide sequence of the 5′ portion of the small intron of the gene in λSG40 differs fairly markedly from the small introns of the β^A and γ genes.

3. Benz EJ Jr, Barker JE, Pierce JE, Turner PA, Nienhuis AW: Hemoglobin switching in sheep: Commitment of erythroid stem cells to expression of the β^{C}-globin gene and accumulation of β^{C}-globin mRNA. Cell 14:733–740, 1978.
4. Nienhuis AW, Croissant RD, Barker JE: Induction of hemoglobin C synthesis in sheep: Characterization of the "switching" stem cell. In Stammatoyannopoulos G, Nienhuis AW (eds): "Cellular and Molecular Regulation of Hemoglobin Switching," New York: Grune and Stratton, 1979, pp 397–420.
5. Barker JE: Hemoglobin switching in sheep: Characteristics of BFU-E-derived colonies from fetal liver. Blood, 56:495–500, 1980.
6. Mears JG, Ramirez F, Leibowitz D, Bank A; Organization of human δ- and β-globin genes in cellular DNA and the presence of intragenic inserts. Cell 15:15–23, 1978.
7. Flavell RA, Kooter JM, DeBoer E, Little PFR, Williamson R: Analysis of the human β-δ-globin gene loci in normal and Hb Lepore DNA: Direct determination of gene linkage and intergene distance. Cell 15:25–41, 1978.
8. Lawn RM, Fritsch EF, Parker RC, Blake G, Maniatis T: The isolation and characterization of linked δ- and β-globin gene from a cloned library of human DNA. Cell 15:1157–1174, 1978.
9. Fritsch EF, Lawn RM, Maniatis T: Molecular cloning and characterization of the human β-like globin gene cluster. Cell 19:959–972, 1980.
10. Kretschmer PJ, Kaufman RE, Coon HC, Chen M-J, Geist CE, Nienhuis AW: Hemoglobin switching in sheep: Cloning and characterization of the β^{A} and β-like embryonic globin genes from genomic DNA. J Biol Chem. 255:3204–3211, 1980.
11. Kretschmer PJ, Coon HC, Davis A, Harrison M, Nienhuis AW: Hemoglobin switching in sheep: Isolation of the fetalγ-globin gene and demonstration that the fetalγ and adult β^{A}-globin genes lie within 8 kilobase segments of homologous DNA. J Biol Chem, in press, 1981.
12. Benz EJ Jr, Kretschmer PJ, Geist CE, Kantor JA, Turner PH, Nienhuis AW: Hemoglobin switching in sheep: Synthesis, cloning, and characterization of DNA sequences coding for the β^{B}-β^{C} and γ-globin mRNAs. J Biol Chem 254:6880–6888, 1979.
13. Southern EM: Detection of specific sequences among DNA fragments separated by gel electrophoresis. J Mol Biol 98:503–517, 1975.
14. Maxam AM, Gilbert W: A new method for sequencing DNA. Proc Natl Acad Sci USA 74:560–564, 1977.
15. Maxam AM, Gilbert W: Sequencing end-labeled DNA with base-specific chemical cleavages. In Grossman L, Moldave K (eds): "Methods in Enzymology," New York: Academic Press, 1980, pp 499–560.
16. Konkel DA, Tilghman SM, Leder P: The sequence of the chromosomal mouse β-globin major gene: Homologies in capping, splicing and Poly(A) sites. Cell 15:1125–1132, 1978.
17. Konkel DA, Maizel JV Jr, Leder P: The evolution and sequence comparison of two recently diverged mouse chromosomal β-globin genes. Cell 18:865–873, 1979.
18. Tiemeier DC, Tilghman SM, Polsky FI, Seidman JG, Leder A, Edgell MH, Leder P: A comparison of two cloned mouse β-globin genes and their surrounding and intervening sequences. Cell 14: 237–245, 1978.
19. Leder P, Tilghman S, Tiemeier D, Konkel D, Seidman JG, Miller HI: The organization of mouse β-globin genes. In Stammatoyannopoulos G, Nienhuis AW (eds): "Cellular and Molecular Regulation of Hemoglobin Switching," New York: Grune and Stratton, 1979, pp 493–500.
20. Kaufman RE, Kretschmer PJ, Adams JW, Coon HC, Anderson WF, Nienhuis AW: Cloning and characterization of DNA sequences surrounding the human γ-, δ- and β-globin genes. Proc Natl Acad Sci USA 77:4229–4233, 1980.
21. Proudfoot NJ, Brownlee GG: 3′ non-coding region sequences in eukaryotic messenger RNA. Nature 263:211–214, 1976.
22. Pribnow D: Nucleotide sequence of an RNA polymerase binding site at an early T7 promoter. Proc Natl Acad Sci USA 72:784–788.
23. Lacy E, Hardison RG, Quon D, Maniatis T: The linkage arrangement of four rabbit β-like globin genes. Cell 18:1273–1283, 1979.
24. Dayhoff MD: "Atlas of Protein Sequence and Structure," Vol. 5, Silver Spring, Maryland: National Biomedical Research Foundation Publication, 1972.
25. Armone A: X-ray diffraction study of binding of 2,3-diphosphoglycerate to human deoxyhemoglobin. Nature 237:146–150, 1972.
26. Bunn HF, Seal US, Scott AF: The role of 2,3-diphosphoglycerate in mediating hemoglobin function of mammalian red cells. Ann NY Acad Sci 241:498–512, 1974.

Organization and Expression of Globin Genes, pages 69–88

The Mouse β Hemoglobin Locus

Marshall H. Edgell, Steven Weaver, Carolyn L. Jahn, Richard W. Padgett, Sandra J. Phillips, Charles F. Voliva, Mary B. Comer, Stephen C. Hardies, Nancy L. Haigwood, Charles H. Langley, Robert R. Racine, and Clyde A. Hutchison III

Department of Bacteriology and Immunology, Curriculum in Genetics, University of North Carolina, Chapel Hill (M.H.E., S.W., C.L.J., R.W.P., S.J.P., C.F.V., M.B.C., S.C.H., N.L.H., C.A.H.), National Institute of Environmental Health and Safety, Research Triangle Park, North Carolina (C.H.L., R.R.R.)

INTRODUCTION

The globin family of genes has for quite some time provided an attractive system in which to explore differential gene regulation. Extensive protein studies of the gene system in many species have indicated that there are nontrivial differences in the regulatory mechanisms acting in the different species. Dramatic variations exist as to the presence of fetal specific and stress globins as well as the number of globins synthesized in the adult. The DNA studies brought together in these proceedings show that there exists considerable variation in the organization of the globin gene systems and suggests that these interspecies studies will be useful in defining interesting components in the regulatory mechanisms. We have found the characterization of the globin genes in mice attractive, not only because it is genetically one of the best characterized of mammals but also because we can explore variations in globin gene organization within a closely related set of well-studied rodents.

We have looked at a few individual mice from various species of Mus to access the intrageneric variation at the β-globin locus. All of the Mus species examined have β-globin genes of sufficient homology with the adult cDNA globin probe from M. musculus to give equally intense hybridization patterns. Within this sample of Mus species we see roughly five DNA patterns or globin DNA restriction "phenotypes" when genomic Eco RI digests are probed (Fig. 1). In the inbred lines of Mus musculus we see three different β-globin DNA phenotypes. They correspond to the three known mouse β-globin haplotypes Hbb^d, Hbb^p, and Hbb^s which were defined from protein studies [1–4].

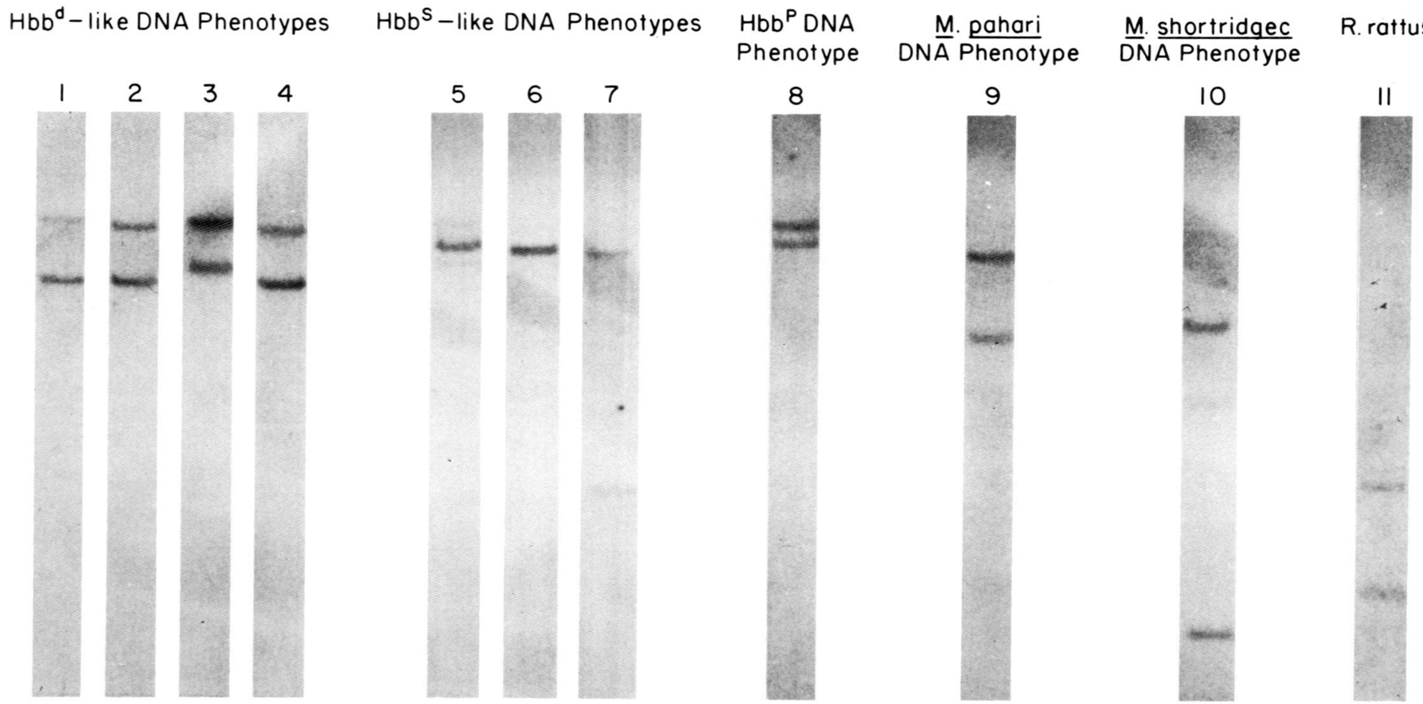

Fig. 1. DNA β-globin phenotypes from selected Mus species and a rat. Four micrograms of genomic DNA were digested to completion with Eco RI and electrophoresed in a 1% agarose gel. The DNA was transferred to nitrocellulose and hybridized with nick-translated adult β-globin cDNA sequences excised from a cDNA clone. Mice with similar band patterns (globin DNA phenotype) have been grouped together in the figure. (Lane 1: M musculus, BALB/c; Lane 2: M molossenus; Lane 3: M cookii; Lane 4: M cervicolor popaeus; Lane 5: M musculus, C57BL; Lane 6: M hispanicus; Lane 7: M caroli; Lane 8: M musculus, Au/SsJ; Lane 9: M pahari; Lane 10: M shortridgei; Lane 11: R rattus).

We have explored in detail two of these β-globin haplotypes, Hbb^d and Hbb^s. The general approach has been to isolate clones containing β-globin genes from DNA libraries in λ Charon 4A which were constructed from size-fractionated partial Eco RI digests of the genomic DNA. The hybridization techniques were sensitive enough to isolate clones even with limited probe homology such as one might expect for the embryonic sequences. From both haplotypes a set of clones was isolated containing a continuous stretch of DNA 65 kilobases (kb) long. Detailed restriction mapping and DNA sequence analyses were then employed to locate the β-like sequences.

THE Hbb^d LOCUS

Two DNA libraries from Hbb^d mice were searched for globin homologous sequences. Both were constructed from partial Eco RI digests of BALB/c genomic DNA. One was made from embryonic DNA by Mark Davis at Cal Tech and the other from adult liver DNA by Norman Arnheim at the State University of New York at Stoney Brook. Screening the library with adult β-globin probe [5] resulted in plaques with two intensities of hybridization (Fig. 2). The strong hybridizing clones contained either a 7 or a 14 kb fragment along with other small nonhybridizing fragments. The hybridizing fragments were the same as those previously cloned and shown by Leder and his colleagues to contain the adult coding regions for β-maj or β-min [6, 7]. The phages that hybridized weakly to the adult β-globin probe all contained multiple small Eco RI fragments. Among the phages isolated, eight different types were represented: These were designated as CE8, CE12, CE14, CE17, CE18, CE19, CA4, and CA11. (The CE or CA designation denotes the source of the cloned DNA: BALB/*c* *e*mbryonic mouse or BALB/*C* *A*DULT LIVER.(Together these clones encompass a continuous stretch of DNA 65 kb in length within the Hbb^d locus. Within this region restriction sites have been mapped in detail and seven β homologous gene-like regions [8] have been located (Fig. 3). Sequence data have been acquired from all of the β globins in the cluster 8–10]. A reference for the extent of the sequence information available from the Hbb^d locus which represents the primary data base from which we draw our comments is shown in Figure 4.

The large-scale features of the locus appear analogous with the other characterized β clusters [11, 12 and these proceedings]. Each of the genes is oriented in the same 5′ to 3′ direction within the cluster and an embryonic gene is the most 5′ structure and an adult gene is the most 3′ structure. There are also more β homologous structures than proteins known to be specified by the locus. The Hbb^d haplotype specifies only four known β-like proteins. These are the two adult betas, βmaj and βmin, and the two embryonic β-like proteins, εy2 and εz.

The Adult Genes βmaj and βmin

Leder and his colleagues have sequenced the entire βmaj and βmin genes [10]. The functioning β-globin genes in each of the species so far characterized share the structural features of a small intervening sequence between residues 30 and 31 and a large intervening sequence between residues 104 and 105 as well as the sequences believed to be necessary for transcription, a "Hogness" box, a CAP site, and a poly (A) attachment site (Fig. 4). The βmaj and βmin genes have almost identical small intervening sequences but their large intervening sequences are nonhomologous

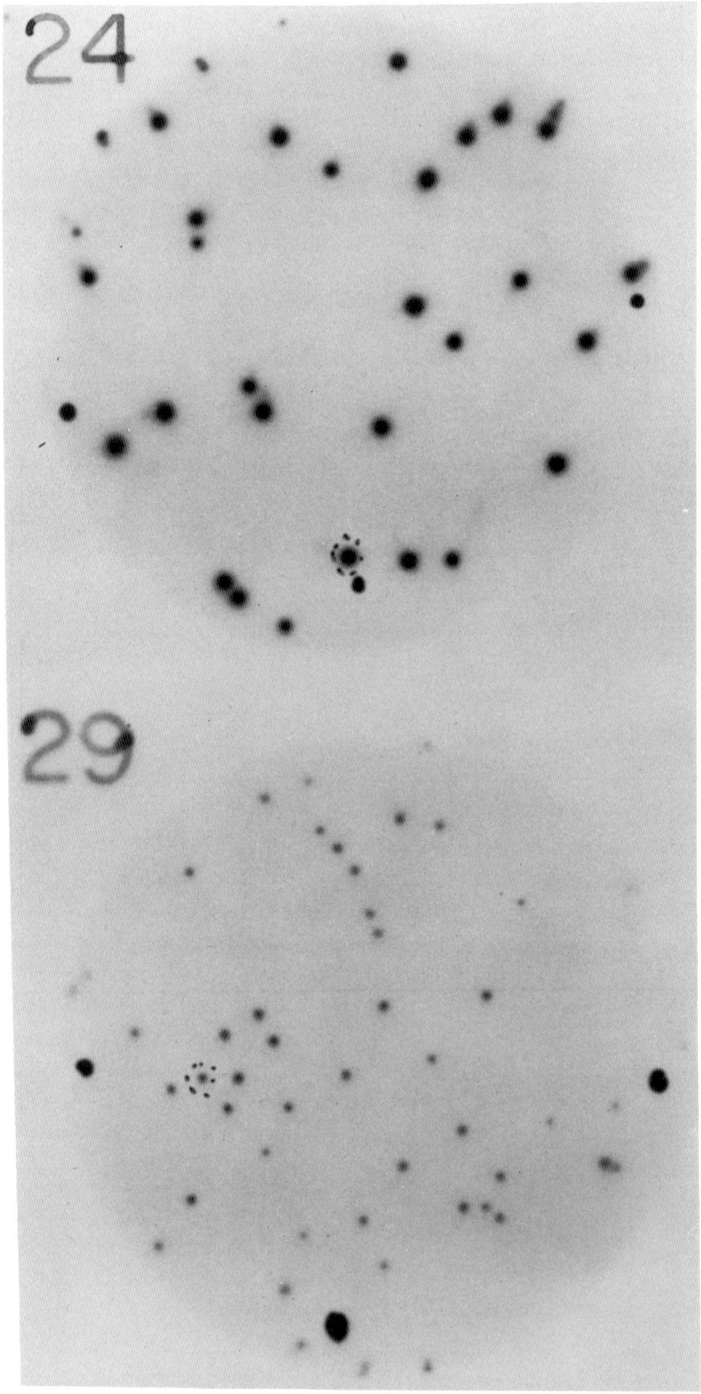

Fig. 2. Plaque filters probed with adult β sequences. Filter transfers were made from petri dishes during the plaque purification of positives identified during a search of the BALB/c partial Eco RI library. These filters are from the third cycle of plaque purification and hence all the plaques represent the same clone. The probe was ^{32}P nick-translated adult cDNA sequences cloned by Rougeon and Mach [5]. Filter 24 shows a strong hybridizing clone with adult sequences. Filter 29 shows a weak hybridizing clone with embryonic sequences.

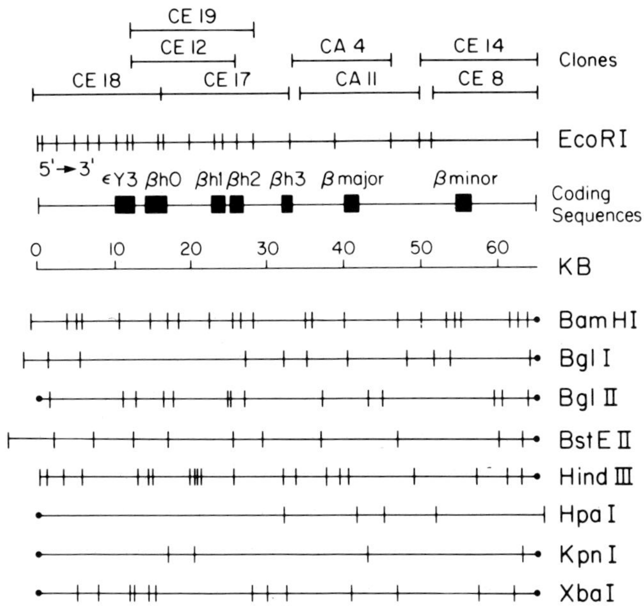

Fig. 3. The map of the Hbb[d] locus. The solid blocks in the coding sequence diagram indicate the locations of β-like structures present in this 65 kb region of the Hbb[d] locus. The dots terminating some of the restriction maps indicate maps for which our data terminate within the cloned regions.

over about 400 np of their 600 np extent [7, 10]. By heteroduplex analysis it has also been shown that the nucleotide environments in which these two adult genes lie are quite different [7]. We have shown [8] that the embryonic and adult genes are linked together in the cluster (Fig. 3) and that the adult genes are ordered 5′-βmaj-βmin-3′. This orientation is opposite to that found for the adult betas in humans in that δ, the minor human globin, is 5′ to the major human gene, β.

The Embryonic Gene, ϵy3

The 5′-most gene in the Hbb[d] cluster preserves the canonical β structure (defined in Fig. 4) within the sequenced portion. This gene codes for one of the two embryonic proteins, ϵy. Translating the nucleic acid sequence data into its equivalent amino acids gives a sequence consistent with the tryptic peptide data of Steinheider [13] and Gilman [14] at all but two codons (positions 41 and 59). We have also shown [8] that this is the only structure in our clones or in the genome which shows strong hybridization to a cDNA clone constructed from embryonic mRNA [15]. Thus, the data indicates that this gene codes for an ϵy-like product which is not exactly the same as the ϵy2 as sequenced by Gilman or Steinheider. We have chosen to call this structure the ϵy3 gene.

The βh0 and βh1 Structures

The βh designation refers to β*h*omologous. We have applied this nomenclature to the β-like elements which specify no known protein. We prefer this nomenclature since the other nomenclature, pseudogene [these proceedings], implies that the

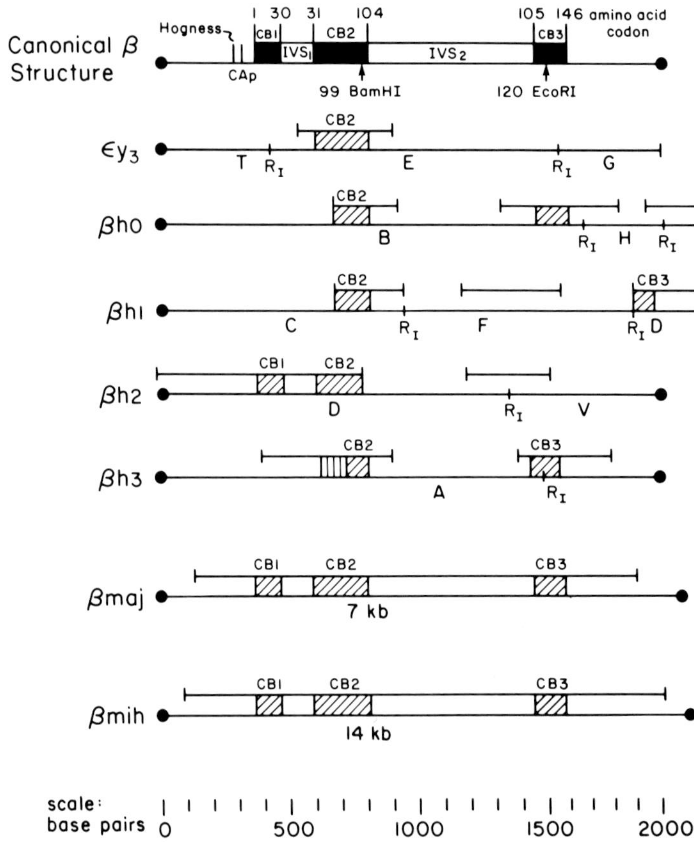

Fig. 4. Sequenced regions in the Hbb^{d} locus. The reference diagram is of a "canonical" β-globin gene which incorporates the features found so far to be in common amongst the globin genes producing proteins in all species. The bottom line in each diagram indicates the Eco RI restriction map. The hatched blocks indicate coding block homology. The upper line indicates the extent of the sequenced regions. The sequenced regions in βmaj and βmin come from Konkel et al, [9, 10].

structure is a nonfunctioning gene unless one makes a strong effort to define this special use of the term pseudo differently from common usage. It should be clear that a sequence can have important functions which are heritable (gene-like) whether or not the sequence specifies a protein and whether or not it even specifies a transcript.

The βh0 and βh1 structures represent a distinct subclass in the same fashion as the human γs represent a subclass of proteins, being recognizably homologous, but distinctly different, from the human embryonic or adult sequences. The βh0 and βh1 sequences are as homologous in their nucleotide sequence to each other as the βmaj and βmin sequences are homologous. They are, however, as different from the adult mouse genes as the embryonic gene differs from the adults. The βh0 and βh1 elements are likewise just as different from the embryonic sequence as are the adult betas. The βh0 and βh1 structures preserve the canonical β form within the sequenced regions.

These sequences do not correspond to any characterized mouse β-like protein sequence. Our translation of their nucleotide sequences gives an amino acid sequence which is slightly more embryonic-like than adult-like. These structures would by analogy to the human γs be suitable candidates for mouse fetal genes. Not only are the sequence relationships to the adults and embryonics appropriate but they occupy the analogous position within the mouse cluster to that of the human γ genes—ie, between the embryonic and adult genes. There is some evidence for a mouse fetal hemoglobin coexpressed with the adult globin in fetal liver-derived [16] or spleen-derived [17] erythrocytes as well as in mouse erythroleukemia cells [18]. However, these proteins have not been fully characterized and their existence has been disputed [19].

The βh2 Structure

The least β-like of the structures that we have found so far which still possesses normal gene structure is βh2 (Fig. 13). Over the sequenced region (Fig. 4) there is only 61% homology with the adult genes. The adult α globin coding blocks for comparison show 55% homology with the adult β-globin coding blocks [25]. Due to this weak homology the βh2 element was not discovered by hybridization with the adult cDNA probe since our normal conditions tolerate only 20%–25% mismatch. Rather, it was found "by accident" while sequencing in this region of the cluster.

The βh3 Structure

Of all the β-like elements in the Hbb^{d} locus, βh3 has the most unusual structure (Fig. 5) [8, Hutchison et al, in preparation]. In describing the various regions within the βh3 structure we will use for convenience the usual coding block and intervening sequence nomenclature to refer to those βh3 regions with sequence homology to the corresponding regions in normal globin genes. This is not meant to imply that those structures in βh3 serve analogous functions. Coding block three in βh3 is 70% homologous with the adult coding block three. This coding block does not, however, have a terminator at residue 146. The closest terminator in the appropriate

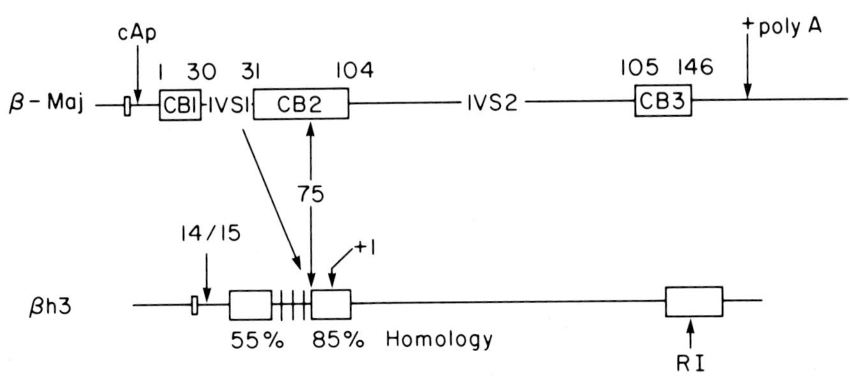

Fig. 5. The βh3 structure. The diagram shows the βh3 structure relative to the adult β-globin βmaj. The region between the blocks showing 55% homology with βmaj CB1 and 85% homology with βmaj CB2 contains "patchy" homology to βmaj CB1, 5′ to residue 75.

reading frame occurs 13 triplets later. The "IVS2" in βh3 is 60 nucleotides longer than the adult βmaj IVS2. The sequence similarities between the adult and βh3 IVS2 region are confined to ten bases of identity at the boundaries with coding blocks two and three. From codon 104 to codon 75 the βh3 coding block two is quite adult-like, with 85% homology, except for a frameshifting additional base in residue 90. At residue 75 this strong in-phase homology becomes "patchy." There appear to be about eight patches of short sequences showing strong homology to the adult sequences, separated by short regions which do not show this strong homology (Figs. 5 and 6).

The βh3 sequence 5′ to this patchy region shows weak homology to globin but to coding block one instead of coding block two. This weak homology, 55% to adult, was detected only by means of computer-assisted analysis. The computer analysis that we employed [Hutchison et al, in preparation] generates a matrix with the structure we want to explore on one axis (ie, βh3 on the vertical axis) and the structure to which we would like to compare it on the other (ie, βmaj on the horizontal axis). The program then takes a sequence of user-defined length (the search argument) from the "vertical" sequence and scans the "horizontal" sequence, placing a dot in the matrix at every occurrence of a match (Fig. 7). The next search argument is generated by moving down one nucleotide in the "vertical" sequence and the horizontal scan is then repeated. The entire process is repeated until the "horizontal" sequence has been scanned with every search argument drawn from the "vertical" sequence. If there are sequences which are organized in the same way in the two structures they will give rise to a group of dots in the matrix clustered together along a diagonal. Random occurrences of matches will not generate clustered dots. Similar matrices have been generated by the computer analysis developed by Jacob Maizels and used to analyze βmaj and βmin [10]. An attractive feature of the analysis is that we can vary the stringency of the analysis by altering the criteria that the computer uses to decide whether a "match" has been found. Both the length of the search argument and the number of tolerated mismatches within the argument can be specified.

Analyses of the βh3 with two different criteria illustrate the utility of the procedure. An 8 out of 10 criterion (Fig. 8) shows a group of dots clustered on a diagonal in the region in which we had previously detected homology (85% to adult). However, the low stringency criterion, 20 out of 40, reveals weak homology, 55% in βh3 to coding block one of adult globin (Fig. 9). This region of weak homology is joined to the region of strong homology via the patchy homology region which spans a 160-nucleotide deletion relative to a canonical β structure. There is also a region starting at the CAP site which contains a match of 14 out of 15 bases with the adult gene.

An appealing model then, for the βh3 structure is that it arose by an illegitimate recombination event, either within a globin gene or between two globin genes. In one such model the recombination would involve an adult-like globin gene giving rise to the region with strong homology (codons 75 and rightward) and a nonadult gene giving rise to the region of weak homology (CB1). The patchy homology between the two regions may reflect the mechanisms of the illegitimate cross-over event since such sequences have been before in other recombinant molecules; SV40 defectives with incorporated host sequences [20].

```
Beta minor  GGCAGCTCACAAGTAGAAGCTGGGTGCTTGGAGACAGAGGTCTGCTTTCCAGCAGGCACTAACTTTCAGTGTCCCCTGTCTATGTTTCCC

Beta h3     CCTGCATCTGCTTCTGACACAGGTTGTTCACTAGCAGCAGTCCTACACAGGGTGTACTGACTGACAAAGAGAAGTTGGCCATCAGTGGCC   90
                 cap homology

             IVS1 :31        35             40             45             50             55
                   Leu Leu Val Val Tyr Pro Trp Thr Gln Arg Tyr Phe Asp Ser Phe Gly Asp Leu Ser Ser Ala Ser Ala Ile Met Gly Asn P
Beta minor  TTTTTAGGCTGCTGGTTGTCTACCCTTGGACCCAGCGGTACTTTGATAGCTTTGGAGACCTATCCTCTGCCTCTGCTATCATGGGTAATC

Beta h3     TATGGGACACAGTAGACATGGGAAACATTGATGGCGAGTCACTGGAGAAGTTGTAATAGACTGCATAAGCAGGACACATGGAAGCTTGGT  180
                                                                                     HindIII

                 60             65             70             75             80             85
            ro Lys Val Lys Ala His Gly Lys Lys Val Ile Thr Ala Phe Asn Glu Gly Leu Lys Asn Leu Asp Asn Leu Lys Gly Thr Phe Ala Ser L
Beta minor  CCAAGGTGAAGGCCCATGGCAAAAAGGTGATAACTGCCTTTAACGAGGGCCTGAAAAACCTGGACAACCTCAAGGGCACCTTTGCCAGCC
                                                             ||||||| |||||| || |||||||||||| ||| |  ||
Beta h3     AGGTGAAGAGAGAGCAGAATGTCTGGCAGGAACTGATTTTCTCTACCCACCTGAAACACCTGGCCATCCTCAAGGGCACTTTTTCTGGCA  270

                90             95             100        104:  IVS2
            eu Ser G lu Leu His Cys Asp Lys Leu His Val Asp Pro Glu Asn Phe Arg
Beta minor  TCAGTG AGCTCCACTGTGACAAGCTGCATGTGGATCCTGAGAACTTCAGGGTGAGTCTGATGGGCACCTCCTGGGTTTCCTTCCCCTGG
            | | || |||| |||||||||||||||||||| ||||||||||||||||||||||||   |   |   |         |   |   |
Beta h3     TGAATGAAGCTGCACTGTGACAAGCTGCATGTAGATCCTGAGAACTTCAGGGTGAGTGGGCCAGATGCTCTGGCTTCTGCTTCTGCTCCT  360
                 | (frameshift)

Beta minor  CTATTCTGCTCAACCTTCCTATCAGAAGGAAAGGGGAAGCGATTCTAGGGAGCAGTTTCGATGATGGTGTGTGGATGTGAATTGTGAG--
                  |  |    |||| | |   |             |                |    ||       |     |   |  | |
Beta h3     GCTCGTTTTTATCACTTCTTTTTCCATTTGCCTTTTTTCCCCACAGTTTCCT-TACTGTTCTGTATTCCTTCACTTAACACATCTTCT--  450

Beta minor  IVS2 = 629 np
Beta h3     IVS2 = 700 np

                                               IVS2  :105          110            115            120
                                                           Leu Leu Gly Asn Ala Ile Val Ile Val Leu Gly His His Leu Gly Lys Asp Phe
Beta minor  -----------TGATGCGTCTTCTGTCTTCCCACAGCTCCTAGGCAATGCGATCGTGATTGTGCTGGGCCACCACCTGGGCAAGGATTTC
                       |         |     |  |||||| || |  || ||    || ||| ||||||||  || |||          || |||
Beta h3     -----------TAAATGTGTGTGCTCTTCTCCACAGTTCTTGGGAAACATAATAGTGTTTGTGCTGTTCCTCCA----------GAATTC 1070
                                                                                                 Eco RI
```

Fig. 6. Partial sequence of βh3. A vertical line between the βh3 and βminor sequences is used to indicate sequence identity. The blocked sequences show the "patches" of homology 5′ to residue 75. The triplets underlined in the βh3 sequence are termination codons.

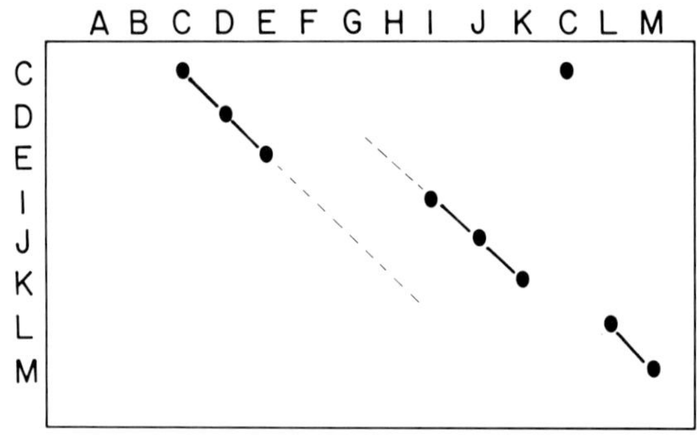

Fig. 7. A diagramatic representation of a homology matrix (see text).

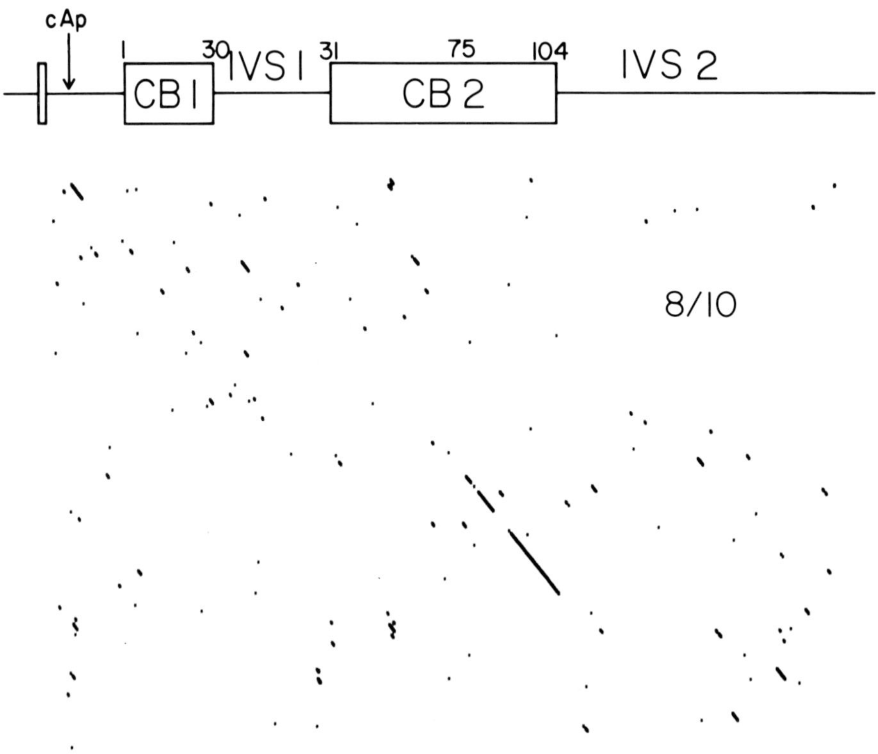

Fig. 8. Homology matrix: βh3 vs βmaj at high stringency. The horizontal axis represents the βmaj sequence; the vertical axis βh3. The criterion to give a point in the matrix requires a sequence match of eight out of ten nucleotides.

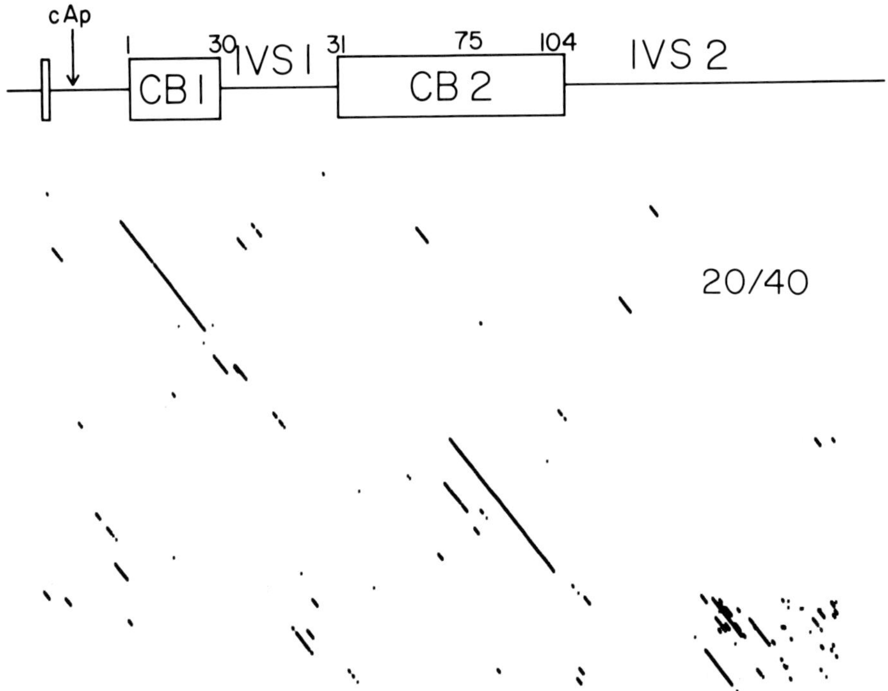

Fig. 9. Homology matrix: βh3 vs βmaj at low stringency. The horizontal axis represents the βmaj sequence; the vertical axis βh3. The criterion to give a point in the matrix requires a sequence match of 20 out of 40 nucleotides.

Distribution of Repetitive Sequences in the Hbb^d β Cluster

As part of our characterization of the β locus we have been looking at the distribution of sequences which are repetitive within the cluster. The fundamental approach is to probe cloned fragments in the cluster with other cloned fragments. We have done this either by probing digests of the clones with purified fragments or by using the cross-blot technique developed by Clyde A. Hutchison [26]. It should be noted that in this approach when two fragments "light up" using the same probe it only means that they each share sequences with the probe. It does not mean that the sequences shared with the probe in the two fragments are the same. Hence, an analysis of this sort generates a lower limit on the repeat complexity within the cluster. A minimum of five repeat families have been identified (Fig. 10) in this fashion [Haigwood et al, submitted, Voliva et al, in preparation]. Two of these repeats have been looked at in more detail.

The second largest HaeIII subfragment from the 7 kb βmaj RI fragment is located 1.3 kb 3′ of βmaj (Fig. 10) and contains the repeat shown by Leder et al [24] to be shared by an adult α clone. It also contains a sequence found near βh2 and βmin [Haigwood et al, submitted]. Neither of those hybridize to the α clones. The βmaj subfragment HaeIII-2 also contains a sequence present in traditional satellite

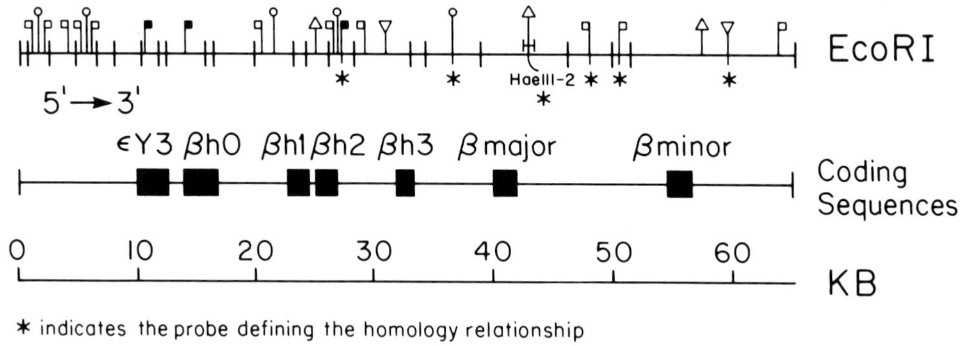

Fig. 10. Repetitive sequences in the Hbb^d locus. The diagram shows the nonglobin homology relationships within the Hbb^d locus. Flags with the same symbol all show homology with the probe defined by the fragment with the same symbol whose staff points to an "*" below the map.

preparations although it appears to not be the simple sequence making up the bulk of such a preparation. The repeated sequences present in this HaeIII fragment are fairly frequent, being found in 85% of the library clones bearing inserts.

The 1.35 kb Eco RI fragment just 5′ to βmin shares sequences with three other Eco RI fragments in the cluster (see Fig. 10). It also is widely dispersed in the mouse genome. Thirty-five percent of the insert bearing clones show strong homology with the 1.35 kb fragment and the remaining 64% show weak homology. The fragment has been sequenced in its entirety [Voliva, personal communication] and the embryonic fragments to which it hybridizes are under sequence analysis to define in detail these repetitive elements.

THE Hbb^s LOCUS

Although mice with the Hbb^s haplotype make only a single protein as adults [1, 2] they contain two adult-like β-globin genes in their DNA [21]. The DNA Eco RI phenotype (genome Eco RI digests probed with adult β sequences) of Hbb^s is also quite different from Hbb^d. We constructed a library in Charon 4A containing a mixture of a partial Eco RI and a partial Eco RI* digest of C57BL DNA. We have isolated 22 independent clones from this library which hybridized to the adult cDNA β sequences [Weaver et al, submitted]. These clones have been designated BA1 through BA22. (The BA designation denotes the source of the cloned DNA: C57*B*L/10, *A*dult liver.)

These clones have been linked together to give a stretch of DNA 60 kb in length within the Hbb^s locus. A restriction map of this region has been constructed and the location of seven β homologous structures has been determined. The Hbb^s locus is very homologous to the Hbb^d locus (compare Figs. 4 and 11). The restriction maps are very similar except in the adult region and the β homologous structures are distributed in an analogous fashion. The 5′-most structure is an embryonic y-like gene and the three 3′-most structures are a βh3-like structure and the two adult genes (Fig. 12).

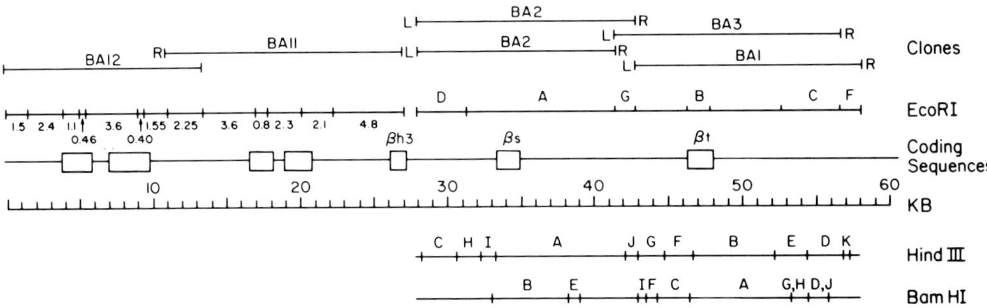

Fig. 11. The map of the Hbb^s locus. The solid blocks in the coding sequence diagram indicate the locations of β-like structures present in the cluster. The dots terminating some of the restriction maps indicate maps for which our data terminate within the cloned regions.

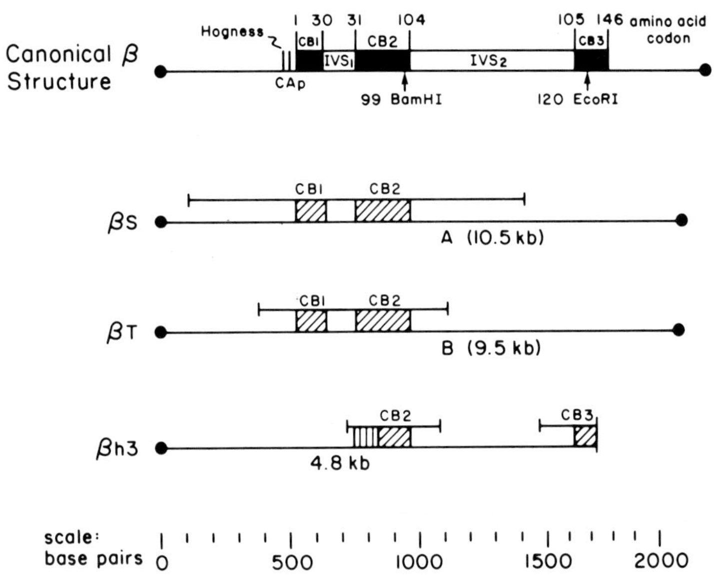

Fig. 12. Sequenced regions in the Hbb^s locus. The reference diagram is of a "canonical" β-globin gene which incorporates the features found so far to be in common amongst the globin genes producing proteins in all species. The bottom line in each diagram indicates the Eco RI restriction map. The hatched blocks indicate coding block homology. The upper line indicates the extent of the sequenced regions.

The εy, βh0, βh1, and βh2 Structures in Hbb^s

The 5′ region of the Hbb^s cluster is very homologous to the same region in Hbb^d. Experiments using the cross-blot procedure of Hutchison [26] show that all of the Eco RI fragments in Hbb^d have very homologous counterparts in Hbb^s. It appears that the embryonic y-like structure in Hbb^s is also in the analogous 5′-most position.

Probing the genome with the embryonic probe shows that the major hybridizing band is the same size as the Hbb^s cloned fragment showing most homology to the Hbb^d embryonic ϵy3.

The βh3 Structure in Hbb^s

A structure exactly analogous to the Hbb^d βh3 structure exists in Hbb^s. We find the strong homology to adult β in the 3′ portion of the structure, the insertion of one base in codon 90, the patchy region of homology around the 160 np deletion at residue 75, and the 5′ region of weak homology to adult β [Hutchison et al, in preparation].

The Hbb^s Adult Genes, βs and βt

Since the Hbb^s haplotype specifies only a single adult protein even though it carries two adult genes in its DNA, the question arises whether these two adult genes are identical or if one has been inactivated in some fashion. Sequence analysis [Weaver et al, in preparation] shows that the coding blocks and intervening sequence portions of the two genes are identical. The large intervening sequences in both βs and βt are quite homologous to βmaj IVS2. There are, however, differences in sequence between the two structures. Three nucleotides differ between the two structures in the region between the CAP region and the initiation codon. Both of these sequences have been seen in transcripts [22] and hence both these adult genes appear to be functioning.

Heteroduplexes between βmaj and βs [Weaver et al, submitted] show that all or most of the 7.2 kb of DNA surrounding βmaj which is included in this heteroduplex is homologous to the DNA surrounding βs (Fig. 13). The restriction maps around βt and βmin are much less similar than those around βs and βmaj (Figs. 4 and 11). Heteroduplexes between βmin and βt demonstrate that the 14 kb of DNA around βmin anneals with the Hbb^s DNA except for two small regions. One of these is a substitution loop indicating a substitution of 1,500 base pairs (bp) in Hbb^d for 100 bp in Hbb^s. The site of the substitution is about 3 kb to the 5′ side of the βmin gene (Fig. 13c).

The second single-stranded feature is a 300 bp symmetrical bubble which maps to the IVS2 of the βmin:βt heteroduplex (Fig. 13A). A similar nonhomology exists between the IVS2 sequences of βs and βt. Therefore, it follows that βs, βt, and βmaj have the same or similar IVS2 sequences, and that the βmin IVS2 is the odd member of the adult large intervening sequences.

The βt and βmin Genes Lie Between Large Inverted Repeats

Single-stranded molecules containing βt or βmin can self-anneal to assume a stem-and-loop configuration which can be visualized in the electron microscope (Fig. 15) [Weaver et al, submitted]. The inverted repeats are 1.5 kb long and 10.5 kb apart (Fig. 14b). The inverted repeats bracket βt; one lying at or near the left end of fragment Eco RI-B and the other near the middle of fragment Eco RI-C (see Fig. 11). In about 30% of the Hbb^s self-annealed molecules, a second, smaller stem and loop formed within the large loop (Fig. 14b). In all cases the stem was about 250 np. However, the secondary loops vary in size as well as in position, suggesting that there is a set of sequence-related short inverted repeats in this region between the

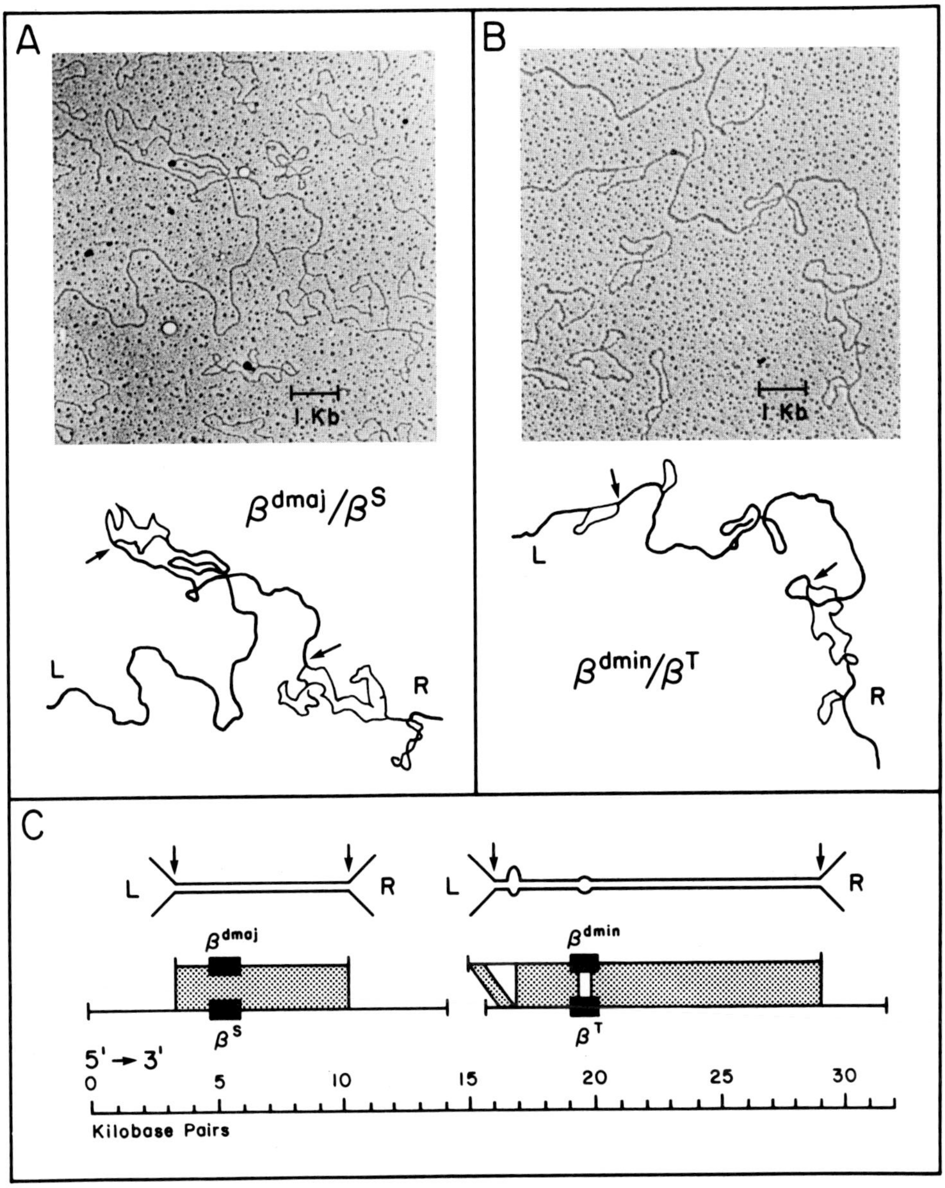

Fig. 13. Heteroduplex analyses of Hbb^{d} and Hbb^{s} adult regions. A. Electron micrograph of the mouse insertions in a heteroduplex between λgtWES bearing a 7.2 kb βmaj Eco RI fragment and a Charon 4A clone (BA2) of 14.3 kb of Hbb^{s} DNA including βs. In the interpretive drawing, thick lines represent double-stranded regions and thin lines represent single strands. Arrows indicate the ends of mouse/mouse strand pairing. L and R indicate the left and right vector arms, respectively. Some of the vector differences can be visualized as single-stranded regions along these arms. B. A heteroduplex between λgtWES bearing a 14 kb βmin Eco RI fragment (λgtWES•MβG3, [7]) and a Charon 4A clone (BA1) of 16 kb of Hbb^{s} DNA containing βt. C. Schematic drawings of the homologous and non-homologous regions between the positionally analogous Hbb^{d} and Hbb^{s} adult clones. The upper drawings are diagrams of the heteroduplexed mouse inserts. In the lower set of diagrams, the shaded regions show positions of insert homology. Positions of the β-globin genes (black boxes) and the Eco RI sites at the boundaries of the mouse inserts in these clones (vertical bars) are indicated.

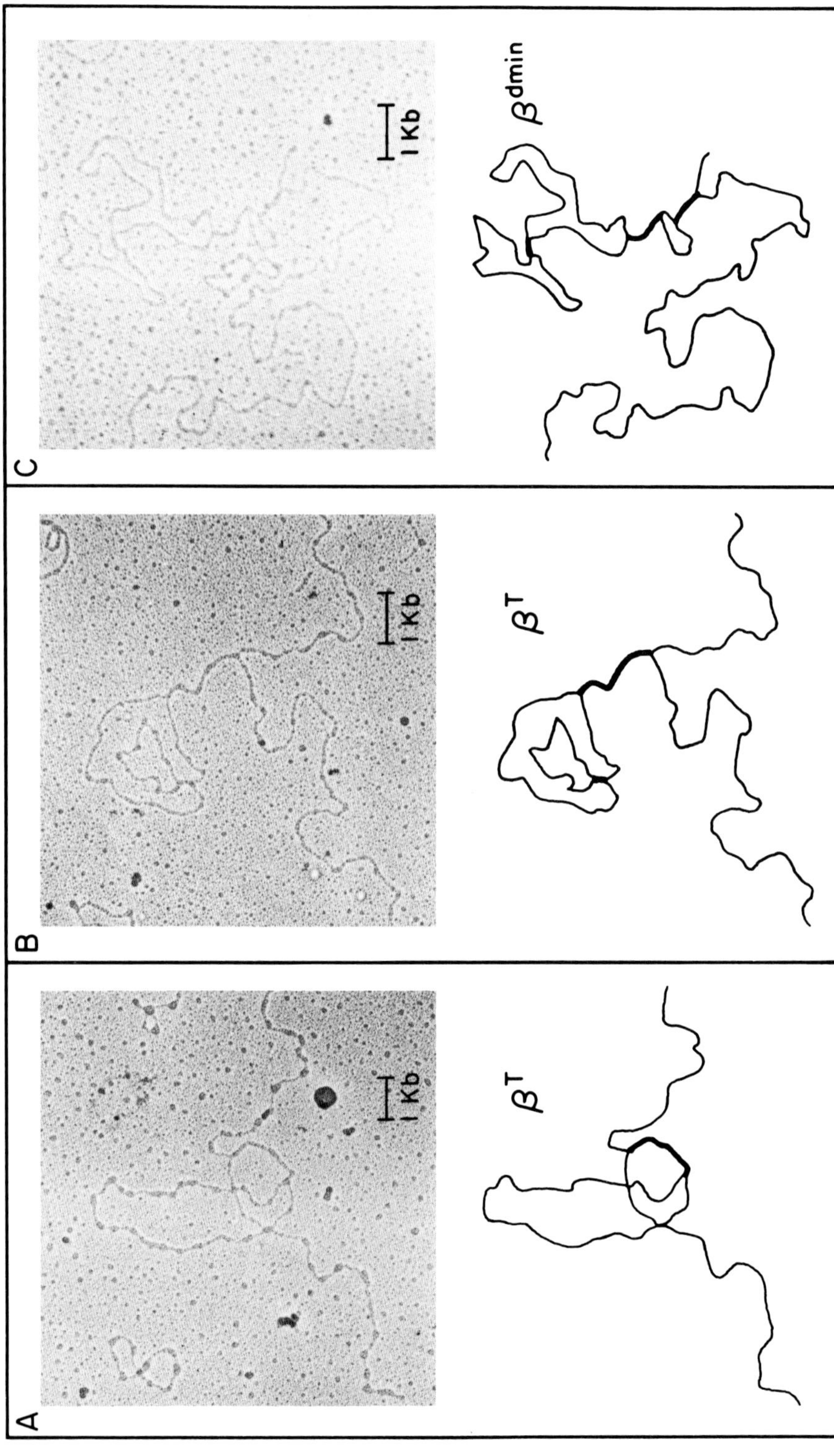

Figure 14

two large inverted repeats. The number and location of these smaller repeats is not yet known.

The stem-and-loop structure seen with molecules bearing βmin are similar but show a substitution feature in the stem (Fig. 14c). This represents a 1,100 bp substitution in one repeat for 150 bp in the other. The 1,100 bp substitution is located on the 5′ side of the βmin gene. The relationship of this substitution to the substitution seen in βmin:βt heteroduplexes is under investigation.

Evolution at the β Locus

Many of the β-globins have a BamHI cleavage site in codons 98–100. This conserved restriction site has allowed us to readily determine the nucleotide differences in the same region of sequence in all of the seven β-globins and hence has provided us with some additional insight into the relationship amongst the β-like structures. It is clear that, within the region we have sequenced, some of these structures are more related than others (Fig. 15).

It has been noted previously that intervening sequences change faster than coding sequences, and that they change by insertions and deletions in addition to nucleotide substitutions, as found in coding regions [10]. This is usually attributed to the selection constraints imposed by the functions of the structure. Of course another possibility for this difference in sequence evolution may be simply that sequences which have once coded for a protein are not good substrates for the enzymology in the cell which gives rise to the observed types of insertions and deletions. Whatever is the case, it is interesting to observe that both the βh2 and βh3

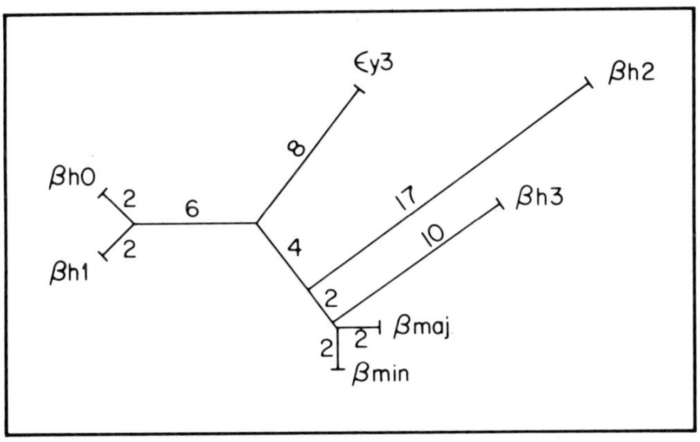

Fig. 15. Nucleotide differences between the β structures. The nucleotide differences are from the 78 nucleotides spanning amino acid residues 76 to 100. Comparisons were done with the frameshift base in βh3 codon 90 removed.

Fig. 14. Electron micrographs of self-annealed βt and βmin clones. Charon 4A clones of genomic mouse DNA containing βt or βmin were alkali denatured and renatured at high dilution and pH 8.5. In the interpretive drawings below the electron micrographs, thick lines represent double-stranded regions and thin lines represent single strands. A. BA1 stem-and-loop structure resulting from annealing of the 1.5 kb inverted repeats bracketing βt. The single-stranded loop is about 10.5 kb long. B. Same as A except a pair of small inverted repeats within the large loop has annealed, forming a smaller stem-and-loop structure. C. CE14 (βmin) stem-and-loop structure showing the annealed 1.5 kb inverted repeat with its substitution loop and a smaller base-paired inverted repeat within the large loop.

structures, which are the two most divergent β structures in the mouse, show this same sort of behavior. For example, the βh3 IVS2 has clearly diverged more rapidly than the coding sequences adjacent to it; only six nucleotides at either boundary of the IVS2 are homologous with adult IVS2. What if anything does this imply about the function of these structures?

One possibility is that these structures arose recently from functioning genes. That would of course explain the origin of the selectional constraints on the sequences as we see them today in βh2 and βh3. These sequences are, however, quite divergent from the other β homologous structures that we have found in Hbb^d and Hbb^s, so that would imply that there are some other quite divergent β-like genes out there in the Mus musculus gene pool. That is not impossible since our current sampling in Mus musculus is only from the pool present in the inbred lines. However, we have looked for the βh3 structure in all of the other Mus species represented in our globin survey (Fig. 1). A βh3-specific probe (a fragment running from the patchy homology to the RI site at residue 121, Fig. 5) "lights up" a single band in all of those animals [Padgett et al, in preparation]. If that band represents a βh3-like structure and not a parent gene then this would be a strong argument against the recent occurrence of the recombination event leading to βh3 in BALB/c.

Another possibility is that the divergent βh2 and βh3 structures have existed for quite some time. If that is the case, then one must accept that these structures have evolved throughout that period in a fashion similar to the genes known to be functioning even when the coding sequence in βh3 has diverged such that it could not give rise to a β-like protein. Certainly one can imagine uses in the cell for β-related sequences which do not depend on the synthesis of a protein. For example, such sequences could serve as a "chelate" at the RNA or DNA level for globin-targeted molecules in order to modulate globin metabolism. That is, imagine a regulatory mechanism which alters the rate of transcription by an end-product feedback mechanism based on globin mRNA. Some molecule would, in such a model, bind to globin mRNA in order to alter the binding of that same molecule to a globin transcriptional control site in the DNA. The production of a globin-related sequence from a structure such as βh3 with an affinity for this regulatory molecule which is significantly higher than the normal globin mRNA would effectively modulate the regulation by lowering the concentration of unbound regulator. That is, it would act as a competitor. Regulatory mechanisms of this complexity are not in fact uncommon. Similar models can be constructed for globin-related sequences interacting as a DNA sequence with a regulatory molecule. It is certainly possible that the selection constraints on such structures would be similar to those of a gene which specifies a protein.

All of these arguments are predicated on the assumption that the observed rapid rate of intervening sequence divergence relative to coding block divergence is the result of the selection constraints imposed on the structures. Generally we assume that these changes come about by a large number of single events. However, another mechanism to accumulate large numbers of changes in a localized region of the DNA is by means of single "catastrophic" or multinucleotide events. Such, for example, occur during gene conversion events, multiple unequal crossing-over events, and transposonlike insertions and deletions. At least two such changes have taken place in this locus which involve globin coding block regions: the events in Hbb^s which gave rise to the two identical adult genes and the illegitimate recombination event giving rise to βh3. If one attributed the set of changes in IVS2 of βmin to such a multinucleotide event then the rate of such changes occurring within the intervening sequences in this locus would not be high relative to such coding block changes. In this context it could be pointed out that for the adult genes all the IVS1s are quite similar and only the IVS2 in βmin differs significantly from the IVS2 sequences present in the other three adult genes, βmaj, βs, and βt.

We do not feel that we can choose between the many possible models for the evolution of this region on the basis of the current data. We would like to believe that the further accumulation of data in this locus will reduce the number of such models.

SUMMARY

We have reviewed here our efforts to characterize the mouse β hemoglobin locus. We describe genomic clones from the Hbb^d and Hbb^s haplotypes containing two adult β-globins and one of the known embryonic β-like globins as well as four other β homologous structures which do not specify any characterized proteins. We have generated extensive restriction maps for both these haplotypes and have collected partial nucleotide sequence data from each of the β-like structures in Hbb^d and from three of the structures in Hbb^s

One of the structures present in both haplotypes, βh3, has an unusual organization for a β-globin gene and appears to be the product of an illegitimate recombination event. Heteroduplex analyses of some of the clones show that the regions containing the adult genes in the two clusters are quite homologous. These studies also show the presence of a set of large inverted repeats surrounding the 3′-most adult gene in each cluster. The distribution of repetitive sequences in the Hbb^d locus has been defined by hybridizing each of the cloned Eco RI fragments in the cluster against the remaining RI fragments. A sampling from other Mus species indicates that there are at least three other β-globin alleles with significantly different patterns of genomic DNA fragments which hybridize to adult globin sequences. Some of the evolutionary possibilities for this locus are also discussed.

ACKNOWLEDGMENTS

This work was supported by grants to Marshall H. Edgell and Clyde H. Hutchison, III, from the USPHS (AI08998 and GM 21313). Support from a training grant to the University of North Carolina Curriculum in Genetics was received by Steven Weaver, Carolyn L. Jahn, Richard W. Padgett, Charles F. Voliva, and Stephen C. Hardies. Marshall H. Edgell would also like to acknowledge the value to this work of the sabbatical year, 1977, spent with Dr. Philip Leder at the National Institute of Health.

REFERENCES

1. Russell ES, McFarland EC: Genetics of mouse hemoglobins. Ann NY Acad Sci 241:25–38, 1974.
2. Popp RA: Sequence of amino acids in the β chain of single hemoglobins from C57B1, SWR and NB mice. Biochim Biophys Acta 303:52–60, 1973.
3. Popp RA, Balift EG: Sequence of amino acids in the major and minor β chains of the diffuse hemoglobin from BALB/c mice. Biochim Biophys Acta 303:61–67, 1973.
4. Gilman JG: Mouse hemoglobin β chains. Biochem J 159:43–53, 1976.

5. Rougeon F, Mach B: Cloning and amplification of α and β mouse globin gene sequences synthesized in vitro. Gen 1:229–239, 1977.
6. Tilghman SM, Tiemeier DC, Polsky F, Edgell MH, Seidman JG, Leder A, Enquist LW, Norman B, Leder P: Cloning specific segments of the mammalian genome: Bacteriophage λ containing mouse globin and surrounding gene sequences. Proc Nat Acad Sci USA 74:4406–4410, 1977.
7. Tiemeier DC, Tilghman SM, Polsky FI, Seidman JG, Leder A, Edgell MH, Leder P: A comparison of two cloned mouse β-globin genes and their surrounding and intervening sequences. Cell 14: 237–245, 1978.
8. Jahn CL, Hutchison CA III, Phillips SJ, Weavers S, Haigwood NL, Voliva CF, Edgell MH: DNA sequence organization of the β-globin complex in the BALB/c mouse. Cell 21:159–168, 1980.
9. Konkel DA, Tilghman SM, and Leder P: The sequence of the chromosomal mouse β-globin major gene: Homologies in capping, splicing and poly (A) sites. Cell 15:1125–1132, 1978.
10. Konkel DA, Maizel JV Jr, and Leder P: The evolution and sequence comparison of two recently diverged mouse chromosomal β-globin genes. Cell 18:865–873, 1979.
11. Lawn RM, Fritsch EF, Parker RC, Blake G, Maniatis T: The isolation and characterization of linked d- and β-globin genes from a cloned library of human DNA. Cell 15:1157–1174, 1978.
12. Lacy E, Hardison RC, Quon D, Maniatis T: The linkage arrangement of four rabbit β-like globin genes. Cell 1273–1283, 1979.
13. Steinheider G, Melderis H, Ostertag W: Embryonic ϵ chains of mice and rabbits. Nature 257: 714–716, 1975.
14. Gilman JG: Mouse haemoglobin beta chains. Biochem J 155:231–241, 1976.
15. Fantoni A, Bozzoni I, Ullu E, Farace MG: Construction of a recombinant bacterial plasmid containing DNA sequences for a mouse embryonic globin chain. Nucleic Acids Res. 6:3505–3517, 1979.
16. Wu N-C, Sikkema DA, Zucker RM: Mouse fetal hemoglobin. Biochim Biophys Acta 536:306–311, 1978.
17. Kraus LM, Rasad A, Ohba Y, Patterson MT: Mouse fetal hemoglobin. Ann NY Acad Sci 241: 683–690, 1974.
18. Wu N-V, Zucker RM: Mouse fetal hemoglobin synthesis in murine erythroleukemia cells. FEBS Lett 99:299–302, 1979.
19. Whitney JB III, McFarland EC, Russell ES: Interconversion of mouse adult hemoglobin and a neonatal spleen fast hemoglobin band. Dev Biol 65:233–237, 1978.
20. Gutai MW, Nathans D: Evolutionary variants of Simian virus 40: Cellular DNA sequences and sequences at recombinant joints of substituted variants. J Mol Biol 126: 1978.
21. Weaver S, Haigwood NL, Hutchison CA III, Edgell MH: DNA fragments of the Mus musculus β-globin haplotypes Hbbs and Hbbd. Proc Nat Acad Sci USA 76:1385–1389, 1979.
22. Baralle F, Brownlee GG: AUG is the only recognisable signal sequence in the 5′ non-colony regions of eukaryotic mRNA. Nature 274:84–87, 1978.
23. Enquist L, Tiemeier D, Leder P, Weisberg R, Sternberg N: Nature 259:596–598, 1976.
24. Leder A, Miller HI, Hamer DH, Seidman JG, Norman B, Sullivan M, Leder P: Comparison of cloned mouse α and β-globin genes: conservation of intervening sequence locations and extragenic homology. Proc Nat Acad Sci USA 75:6187–6191, 1978.
25. Nishioka Y, Leder P: The complete sequence of a chromosomal mouse α-globin gene reveals elements conserved throughout vertebrate evolution. Cell 18:875–882, 1979.
26. Sato S, Hutchison CA III, Harris JI: A thermalstable sequence-specific endonuclease from Thermus aquaticus. Proc Nat Acad Sci USA 74:542–546, 1977.

Organization and Expression of Globin Genes, pages 89–99

The β-Like Globin Gene Family of Rabbits

Ross C. Hardison, Elizabeth Lacy, Che-Kun James Shen, Eugene T. Butler III, and Tom Maniatis

Department of Microbiology, Cell Biology, Biochemistry and Biophysics, The Pennsylvania State University, University Park (R.C.H.), Department of Zoology, University of Oxford, Oxford, England (E.L.), Division of Biology, California Institute of Technology, Pasadena (C-K.J.S., E.T.B., T.M.)

The pattern of hemoglobin synthesis changes during ontogeny of the rabbit. Early in the 32-day gestational period, six chromatographic variants are produced in the yolk sac-derived nucleated erythrocytes [1]. The hemoglobin tetramers at day 14 of development contain a pair of either of two embryonic β-like ϵ-globins, ϵ_y or ϵ_z, with a pair of either the embryonic α-like χ-globin or the adult α-globin polypeptides [2, 3]. These early hemoglobins are replaced by the adult hemoglobin, $\alpha_2\beta_2$ around days 14–16 [4], which is the time of the shift to fetal liver erythropoiesis [5]. No hemoglobin produced exclusively by the fetal liver-derived erythrocytes has been reported for rabbits [6]. Thus the β-like globin gene family has at least three members, ϵ_y, ϵ_z, and β, with a shift in expression from ϵ_y and ϵ_z to β occurring around days 14–16 of embryonic development. The α-like globin gene family has a minimum of two members, χ and α. The synthesis of χ-globin declines before the 16th day of gestation, and α-globin is synthesized through the remainder of fetal and adult life.

In order to study the events mediating this differential gene expression we have isolated the rabbit β-like globin gene family from a library of genomic DNA fragments carried in recombinant bacteriophage [7, 8]. The gene linkage arrangement [9], gene structure [10, 11], and distribution of repeated sequence elements among the globin genes [12] have been described, and we are analyzing the pattern of transcription from both globin and nonglobin regions within the β-like globin locus [10]. This paper reviews our studies on the β-like globin gene family.

STRUCTURE AND LINKAGE ARRANGEMENT OF THE GENES

The rabbit β-like gene family is comprised of four members (genes $\beta1$–$\beta4$) linked together on a 44-kilobase (kb) chromosomal segment (Fig. 1). Gene $\beta1$ encodes the adult β-globin polypeptide, whereas gene $\beta2$ cannot encode a functional globin polypeptide and is designated a pseudogene, $\psi\beta2$. Genes $\beta3$ and $\beta4$ are both transcribed in embryonic erythrocytes and probably encode the embryonic ϵ-globin

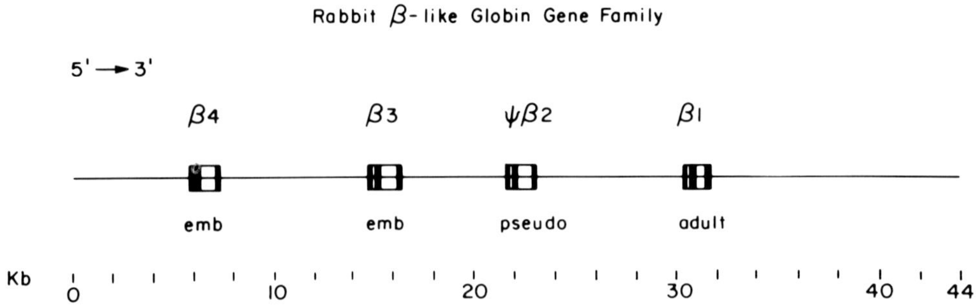

Fig. 1. The organization of the rabbit β-like globin gene family. The four genes are linked on a 44 kilobase segment of the chromosome. Black boxes designate mRNA-coding regions and white boxes are the intervening sequences. Gene β1 is the adult β-globin, ψβ2 is a pseudogene, and β3 and β4 are embryonic β-like globin genes. The arrow indicates the 5′→3′ orientation of all four genes. Distance is in kilobase pairs.

polypeptides. The 5′→3′ transcriptional orientation of all the genes is the same, with the spacing between them varying from 5 to 8 kb.

The rabbit β-like globin genes have common structural features. All four are interrupted by two noncoding, intervening sequences—a small one between codons 30 and 31 and a large one between codons 104 and 105 (Fig. 2). The large intervening sequence (IVS) in the adult β-globin gene was first seen by a blot-hybridization analysis of genomic DNA [13] concurrently with the discovery of a large IVS in the mouse adult β-globin gene by R-loop analysis of a cloned genomic DNA fragment [14]. Nucleotide sequence analysis of cloned DNA revealed a second small IVS and established the precise location of both intervening sequences in the rabbit [10, 15] and mouse [16] β-globin genes. The nucleotide sequence of gene ψβ2 [11] and partial sequence analysis of genes β3 and β4 [10, unpublished results] show that the positions of IVS I and II are conserved in all four rabbit β-like genes. Two IVSs are found in equivalent positions in α-like and β-like globin genes from all species examined [see 17 and 18 for review].

However, the sizes and sequences of the IVSs vary among the genes. The large IVS ranges in size from 573 base pairs (bp) for β1 to about 800 bp for ψβ2, β3, and β4. No homology, aside from short regions at the coding/IVS boundary, has been found between the large IVSs of the four genes [10, 11]. Although IVS I of ψβ2 is 26 bp shorter than its counterpart in β1 as a result of a deletion in the 3′ end of the IVS, substantial sequence homology (about 70%) is still observed between the two small IVSs. The greater conservation of IVS I than IVS II has been found for several recently diverged globin gene pairs, ie, human $^{G}\gamma/^{A}\gamma$, β/δ, $^{G}\gamma/\epsilon$, and mouse β^{maj}/β^{min}. Two possibilities have been suggested to account for these data: (1) Either the 5′ regions of these gene pairs have been corrected against each other or (2) the 5′ regions of the genes are subject to greater selective pressure [18].

The adult β-globin gene isolated from our rabbit library [7] is one of the two codominant alleles expressed in rabbits [19, 20, 21]. The gene isolated by van den Berg et al [15] is another adult β-globin allele, as shown by the complete nucleotide sequence of both genes [22, 10]. The four amino acid replacements in the two alleles each result from single base substitutions; however, no silent site substitutions and only two substitutions in the noncoding regions are found. This is in marked

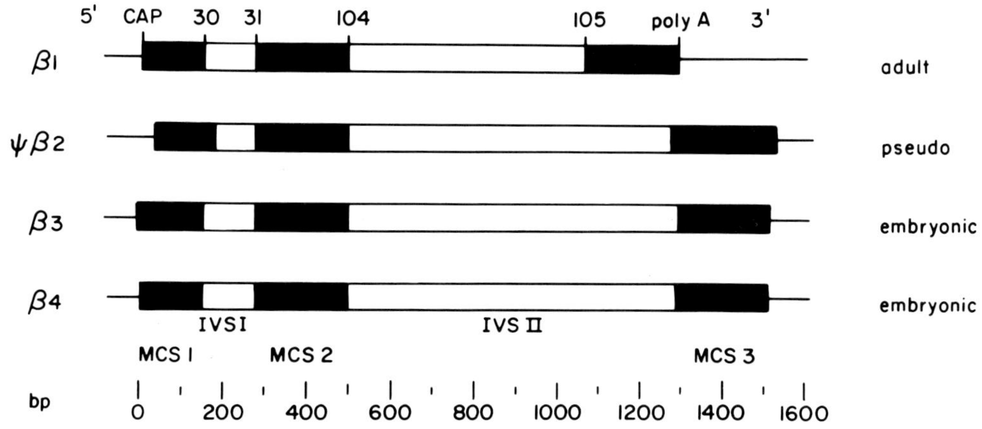

Fig. 2. A comparison of the general structures of the rabbit β-like globin genes. The four genes were aligned at codon 104. The black regions encode mRNA (translated and untranslated) and the white regions are intervening sequences. Codon positions are indicated by the numbers at the top. IVS: Intervening sequence, MCS: mRNA coding sequence. Distance is in base pairs.

contrast to the comparison of two human $^{A}\gamma$-globin alleles, in which 15 base substitutions and three gaps are seen in the noncoding regions but no substitutions (silent or replacement) are found in the protein coding regions [23]. Also, two alleles of human preproinsulin differ by four base substitutions in the intervening and 3′ untranslated sequences, but are identical in the protein coding regions [24]. The amino acid replacements in the rabbit β-globin alleles were fixed in the population by positive selection, not neutral drift [25]. This could have occurred either by a saltatory event in the recent past (about 0.5 million years ago), in which case substitutions in noncoding regions have not had time to accumulate, or the replacement substitutions occurred gradually over the past 12 to 15 million years ago with a selection against divergence in noncoding regions. If the latter were true, such a hypothetical selection did not operate on the noncoding regions of the human $^{A}\gamma$-globin or preproinsulin alleles.

TRANSCRIPTION OF THE GENES

Gene β1 is transcribed into a presumptive precursor RNA of about 1,400 nucleotides (nt) (including the poly A) which is colinear with the gene, including the intervening sequences [10, 26]. The 5′ and 3′ ends of this precursor match the ends of the mature mRNA, so it does not contain any extragenic flanking sequences [27]. This latter study, employing the S1 mapping procedure of Berk and Sharp [28], also provides evidence for the ordered removal of IVS I and II, with at least two processing steps for each IVS [see Flavell et al, this volume]. The presence of intervening sequences in the precursor RNA was first shown for the mouse β-globin gene [29, 30, 31] but the order and number of processing steps may differ from that seen with rabbit β-globin [32]. No direct evidence, however, has been presented that the observed processing intermediates are obligatory for correct expression [27], nor is it known that the predominant 1,400 nt precursor is the primary transcript.

Mature mRNAs from genes $\beta3$ and $\beta4$ are abundant in 12-day rabbit embryos but not in adult bone marrow. A full-length cDNA copy of $\beta4$ mRNA and a partial copy of $\beta3$ mRNA have been cloned in the plasmid pBR322. The nucleotide sequence of the $\beta4$ cDNA plasmid aligns with the genomic DNA sequence, showing that the mRNA indeed came from $\beta4$ and not another closely homologous gene [10]. Gene $\beta3$ shows only weak homology to the other genes and does not cross react with other genomic DNA fragments, strongly suggesting that the mRNA seen with the $\beta3$ probe is indeed a transcript of gene $\beta3$.

We have recently extended our analysis of β-like globin gene transcription during embyronic development. Transcripts of $\beta3$ and $\beta4$ are detectable at day 11 of gestation (four days after implantation), are very abundant at days 12 to 14 (just preceding the switch to fetal liver erythropoiesis), but are much reduced in the 22-day fetal liver (about 8–10 days prior to birth). In contrast, gene $\beta1$ transcripts are rare at days 11 and 12, are increased by day 14, and are abundant in the 22-day fetal liver and adult bone marrow [R.H., unpublished observations]. These data are consistent with the isoelectric focusing analysis of rabbit hemoglobins that showed only adult hemoglobin to be present at day 16 of gestation [4]. The results taken together indicate that genes $\beta3$ and $\beta4$ should be considered embryonic β-like globin genes. We have not yet, however, directly demonstrated that genes $\beta3$ and $\beta4$ encode the ϵ_y and ϵ_z polypeptides partially characterized by Steinheider et al [1, 3].

The expression of $\beta3$ and $\beta4$ is coordinate, since $\beta3$ RNA is always present with $\beta4$ RNA, although the time of maximal expression of the two genes may differ. The switch to $\beta1$ production is reciprocal, with $\beta3$ and $\beta4$ RNA synthesis declining while $\beta1$ synthesis increases. We observe α-globin polypeptide in abundance at the earliest developmental stages examined (day 11; data not shown), which is clearly before the switch to fetal liver erythropoiesis and before the switch to β-globin production. However, the precise temporal pattern of embryonic and adult globin expression during embryonic development is still being investigated.

PSEUDOGENE $\psi\beta2$

No transcript has been found for gene $\beta2$ by a blot-hybridization analysis [33] of RNA from either adult bone marrow or primitive erythrocytes from 12-day embryos [10]. A more rigorous analysis using the S1 nuclease mapping techniques [28] failed to detect any $\beta2$ RNA the size of either precursor or mature mRNA [E.L., unpublished results]. The latter assay can detect less than ten copies of a transcript per cell, so the negative results strongly suggest that $\beta2$ is not expressed in the tissues examined.

Comparison of the nucleotide sequences of $\beta2$ and $\beta1$ revealed two frameshift mutations and two in-phase translation termination codons, as well as mutations at several sites highly conserved in other globin genes (Fig. 3) [11]. These alterations render $\beta2$ nonfunctional as a globin gene, and it is therefore referred to as a pseudogene, $\psi\beta2$.

An extensive sequence comparison between $\psi\beta2$ and $\beta1$ is presented by Lacy and Maniatis [11], but three observations bear repeating here. 1) The junctions between mRNA coding and intervening sequences in $\psi\beta2$ have diverged significantly; for example, the G of the highly conserved AG at the 3′ end of IVS II has been deleted (Fig. 3). Thus, if $\psi\beta2$ were ever transcribed, the precursor RNA could not be spliced in the same way as the $\beta1$ transcript. 2) The 5′ flanking and untranslated regions are very similar between $\psi\beta2$ and $\beta1$. However, the "CCAAT" box of homology, found 80 nucleotides before the CAP site of many eukaryotic genes [34, 18], has been cleanly deleted in $\psi\beta2$. This deletion does not prevent transcription of $\psi\beta2$ in vitro [N. Proudfoot and M. Shander, unpublished observations]. 3) By comparing the nucleotide sequences of the protein coding regions of $\psi\beta2$ and $\psi1$, one can calculate the time since the genes started diverging [11, 18]. This analysis indicates that $\psi\beta2$ diverged from $\beta1$ about 55 million years ago, about the same time that human β and δ diverged. Furthermore, $\psi\beta2$ apparently diverged as a functional globin gene for 22 million years and then diverged as a nonfunctional gene for the past 33 million years.

REPEATED SEQUENCES IN THE GENE CLUSTER

The distribution of repeated sequences throughout the β-globin gene cluster has been analyzed by Shen and Maniatis [12]. Eco RI fragments containing repeated sequence elements were initially mapped by two-dimensional blot-hybridization [35]. Electron microscopy of heteroduplexes and homoduplexes of the several isolated genomic clones mapped the position, size, and orientation of many of the repeated sequences. At least five families of interspersed repeats could be defined (Fig. 4). Two of the families (C and D) have different lengths of repeated elements at some positions in the gene cluster (designated by the lengths of the arrows in Fig. 4). No one family is interspersed among all four genes, but each gene has at least one pair of inverted repeats around it. A 1,400 bp inverted repeat of the D family brackets the entire gene family as isolated. These repeated elements appear to be highly reiterated in the genome, as evidenced by the multitude of bands seen when restriction fragments bearing the repeats are used as probes in genomic blots.

At least two of the gene families (C and D) are homologous to transcripts generated in vivo or in vitro. One of us [E.T.B., unpublished results] has isolated a cDNA plasmid (called h29) synthesized from 12-day-embryo RNA that hybridizes to the restriction fragments containing the D family of repeats. However, because h29 DNA is highly repeated in the genome we do not know if the RNA from which the plasmid was derived actually originated from the β-like gene locus.

The C family of repeats are transcription units for RNA polymerase III in vitro [C-K. J.S., unpublished results]. At a low stringency, these DNA fragments will hybridize to the "Alu family" sequences that are highly repeated in the human genome and are interspersed among the human β-like globin genes [36, 37]. Some of the Alu family repeats around the human β-globin genes promote transcription by RNA polymerase III in vitro [38]. Questions about the existence of these transcripts in vivo and their role (if any) in structural gene expression or replication remain unanswered. The fact that they are partially conserved between rabbits and humans suggests a function of some importance.

The rabbit β-like globin gene cluster contains nonglobin sequences which are represented in embryonic poly A-containing RNA as shown by hybridization of the cloned DNA fragments to labeled embryonic cDNA [R.C.H. and E.T.B., unpublished results]. Since all of the hybridizing fragments contain repetitive sequences, it is not known whether any of the transcripts originate from the β-globin gene cluster.

```
ψβ2  TCAGACTCATAAGCCAGTGCCAGAGCTGACCATGACCAGTGTTCTGTCACTCTTCTGGCCTCACCCTGGCATTGGTGACTCACCAGGGTAGGCAGTGCAGGGGCCACTACTGGGCATAAAAG

ψβ2  GCAGAGTGGGACAGCTGCTGCTTATGCTTGTATTTGACACAACTGTGTTTACTAGCAATATTCCAAAT———AGAATG GTG CAT CTC TCT GGT GAA GAG AAG GAT ACA TTC
β1                         CapACACTTGCTTTTGACACAACTGTGTTTACTTGCAATCCCCCAAAACAGACAGAATG GTG CAT CTG TCC AGT GAG GAG AAG TCT GCG GTC
                                                                                       1               5                        10
                                                                                 ↓IVS1
ψβ2  ACT GCC CTG TGG GGC AAG GTT GAT GT- GAA GAA GTT GGT GAT GAG GCC CTG AGA GGG TTGCTTTTCAAGTTATAGGGCAGTCTATTGAGGCAGATGGAAGCTGG
β1   ACT GCC CTG TGG GGC AAG GTG AAT GTG GAA GAA GTT GGT GGT GAG GCC CTG GGC AGG TTGGTATCCTTTTTACAGCACAACTTAATGAGACAGATAGAAACTGG
                 15                  20                  25                  30
                                                                                                  ↓IVS1
ψβ2  GCTTGTGTAGACAGAGCTGGCCCCAGGGTTTCTGCCAGG—C—ACT———GAATCCCCCA———————————————————— GTG CTG ATT GTC TAC CCA AGA ACC CAG AGG
β1   TCTTGTAGAAACAGAGTAGTCGCCTGCTTTTCTGCCAGGTGCTGACTTCTC——TCCCCTGGGCTGTTTCATTTTCTCAGG CTG CTG GTT GTC TAC CCA TGG ACC CAG AGG
                                                                                      31              35                  40

ψβ2  TTC TTT ATA CCG TTT GGG GAC CTG TCC TCT GAA CCT GCT GTT ATG AGT AAT GTC AAG GTG ACG GCC CAT TGC AAG AAG GTG CTG GCT ACC TTC
β1   TTC TTC GAG TCC TTT GGG GAC CTG TCC TCT GCA CAT GCT GTT ATG AGC AAT CCT AAG GTG AAG GCT CAT GGC AAG AAG GTG CTG GCT GCC TTC
                 45                  50                  55                  60                  65                  70

ψβ2  AGT GAG GGT CTG AAT CAC CTG GAC AAC CTC GGG GAA AAT TTT GTC CAG CTG AGT GAG CTG CAC TGT CAC AAG CTC CAT GTG GAT CCT GAG AAC
β1   AGT GAG GGT CTG AAT CAC CTG GAC AAC CTC AAA GGC ACC TTT GCT AAG CTG AGT GAA CTG CAC TGT GAC AAG CTG CAC GTG GAT CCT GAG AAC
                 75                  80                  85                  90                  95                  100
       ↓IVS2
ψβ2  TTC AGT GTTAGTCTAGGCAATGTTCCACTTTTTTCTTTTCACATTCTCATCTTCTTTTTTTGTTGTTTGTTTGTTAAGCAACCTGTCTTTGTCAACATTTCTATCTAATTCCCTATATTTC
β1   TTC AGG
         104
```

```
     580       590       600       610       620       630    ∽324 bp   970       980       990       1000      1010
ψβ2  GTCATTTATTGCTTTTCAAAAATTTTCATTTTTCTTTAGTATTTCGTTTTTTCATTCCTCAC————GGACATAGCAAAGGCACATTGTAGGTGGGGTTGCAGATGTTGCTAAGAAGCAA

     1020      1030      1040      1050      1060      1070      1080      1090      1100      1110      1120      1130
ψβ2  CCACTATTACATCAATTAAGGGATTAATTAGTAATATTATTATATATGCATATACATGTACATACCTATGTGATTATACACTTATATATATATTGGTTGTTTTAAATTGAGACTCTATTTAGA
                                                                                                          ↓IVS2
     1140      1150      1160      1170      1180      1190      1200      1210      1220      1230      1240       1250
ψβ2  TTAAGATGTAGGAAGAAAATAATTAGTGTTGGCCCAATTTCAGAATACAAGCTTGATTTCTGTGCTAACCATACGTAAAATATCTCCATAGCTCTTTCTCACA CTC CAG GGC AAT GGG
                                                                                                          ||| | | ||| ||  | |
β1                                                                                                        CTC CTG GGC AAC GTG
                                                                                                          105

     1260            1270            1280            1290            1300            1310            1320            1330            1340
ψβ2  CTG GTG ATT GTG CTG GCC TTC TAC TTC GGC AAA GAA TTT ACC CCA TAG CTG AAG G–T GCC TAT CAC AGA GTG GCA GCT GGT GTG GCC AAT GCC
     ||| ||   || ||| |||  |        |  || ||| ||| ||| ||  ||  ||   ||  ||  ||  | | ||| ||| ||  |   |||  |  ||| ||| ||| ||| ||| |||
β1   CTG GTT GTT GTG CTG TCT CAT CAT TTT GGC AAA GAA TTC ACT CCT CAG GTG CAG GCT GCC TAT CAG AAG GTG GTG GCT GGT GTG GCC AAT GCC
     110                 115                 120                 125                 130                 135                 140

     1350            1360            1370      1380      1390      1400      1410      1420      1430      1440      1450      1460
ψβ2  TGG GCT CAC AAG TAC CCC TAAGATCTTGACTTGTTCTCTACTTACCCTGTTCCATGGAAAATCTTTTCTATATTTCATCTATTG–AACATGAAGACATAATACCCAACCTAAGGGTG
      |  ||| ||| ||  ||| | | | |||||||                    || |||| | |||   |||||                     ||| |  |||| | || |     |||
β1   CTG GCT CAC AAA TAC CAC TGAGATCTT————————————TTTCCCTCTGCCA——AAAAT——————————————TATGGGGACATCATGA–AG——CCCCT————————
                 145

     1470                 1480
ψβ2  TGAGCC——————————TAATAAAAGATCA
     |||||            ||||||| ||
β1   TGAGCATCTGACTTCTGGCTAATAAAGGAAATTTATTTTCATTGC
```

Fig. 3. The nucleotide sequence of gene $\psi\beta 2$. The sequence has been aligned with the nucleotide sequence of gene $\beta 1$. Vertical lines are drawn between homologous nucleotides in the two genes, and dashes indicate deletions. Nucleotides in $\psi\beta 2$ are numbered above the line beginning (+1) with the nucleotide homologous to the Cap site in $\beta 1$. Numbers below the line designate the codon in gene $\beta 1$. The translation initiation codon (ATG) in $\beta 1$ is underlined, as is the AATAAA found before the 3′ poly A addition site in most eukaryotic mRNAs. Reprinted with permission from Cell (1980).

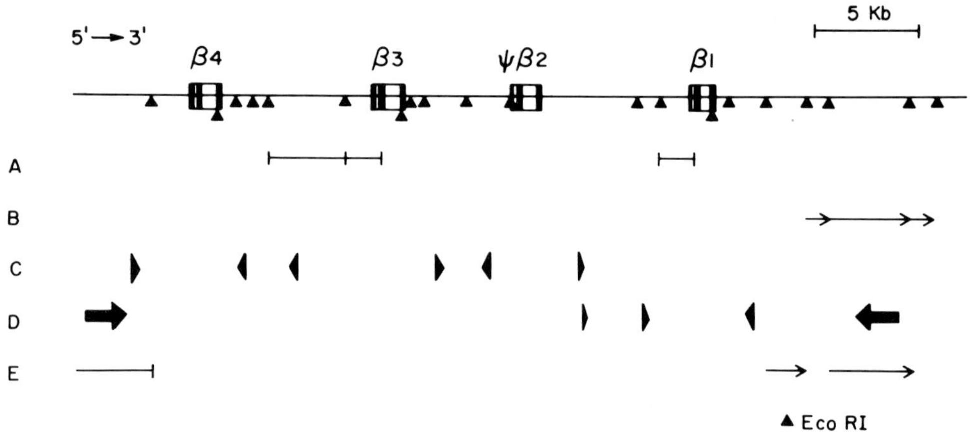

Fig. 4. Repetitive elements interspersed among rabbit β-like globin genes. The bars and arrows indicate the position of each member of a family of repeated sequences. The direction of each arrow indicates the relative orientation of the repeat; the length of the arrow shows the length of the repeat. The bars indicate that a member of a repeat family is found in that Eco RI restriction fragment. The triangles are Eco RI cleavage sites. Although families B and E have not been mapped by electron microscopy, the members located 3′ to β1 probably are in the same orientation (direct repeats) because no snap-backs have been observed in homoduplexes of this region.

DISCUSSION

A comparison of globin gene structure and arrangement in different species reveals some interesting similarities and differences. The embryonic and fetal genes of rabbit β [9], human β [37], human α [39], and mouse β [40] globin families lie to the 5′ side of the adult β-globin genes; that is, the genes are arranged 5′→3′ in the order of their expression. This is apparently not true in chickens [41]. The genes in all these families are usually within 8–16 kb of each other, yet the sheep β-like globin genes appear to be much further apart [42]. These variations in chromosomal organization might be better understood in terms of chromosomal activation. For example, if the boundaries of the activated chromosomal domains were mapped [43], perhaps the genes would fit into more obvious functional units.

Many coordinately expressed globin genes are arranged in pairs, but the extent of sequence homology between the two members varies considerably. The human α_1 and α_2-globin genes share extensive homology [39], as do the human $^{G}\gamma$ and $^{A}\gamma$-globin genes [23]. This homology includes coding, intervening, and flanking sequences and is thought to be the result of a gene correction mechanism which acts during evolution [23, 39, 44]. Although the adult human δ and β-globin genes appear to have been corrected against each other at one time in their evolutionary history [18], their intervening and flanking sequences have diverged extensively [45, 18]. Similarly, the rabbit β3 and β4 genes are coordinately expressed but the only detectable homology is in the middle coding region of the two genes [10]. At the extreme end of the divergence spectrum are the pseudogenes which do not encode a functional globin polypeptide.

Some aspects of chromosomal organization are common to the globin gene clusters in different species. The conservation of number and positions of intervening sequences has already been mentioned. Pseudogenes have been found in most globin gene families: rabbit β [10, 11] human α [46], mouse β [40], mouse α [47, 48], and perhaps human β [37] and goat β [49]. In the rabbit β, human α, and mouse α-globin families, the pseudogene is located between the embryonic/fetal genes and the adult genes. The pseudogenes may simply be by-products of the gene duplication and divergence process in the evolution of multigene families. Although this does not preclude some role in mediating expression of functional globin genes, it is apparent that some pseudogenes arose after the divergence of adult and embryonic/fetal genes. Thus the phenomenon of developmental regulation of globin gene expression predates the appearance of some pseudogenes [11, 46].

Repeated sequence elements are interspersed among globin genes in the rabbit β [12], human β [37], human α [39], and chicken β [41] families. Some of these repeats in the human β [38], rabbit β, and human α [C-K.J.S., unpublished results] gene families are transcribed in vitro by RNA polymerase III. It has been suggested that this family of repeats in the human genome, the Alu family, may act as replication origins or may be involved in the production of transcripts from structural genes [36]. Further studies of all the transcription units within the globin gene families should help elucidate mechanisms for their developmentally regulated expression.

SUMMARY

The set of four differentially expressed β-like globin genes from rabbits is linked together in a 44-kilobase chromosomal segment in the order 5′-β4-(8 kb)-β3-(5 kb)-β2-(7 kb)-β1-3′. All four genes are interrupted by two intervening sequences (IVSs)—a small one between codons 30 and 31 and a large one between codons 104 and 105. Gene β1 encodes the adult β-globin polypeptide, whereas gene β2 cannot encode a functional globin polypeptide and is designated a pseudogene, $\psi\beta$2. The two codominant alleles of adult β-globin differ by four nucleotide substitutions that result in four amino acid replacements; but except for two nucleotide substitutions in the large IVS, the flanking and intervening sequences of the two alleles are identical. Genes β3 and β4 are coordinately transcribed in embryonic erythrocytes and probably encode the embryonic ϵ-globin polypeptides. Transcripts of β3 and β4 are replaced by β1 transcripts at the time of the shift to fetal liver erythropoiesis. At least five families of repeated sequences are interspersed among the β-like globin genes. These sequence elements are repeated elsewhere in the genome. One of the families is transcribed in vivo and another promotes transcription by RNA polymerase III in vitro.

ACKNOWLEDGMENTS

This research was supported by a grant from the National Science Foundation to T.M., an NIH graduate training grant to the California Institute of Technology (E.L.), an NIH postdoctoral training grant to the California Institute of Technology (C.-K. J.S., and E.T.B.), postdoctoral fellowships from the Jane Coffin Childs Memorial Fund for Medical Research and from the NIH (R.C.H.), and a Rita Allen Foundation Career Development Award (T.M.).

REFERENCES

1. Steinheider G, Melderis H, Ostertag W: Mammalian embryonic hemoglobins. In Martin H, Novicke L (eds): "International Symposium on the Synthesis, Structure and Function of Hemoglobin," Lehmans, Munich: Germany, 1972, pp 222–235.
2. Melderis H, Steinheider G, Ostertag W: Evidence for a unique kind of α-type globin chain in early mammalian embryos. Nature 250:774–776, 1974.
3. Steinheider G, Melderis H, Ostertag W: Embryonic ϵ chains of mice and rabbits. Nature 257:714–716, 1975.
4. Jelkmann W, Bauer C: Oxygen affinity and phosphate compounds of red blood cells during intrauterine development of rabbits. Pflügers Arch 372:149–156, 1977.
5. Sorenson G: Hepatic hematocytopoiesis in the fetal rabbit: A light and electron microscopic study. Ann NY Acad Sci 111:44–69, 1963.
6. Kitchen H, Brett I: Embryonic and fetal homoglobin in animals. Ann NY Acad Sci 241:653–670, 1974.
7. Maniatis T, Hardison R, Lacy E, Lauer J, O'Connell C, Quon D, Sim GK, Efstratiadis A: The isolation of structural genes from libraries of eukaryotic DNA. Cell 15:687–701, 1978.
8. Lacy E, Lawn R, Fritsch E, Hardison R, Parker T, Maniatis T: Isolation and characterization of mammalian globin genes. In Stamatoyannopoulos G, Nienhuis A (eds): Cellular and Molecular Regulation of Hemoglobin Switching. New York: Grune and Stratton, 1979, pp 501–519.
9. Lacy E, Hardison R, Quon D, Maniatis T: The linkage arrangement of four rabbit β-like globin genes. Cell 18:1273–1283, 1979.
10. Hardison R, Butler E III, Lacy E, Maniatis T, Rosenthal N, Efstratiadis A: The structure and transcription of four linked rabbit β-like globin genes. Cell 18:1285–1297, 1979.
11. Lacy E, Maniatis T: The nucleotide sequence of a rabbit β-globin pseudogene. Cell 21:545–553, 1980.
12. Shen C-KJ, Maniatis T: The organization of repetitive sequences in a cluster of rabbit β-like globin genes. Cell 19:379–391, 1980.
13. Jeffreys AJ, Flavell RA: The rabbit β-globin gene contains a large insert in the coding sequence. Cell 12:1097–1108, 1977.
14. Tilghman S, Tiemeier D, Seidman J, Peterlin M, Sullivan M, Maizel J, Leder P: Intervening sequence of DNA identified in the structural portion of mouse β-globin gene. Proc Natl Acad Sci USA 75:725–729, 1978.
15. van den Berg J, van Ooyen A, Mantei N, Schamböck A, Grosveld G, Flavell R, Weissman C: Comparison of cloned rabbit and mouse β-globin showing strong evolutionary divergence of two homologous pairs of introns. Nature 276:37–44, 1978.
16. Konkel D, Tilghman S, Leder P: The sequence of the chromosomal mouse β-globin major gene: Homologies in capping, splicing and poly (A) sites. Cell 15:1125–1132, 1978.
17. Maniatis T, Fritsch E, Lauer J, Lawn T: The molecular genetics of human hemoglobins. Ann Rev Biochem, in press, 1980.
18. Efstratiadis A, Posakony J, Maniatis T, Lawn R, O'Connell C, Spritz R, DeRiel J, Forget B, Weissman S, Slightom J, Blechl A, Smithies O, Baralle F, Shoulders C, Proudfoot N: The structure and evolution of the human β-globin gene family. Cell 21:653–668, 1980.
19. Galizzi A: The amino acid sequence of the β-chains of rabbit hemoglobin. Eur J Biochem 17:49–55, 1970.
20. Gallizi A: Polymorphic alleles of the gene for the β-chain of rabbit hemoglobin. Nature 229:142–143, 1971.
21. Bricker J, Garrick MD: An isoleucine-valine substitution in the β-chain of rabbit hemoglobin. Biochim Biophys Acta 351:437–441, 1974.
22. van Ooyen A, van den Berg J, Mantei N, Weissman C: Comparison of total sequence of a cloned rabbit β-globin gene and its flanking regions with a homologous mouse sequence. Science 206:337–344, 1979.
23. Slightom J, Blechl A, Smithies O: Human fetal $^{G}\gamma$ and $^{A}\gamma$ globin genes: Complete nucleotide sequences suggest that DNA can be exchanged between these duplicated genes. Cell 21:627–638, 1980.

24. Ullrich A, Dull T, Gray A, Brosius J, Sures I: Genetic variation in the human insulin gene. Science 209:612–615, 1980.
25. Perler F, Efstratiadis A, Lomedico P, Gilbert W, Kolodner R, Dodgson J: The evolution of genes: The chicken preproinsulin gene. Cell 20:555–566, 1980.
26. Flavell RA, Bernards R, Grosveld GC, Hoeijmakers-Van Dommelen HAM, Kooter JM, DeBoer E, Little PFR: The structure and expression of globin genes in rabbit and man. In Axel R, Maniatis T, Fox CF (eds): "Eukaryotic Gene Regulation. ICN-UCLA Symposium on Molecular and Cellular Biology," Vol XIV, New York: Academic Press, 1979, pp 335–354.
27. Grosveld GC, Koster A, Flavell R: A transcription map for the rabbit β-globin gene. Cell, in press, 1981.
28. Berk A, Sharp P: Sizing and mapping of early adenovirus mRNAs by gel electrophoresis of S1-endonuclease digested hybrids. Cell 12:712–732, 1977.
29. Tilghman S, Curtis P, Tiemeier D, Leder P, Weissmann C: The intervening sequence of a mouse β-glb gene is transcribed within the 15S β-glb mRNA precursor. Proc Natl Acad Sci USA 75:1309–1313, 1978.
30. Smith K, Lingrel JB: Sequence organization of the β-globin mRNA precursor. Nucleic Acids Res 5:3295–3301, 1978.
31. Kinniburgh A, Mertz J, Ross J: The precursor of mouse β-globin messenger RNA contains two intervening RNA sequences. Cell 14:681–693, 1978.
32. Kinniburgh A, Ross J: Processing of the mouse β-globin mRNA precursor. Cell 17:915–921, 1979.
33. Alwine J, Kemp D, Stark G: Method for detection of specific RNAs in agarose gels by transfer to diazobenzyloxymethyl-paper and hybridization with DNA probes. Proc Natl Acad Sci USA 74:5350–5354, 1977.
34. Benoist C, O'Hare K, Breathnach R, Chambon P: The ovalbumin gene: Sequence of putative control regions. Nucleic Acids Res 8:127–142.
35. Sato S, Hutchison CA III, Harris JI: A thermostable sequence-specific endonuclease from Thermus aquaticus. Proc Natl Acad Sci USA 74:542–546, 1977.
36. Jelinek W, Toomey T, Leinwand L, Duncan C, Biro P, Choudary P, Weissman S, Rubin C, Houck C, Deininger P, Schmid C: Ubiquitous, interspersed repeated sequences in mammalian genomes. Proc Natl Acad Sci USA 77:1398–1402, 1980.
37. Fritsch EF, Lawn RM, Maniatis T: Molecular cloning and characterization of the human β-like globin gene cluster. Cell 19:959–972, 1980.
38. Duncan C, Biro P, Choudary P, Elder J, Wang R, Forget B, DeRiel J, Weissman S: Rna polymerase III transcriptional units are interspersed among human non-α-globin genes. Proc Natl Acad Sci USA 76:5095–5099, 1979.
39. Lauer J, Shen C-KJ, Maniatis T: The chromosomal arrangement of human α-like globin genes: Sequence homology and α-globin gene deletions. Cell 20:119–130, 1980.
40. Jahn CL, Hutchinson CA III, Phillips SJ, Weaver S, Haigwood NL, Voliva CF, Edgell MH: DNA sequence organization of the β-globin complex in the BALB/c mouse. Cell 21:159–168, 1980.
41. Dodgson J, Strommer J, Engel JD: Isolation of the chicken β-globin gene and a linked embryonic β-like globin gene from a chicken DNA recombinant library. Cell 17:879–887, 1979.
42. Kretschmer PJ, Coon H, Davis A, Harrison M, Nienhuis AW: The structure and organization of the sheep globin genes, in press, this volume, 1980.
43. Stalder J, Larsen A, Engel J, Weintraub H: Tissue-specific DNA cleavages in the globin chromatin domain introduced by DNase I. Cell 20:451–460, 1980.
44. Zimmer E, Martin S, Beverly S, Kan Y, Wilson A: Rapid duplication and loss of genes coding for the α chains of hemoglobin. Proc Natl Acad Sci USA 77:2158–2162, 1980.
45. Lawn R, Fritsch E, Parker R, Blake G, Maniatis T: The isolation and characterization of linked δ- and β-globin genes from a cloned library of human DNA. Cell 15:1157–1174, 1978.
46. Proudfoot N, Maniatis T: Nucleotide sequence of a human α-globin pseudogene. Cell 21:537–544, 1980.
47. Nishioka Y, Leder A, Leder P: Unusual α-globin-like gene that has cleanly lost both globin intervening sequences. Proc Natl Acad Sci USA 77:2806–2809, 1980.
48. Vanin ES, Goldberg GI, Tucker PW, Smithies O: A mouse α-globin-related pseudogene lacking intervening sequences. Nature 286:222–226, 1980.
49. Lingrel JB, Haynes JR, Schon EA, Clearly ML, Gallagher PM: Structural organization of the normal and defective globin genes of the goat, in press, this volume, 1980.

Co-Evolution and Control of Globin Genes

Oliver Smithies, Ann E. Blechl, Sixiang Shen, Jerry L. Slightom, and Elio F. Vanin

Laboratory of Genetics, University of Wisconsin, Madison

INTRODUCTION

Recent successes in cloning and sequencing long stretches of DNA have resulted in a great deal of detailed information about the mammalian globin genes at the nucleotide level. From this information one can draw a number of interesting new inferences concerning the evolution and co-evolution of these genes. One can also begin to see indications of unexpected genetic elements within the globin system which may be important in controlling the expression of the globin structural genes at different development stages. The description of these various new findings, and the ideas which stem from them, will be drawn from data obtained in our laboratory, and from the work of many other groups. (See paper by Maniatis et al in this volume.)

ANATOMY OF THE GLOBIN GENE CLUSTERS AT THE DNA LEVEL

Figure 1 (modified from Efstradiatis et al [1]) summarizes the general disposition of the human α and β globin gene clusters on chromosomes 16 and 11 respectively. Each cluster contains several genes. They are all relatively close together on their respective chromosomes (the scale is such that the line representing the DNA of the β region is about 60,000 base pairs (60 kbp); the α cluster is about 40 kbp). The genes are all oriented in the same direction relative to transcription with 5′ to the left and 3′ to the right, so that transcription is left to right. All the human globin polypeptides (α,β,γ,δ,ϵ, and ζ) are accounted for, and the α, γ, and ζ are coded for by duplicate genes.

In addition to the genes coding for known polypeptides, there are some extra genes in these clusters, which I shall refer to as pseudogenes: one is known in the α cluster and two are known in the β cluster. Between many, but not all, of the genes there are repeated sequences of DNA. The genes are arranged on the chromosome in the order in which they are used developmentally; for example, in the β globin gene cluster the embryonic globin gene, ϵ, is 5′ to the fetal globin genes, $^{G}\gamma$ and $^{A}\gamma$,

A closely related version of this paper is published in the 39th Symposium of the Society for Developmental Biology, Subtelny S, Abbott UK (eds): "Levels of Genetic Control in Development." New York: Alan R. Liss, 1981, in press.

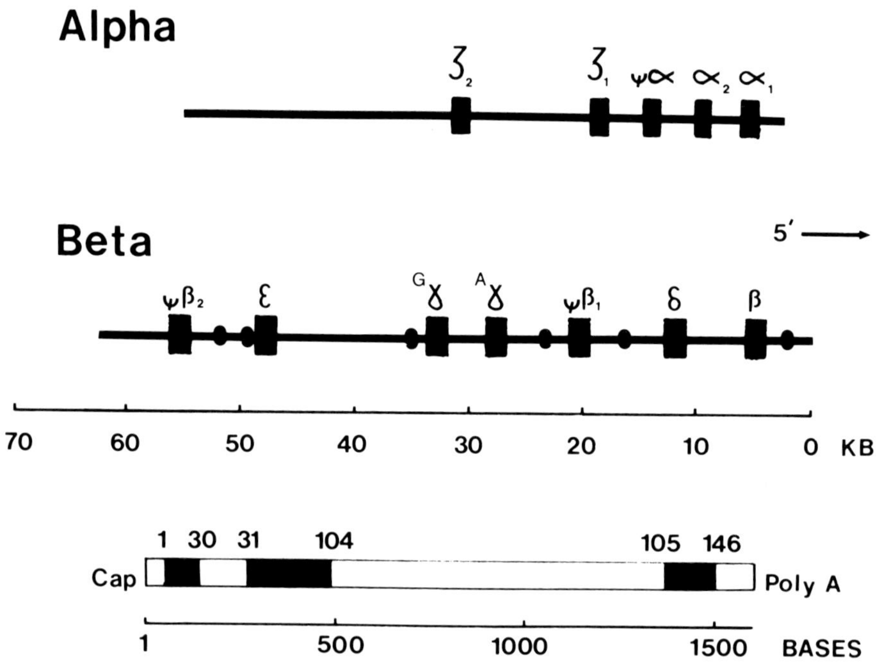

Fig. 1. The human α and β globin clusters (modified from Efstradiatis et al [1]). Genes in the clusters are shown as black bars. The solid circles indicate the distribution of some repeated DNA sequences in the β gene clusters (Fritsch et al [16]). The arrangement of coding and intervening sequences in the $^{G}\gamma$ fetal globin gene is shown on an enlarged scale; the numbers refer to amino acid residues.

and these are 5′ to the adult globin genes δ and β; similarly the embryonic ζ genes are 5′ to the adult α genes.

EVOLUTION OF GLOBIN GENES

Amino acid sequence data available for many years [2] have established that the globins evolved from a common ancestor by a process of gene duplication and subsequent evolution. Consequently it is not surprising that the recent data at the DNA level show that all of the genes which yield a globin product (ie, excluding the pseudogenes) have the same general anatomy, which is illustrated in the case of the $^{G}\gamma$ gene in the lower part of Figure 1. For example, all the productive β-like genes have two intervening sequences (or introns): the first between the codons for amino acid residues 30 and 31 and the second between residues 104 and 105. Both the α- and β-type globin genes have their intervening sequences in homologous positions, which leads to the likely inference that the common ancestor of the α and β genes had already acquired intervening sequences, as was originally pointed out by Leder et al [3].

The genes for the α- and β-type globins are located on separate chromosomes in most species, and it is not surprising that although they once had a common ancestor they have now diverged considerably. An example of this divergence at the DNA level is seen in the size of the second intervening sequence. In all

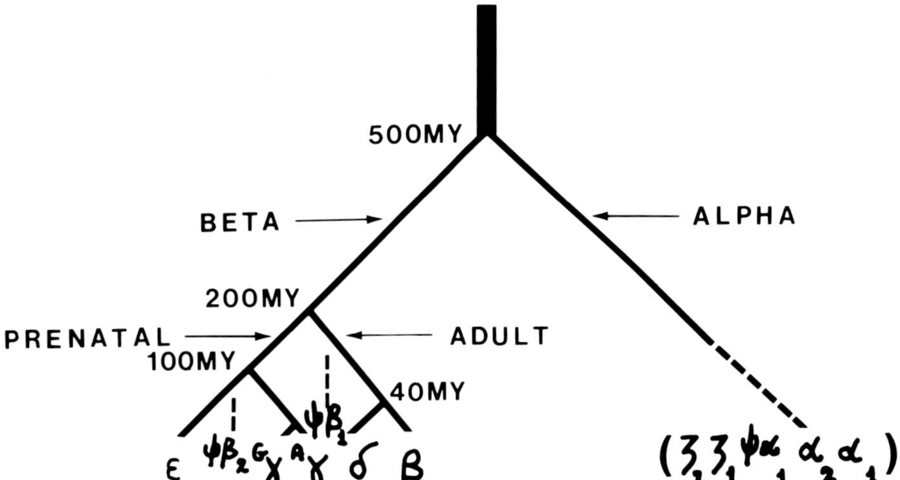

Fig. 2. An evolutionary tree for the human globin genes (modified from Efstradiatis et al [1]). The tree was constructed by direct comparisons of the nucleotide sequences of the coding regions of the respective globin genes excluding silent bases and nucleotides in the intervening sequences. The dotted lines indicate parts of the tree that have not yet been ascertained. MY: Millions of years.

sequenced β-type globin genes the second intervening sequence is many times larger than the first, while in the α-type globin genes the two are of similar size [1, 4, 5].

The complete DNA sequences of all the human β-type globin genes are now available, and the laboratories involved in sequencing these genes have collaborated in a joint paper comparing the sequences [1]. A rather accurate and detailed evolutionary tree was constructed from these sequences by comparing nucleotides which are important in coding, but excluding silent nucleotides (such as nucleotides in the introns and many of the third bases in coding triplets). This tree is reproduced in Figure 2, with a few changes mainly to indicate the presence of the pseudogenes. As yet there is insufficient evidence to allow us to place the pseudogenes accurately—but putting them into the tree in a preliminary way is nevertheless a helpful reminder that their presence must be taken into account.

GENERAL ASPECTS OF GENE DUPLICATIONS

The evolutionary tree (Fig. 2) leads to the major consideration of the evolution of the globin genes. For many years one of us (O.S.) has been interested in what goes on at the branch points of trees such as this—ie, what is the nature of the event which leads from a single copy of a gene to two copies? Another example of such a gene duplication is that of the human haptoglobin genes (reviewed in Smithies [6]). If we take the simplest interpretation of the globin evolutionary tree, it shows that for the β-type globins the first surviving duplication event occurred about 200 million years ago.

Let us consider this event in general terms. Figure 3A diagrams the series of events classically thought to lead to a de novo duplication. You will see that the

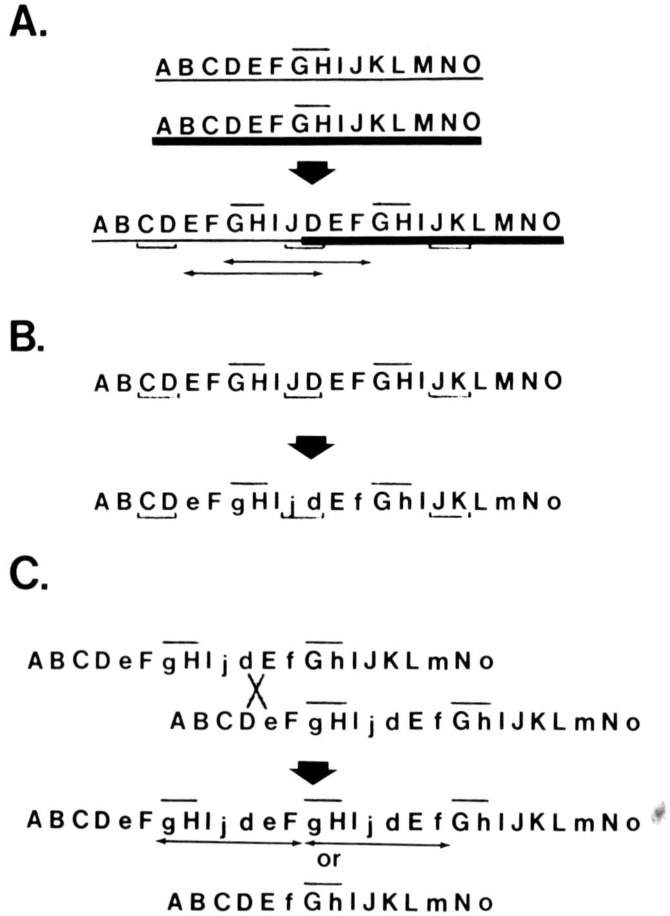

Fig. 3. De novo gene duplications and their subsequent evolution. A. A schematic representation of the de novo formation of a duplication of a gene "GH" by the breakage and re-union of identical DNA molecules at nonhomologous positions, CD and JK. The brackets indicate the regions which require study in order to determine the nature of the breakage and re-union event. The horizontal arrows illustrate that the length of the duplicate regions can be determined from the distance between many pairs of homologous positions (eg, from the E through D distance). B. The divergence of duplicated genes by mutations subsequent to the initial duplication is illustrated. A change from large to small letters represents observable sequence differences which do not obscure the relatedness of two regions. Note that comparisons of the CD, jd, and JK sequences can still show the site of breakage and re-union. C. Unequal but homologous crossing over can lead to gene triplication or to gene loss. The horizontal arrows show that the spacing of the triplicate genes is the same as between the duplicates.

extent of a duplication is likely to be considerably larger than the region actually coding for the specific gene product being studied. This means that duplicated genes will usually be surrounded by regions of DNA which are also duplicated. Notice that the true length of the duplication is given by the distance between homologous points (eg, the distance from E through D) rather than by the intergenic distance (I through F). Figure 3A also shows that it may be possible to deduce the type of event which led to a given duplication by examining the DNA sequences from three particular regions. In the duplication of Figure 3A these three regions are the CD region (which existed prior to the duplication), the JD region (which is the "novel joint"), and the other preexisting region, JK.

Once a duplication exists in the genome, mutations will occur which will cause the initial duplicates to diverge; consequently it may also be possible to deduce what has happened to a duplication since its initial formation by comparing the divergence of the duplicated regions of DNA which were identical when the duplication first arose (Fig. 3B). In addition, secondary unequal but homologous crossover events may be expected to lead to an increase (or decrease) in gene number without the need to invoke anything more unusual than recombination between homologous (duplicated) DNA sequences; see Fig. 3C. Products of this secondary increase in gene number have some quite characteristic features—such as the fact that the now triplicated genes are uniformly spaced and that each will have very similar surroundings. This last feature should be useful in distinguishing genes which have arisen by secondary events from those which have arisen by a de novo duplication. Thus, in Figure 3C, the triplicated genes have between them the two very similar sequences IjdeF and IjdEf, but outside the triplicated region two quite different sequences occur: ABCDeF and IJKLmNo.

DUPLICATION OF HUMAN FETAL GLOBIN GENES

One of the aims of our work is to describe the evolution of the β-globin gene cluster in terms of de novo duplications and subsequent secondary events. But you will realize from the complexity of the region that its evolution is not likely to be simple. Not only do we have to account for the nature and distribution of seven globin genes (the pseudogenes are included in this count), but in addition we have to explain the positions in the region of at least six copies of a repeated DNA sequence. We have tried to assemble a detailed evolutionary path compatible with all the details of the known map but have not yet been successful because there are insufficient data concerning the similarities and differences within the intergenic regions. However, we did much better when we confined our attention to the fetal globin gene region which we have now almost completely sequenced.

The length of the duplication which created the two human fetal globin genes is approximately 5 kbp; we know this because the distance between homologous points in the two genes is 5 kbp (see Fig. 3A). A complete comparison of the duplicate regions would therefore need more than 10,000 bp of DNA sequence data. Figure 4 presents a comparison of the data we have so far accumulated—over 7,000 bp of sequence. The vertical bars show base pair differences per 50 base pairs plotted as a function of their position along the duplication.

The results of the comparison are to a large extent as predicted. The general location of the 5′ end of the duplication is easily identified. To the left (5′) of the point of duplication, the sequence of the $^{G}\gamma$ gene (BC in Fig. 3) deviates completely (> 70% difference) from the sequence in the comparable position 5′ to the $^{A}\gamma$ (Ij in Fig. 3). Yet to the right or 3′ side of the point of duplication the sequences of the $^{G}\gamma$ and $^{A}\gamma$ genes are very homologous; thus the $^{G}\gamma$ sequence (De in Fig. 3) is identical in sequence to the $^{A}\gamma$ sequence (dE in Fig. 3) in more than 70% of its nucleotides.

We are, of course, very interested in finding whether there is anything unusual about the DNA in this region which might explain why the duplication end-points

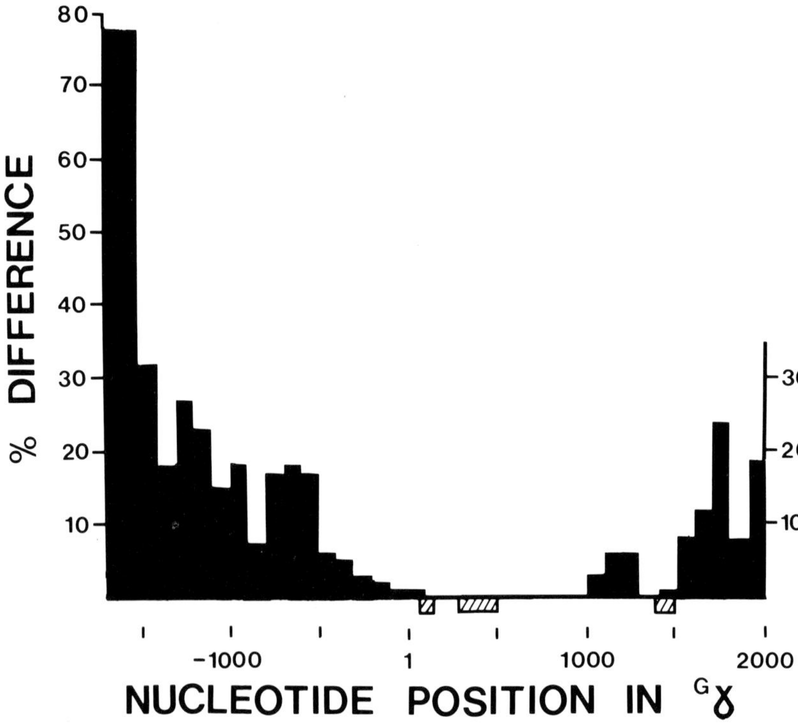

Fig. 4. A comparison of the duplicated $^G\gamma$ and $^A\gamma$ fetal genes and their surroundings. Vertical bars show the differences for every 50 base pairs in the nucleotide sequences of the $^G\gamma$ and $^A\gamma$ duplicate genes from one chromosome plotted against their position along the duplication. Position 1 corresponds to the first nucleotide of the γ-globin mRNAs. The coding regions are indicated by the hatched black bars.

are in these places. For example, we want to know whether the sequence CD has some unusual feature in common with the sequence JK which might account for the breakage and reunion occurring at these particular places. Until we have located and sequenced the right end of the duplication (JK) we cannot answer this question. Nevertheless we can say, even at this stage of our investigation when we have only sequenced CD and Jd but not yet JK, that the region near the break point is AT-rich (> 75% AT). Our current supposition is that the two fetal globin genes arose from a de novo duplication, that the two nonhomologous breaks giving rise to the duplication occurred in AT-rich regions, and that the two AT-rich regions were sufficiently similar in sequence to permit the broken DNA molecules to rejoin via their nonhomologous ends. However, we will not be able to prove the correctness of this supposition until we have the data for JK.

DNA EXCHANGES BETWEEN DUPLICATED GENES

A less expected feature of the comparison between the duplicated genes is a 1,200 bp stretch of DNA over which the $^G\gamma$ and $^A\gamma$ genes have only two base pair differences! This virtual identity contrasts markedly with the level of similarity (about 70% to 80%) between the other parts of the duplication. Clearly there must either be some strong selective factor which has prevented this region from diverging at the same rate as the remaining part of the duplication, or some type of molecular mechanism exists which has kept this region identical in the duplicated part on this chromosome (the region is not completely identical on the other chromosome from the same individual).

Once again in order to look for a possible explanation at the molecular level for the existence of these nondivergent regions we examined the DNA sequences at their boundaries. It was immediately apparent that the 3′ boundary has some unusual features. In Figure 5 we have diagrammed the DNA sequences of the duplicated genes at and on either side of this boundary. The unusual feature is a string of TG dinucleotides (about two thirds of the way into the larger intron) which form a stretch of simple sequence DNA. The length of this stretch of simple sequence DNA differs in the two duplicate genes in a way which looks like the result of an unequal but homologous crossing-over event. To the right of the simple sequence DNA the larger introns of the two genes differ in 12/269 bases; to the left, they are identical over a length of 571 bases. We have therefore proposed that this simple sequence DNA is a "hot spot" for initiating an intrachromosomal recombinational event, and that this event has led to an intergenic exchange of DNA [7]. The result, in this instance, is that a large part of the $^{A}\gamma$ globin has been converted to be identical in sequence to that of the $^{G}\gamma$ globin gene. There is an indication that the conversion event shows some polarity, ie, it extends 1,200 base pairs to the 5′ side of the "hot spot," and there is also an indication that an earlier more extensive conversion event occurred which extends 300 base pairs further to the 5′ side.

You will realize that the existence of gene conversion events can cause evolutionary trees, such as that presented in Figure 2, to be misleading. In many cases this type of tree may not represent the history of the duplicational events so much as the history of the occurrences of gene conversions!

The type of event we think most likely to account for the intergenic exchange of DNA within a single chromosome in the human fetal globin genes is a sister chromatid exchange. (DNA sequence data from the other chromosome exclude its participation in this particular event.) Sister chromatid exchanges between duplicated genes can lead to daughters which contain triplicate genes and a single gene, or they can lead to an intergenic exchange of DNA with no change in the number of genes in the two daughters. Figure 6 (from Slightom et al [7]) illustrates the basis for this statement using intermediates in the recombinational event of the types currently proposed by workers in recombination (see Radding [8] for a review). The figure accounts quite neatly for the data.

PSEUDOGENES IN THE GLOBIN CLUSTERS

At present the DNA sequences of the human pseudogenes have not been published—although their orientation relative to the expressed genes is known. However, the sequence of a mouse α globin-related pseudogene has been determined. About two years ago we described the isolation from the genome of outbred CD1 Swiss mice of a clone, 30.5, which appeared to contain a gene related to the α globin gene family [9, 10]. We have recently completed sequencing this gene (Vanin et al [11]), and Nishioka et al [12] have sequenced the comparable gene from Balb/c mice.

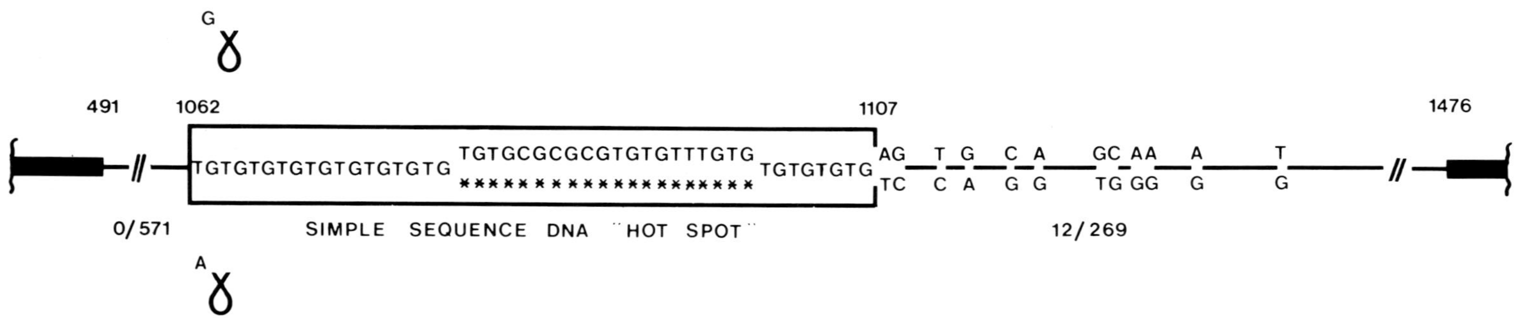

Fig. 5. A comparison of the DNA sequences of the larger introns of the duplicate ${}^{G}\gamma$ and ${}^{A}\gamma$ globin genes from one chromosome. The introns extend from nucleotide positions 491 to 1,476 [7]. The nucleotide sequences of the simple sequence DNA "hot spot" (see text) are shown in the box, with the asterisks indicating where the ${}^{A}\gamma$ gene is shorter. On the 5′ side of the box, the two introns are identical in sequence for 571 bases. On the 3′ side, they differ in the 12 (out of 269) positions shown.

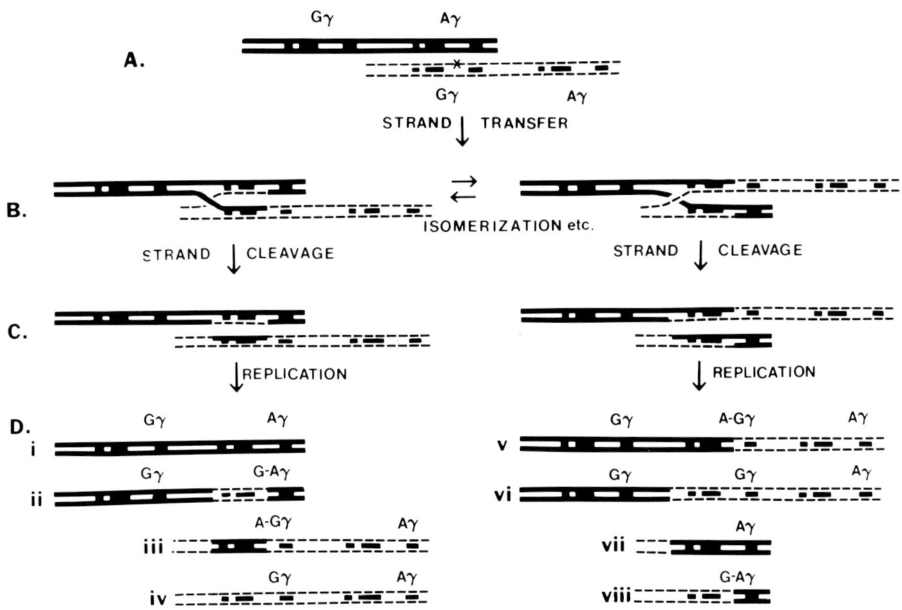

Fig. 6. Diagramatic representation and possible outcomes of a recombinational event between a $^{G}\gamma$ globin gene aligned with an $^{A}\gamma$ globin gene (from Slightom et al [7]). The participating DNA molecules are identical but are distinguished for illustrative purposes by light and heavy paired lines. Coding regions are shown by black bars between the paired lines. A. Initial alignment when the recombinational event was initiated at a simple sequence "hot spot" (X) in the larger intron. B. Strand transfer, isomerization, branch migration, and ligation yield the isomers. C. Strand cleavage leads to the intermediate heteroduplex products. D. Replication leads to the final products. Note that products ii and iii show "conversion."

This α globin-related gene, 30.5, has several interesting features. First, the gene is clearly more related to the mouse α gene (84% identical over the coding region) than to the mouse β genes (54% identity), and it is more closely related to the mouse than to the human or rabbit α globin genes. Secondly, it completely lacks both intervening sequences; they are as cleanly absent as if they had never been there (Fig. 7). Thirdly, the gene could not code for a globin polypeptide, since it has frame shifts which produce missense amino acids and finally stop codons long before a normal globin could be produced (Fig. 8). Because of this third feature we call it a pseudogene, $\psi\alpha 30.5$. Its second unusual feature also would have made it a pseudogene, because there is now a considerable body of data showing that if RNA is transcribed from globin (and other) genes without introns, then the RNA will not yield a stable cytoplasmic message [13]. Thus the mouse α globin-related gene $\psi\alpha 30.5$ is a nonproductive pseudogene in at least two respects.

How did this pseudogene evolve, and what is its function, if any? There are three general types of evolutionary history to be considered, none of which can be eliminated at present. The first evolutionary possibility is that the pseudogene is the modern descendant of an old precursor globin gene which lacked introns. This possibility is unlikely because the gene is so obviously related to the modern mouse

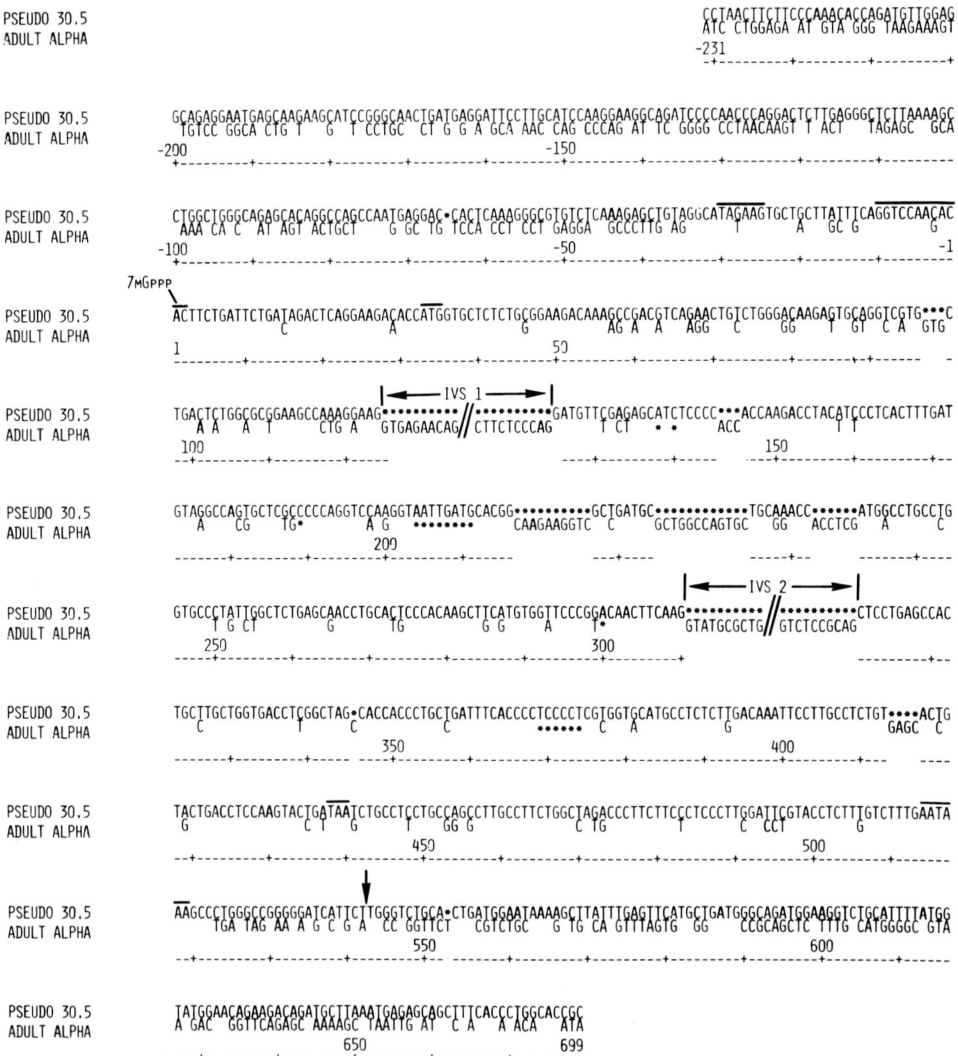

Fig. 7. A comparison of the sequences of a mouse pseudogene, ψα30.5, and an adult α globin gene [4] (modified from Vanin et al [11]). The nucleotide sequence of ψα30.5 (upper lines) is compared to the sequence of a productive adult α globin gene (lower lines). The sequence of ψα30.5 is listed in full. Where the sequence of the adult α globin gene differs from ψα30.5 the differing bases are shown; where they are identical only the sequence of ψα30.5 is shown. Asterisks indicate that no bases are present at the indicated positions. The numbering system on the counting scale refers to ψα30.5, with number 1 corresponding to the nucleotide homologous to the first nucleotide of α globin mRNA. The intervening sequences, IVS 1 and IVS 2 of the adult α globin gene are shown. The position of the cap of adult α globin mRNA is shown by 7mGppp. Potentially important signals are overlined.

```
1..........5............10............15...............21/23.......26
(Met)ValLeuSerAlaGluAspLysAlaAspValArgThrValTrpAspLysSerAlaGlyArgAlaAspSerGlyAla

27....................35
GluAlaLysGlyArgMetPheGlySerIleSerProProArgProThrSerLeuThrLeuMet * AlaSerAlaArg

54........57
ProGlnValGlnGlyAsn * CysThrGly * CysCysLysProTrpProAlaTrpCysProIleGlySerGluGln

ProAlaLeuProGlnAlaSerCysGlySerArgThrThrSerSerSer * AlaThrAlaCysTrp * ProArgLeu

                                                                    134...136
AlaProProCys * PheTyrProSerProArgGlyAlaCysLeuSer * GlnIleProCysLeuCysThrValLeu

137......140
ThrSerLysTyr * *
```

Fig. 8. The amino acid sequence corresponding to the nucleotide sequence of $\psi\alpha$30.5 [11]. The amino acid sequence is that which would be obtained by translating $\psi\alpha$30.5 from the same initiator codon as the productive α globin gene. The numbered scales show amino acid residue numbers in α globins. Underlined residues have been found in a α globins. Italized residues are out of phase or have not been found in α globins. Asterisks indicate stop codons.

adult α globin gene in the coding regions, but we cannot exclude it in view of our recent data showing that genes can exchange DNA along chromosomes. The second evolutionary possibility is that the introns were recently lost at the DNA level. We prefer this possibility but we cannot prove that it is the best. A third possibility is that the pseudogene is derived from globin mRNA via the action of reverse transcriptase. Nishioka et al [12] favor this possibility. Whichever evolutionary scheme is correct, there remains the even more interesting question of whether pseudogenes have a function.

Could the pseudogene 30.5 be a "dead" gene or could it have a function? We have no direct answer to this question at present, but a good case can be made for the importance of pseudogenes in general by noting their widespread occurrence in other species. Maniatis and his collaborators [5, 14–16] have isolated a number of globin-related DNA sequences which cannot be identified with known polypeptides. In rabbits the gene β_2, located between the fetal and adult β globin genes, is a pseudogene in that it could not code for a functional globin polypeptide [Lacy and Maniatis, personal communication]. In humans there is a pseudogene, $\psi\alpha_1$, between the adult and fetal α globin genes which could not code for a functional globin polypeptide [Proudfoot and Maniatis, personal communication]. Both the rabbit pseudogene, β_2, and the human pseudogene, $\psi\alpha_1$, have intervening sequences, but their DNA sequences are abnormal in ways which would prevent the intervening sequences from being spliced out of the RNA in the usual manner [Lacy, Proudfoot, and Maniatis, personal communication]. Thus all three sequenced pseudogenes share the same basic faults: They could not code for functional globins, and they have abnormal or absent intervening sequences. When we add to this the fact that there are similar DNA sequences, which look as if they will prove to be pseudogenes, between the adult and prenatal β globin genes in both mouse [17] and human [16], then a pattern begins to emerge. It looks as if there are pseudogenes between the prenatal and adult globin genes in all species that have been studied at the level required to characterize pseudogenes. We consequently think that it is unlikely that they are "dead" genes without function.

If they are not "dead" genes what could their function be? An attractive possibility is that they could act to modulate the expression of the productive genes. There is considerable evidence indicating that globin genes may be "open" or "closed" at the chromatin level in the way originally suggested by Weintraub's pioneering work [18]. Although the "open" configuration is a necessary condition for a productive gene, it does not appear to be a sufficient condition. Several investigators have found that globin genes can be in the "open" configuration even though the corresponding globins are not being synthesized [19, 20]. Consequently we are considering the possibility that production versus nonproduction of globins may be controlled by which gene within an "open" region is selected for transcription. If a pseudogene is selected for transcription within an "open" region, then the transcript is diverted into a degradative pathway and no globin product is formed. If one of the normal genes is selected, then the nuclear transcript can be processed and transported to yield a cytoplasmic mRNA on which productive translation can occur. Thus the pseudogenes, on this hypothesis, may be acting as "diverting" genes which modulate the expression of the "open" globin regions by diverting the transcriptional machinery into a nonproductive pathway. There are partial precedents for the thought that mechanisms exist for selecting one of a number of products from a single transcriptional unit. In the adenovirus type 2 system, different portions of the late transcript are used in different mRNAs [21]. In immunoglobulin heavy chains, the secreted and membrane forms of the μ-chain are probably the result of using different portions of the transcript from one region of DNA [22]. Whether or not the "diverting" gene hypothesis proves to be correct, the finding of these pseudogenes raises many questions regarding their possible function in the globin gene clusters.

REPEATED SEQUENCES IN THE β GLOBIN GENE CLUSTER

Within the globin gene clusters are distributed some repeated DNA sequences. Three independent sets of observations have pointed to their existence in the β gene cluster.

The first observation (Duncan et al [23]) was the location of two places in the β gene cluster which could be transcribed in vitro with RNA polymerase III. (This is the polymerase which transcribes the DNA for 5S RNA, tRNA, and some other small molecular weight RNAs; it is not thought to be required for the transcription of protein structural genes.) These polymerase III sites were located 5′ to the fetal $^{G}\gamma$ gene and 5′ to the adult δ gene. The transcript from the site 5′ to $^{G}\gamma$ hybridizes to the site 5′ to the adult δ gene, and also to another site 3′ to the whole β globin cluster. The second observation was made by Fritsch et al [16], who described the presence in the β globin gene cluster of six regions of DNA, all able to hybridize to a probe made from one of them (see Fig. 1). The six regions included all three regions discovered by Duncan et al [23]. The third observation, described in a paper by Jelinek et al [24], hints at the possible nature of these repeat sequences. Included in these sequences is that derived from a family of repeated DNA which is interspersed throughout the human and hamster genome approximately 300,000 times. This family of interspersed repeated DNA is referred to as the Alu family, because many of its members are cut with the restriction enzyme Alu I. The polymerase III site 5′ to the $^{G}\gamma$ gene proves to be a member of this Alu family on the basis of its DNA sequence. This suggests, although it does not prove, that the other five repeated sequences in the β cluster are also members of this family.

Why is this possibility exciting? The answer is that Jelinek et al [24] showed that the sequence common to the Alu family also has much in common with the known origins of DNA replication of several DNA viruses. There is consequently a real possibility that for the first time we will soon have some knowledge of the relationships between the location of origins of DNA replication and the location of structural genes which are expressed at different times during development. Such a correlation would be of great interest, since many developmental biologists suspect that determination and differentiation are intimately related to DNA replication.

The Alu family sequence occurs also near to the pseudogene ψα30.5. Figure 9 compares five DNA sequences: the sequence reported by Jelinek et al [24] for the Alu family in general, the sequences of two cloned members of the Alu family (also

```
ALU FAMILY                       CXACTCGGGAGGCTGAGGCAGGAGAATCGCATGAACC
BLUR 8 CLONE        CCTGGAATCCCAGCTACTTAGGAGGCTCAGACAGAAGAATCCCTTAAACC
CHO CLONE             TTTGCCCCAGXTACTCCAGAGGCAGAGGCCGCCACATTATGT
Gγ GENE             CCTGTAATCCCAGCTACTTGGGAGGCTGAGGAAGGAGAATTGCTTGAACC
PSEUDOGENE          TGGGGAACTTCAGGT*CTTGGCAGGCTGAGGCAGGAGGATTGCTTGCGAG

CONSENSUS SEQUENCE  cctg aa ccCAGctaCT  ggAGGCtgAGgcaG ag AT  c Tgaacc
                              ALU I
                              SITE

                                             41/50    ≤1 DIFFERENCE

                                             15/50    NO DIFFERENCES
```

Fig. 9. A comparison of several members of the "Alu family" of repeated DNA. Five nucleotide sequences are compared from: the human "Alu family" sequence [24]; a clone of one member of the human Alu family, BLUR8 [24]; a Chinese hamster cell (CHO)-cloned genomic DNA fragment [24]; a region on the 5′ side of the human $^{G}\gamma$ fetal globin gene [our unpublished data]; a region on the 5′ side of the mouse pseudogene, ψα30.5 [our unpublished data]. A concensus sequence is given below where not more than one nucleotide in the compared sequences is different (small capitals), or where there are no differences (large capitals). The Alu I site, AGCT, from which the family name is derived is shown.

from Jelinek et al [24], our sequence of the member of the Alu family which is 5′ to the ${}^{G}\gamma$ gene, and our sequence from the 5′ side of the pseudogene $\psi\alpha30.5$. The fidelity of preservation of these sequences is remarkable in that, out of a total of 50 nucleotides, there are 41 positions where there is not more than one difference among the five sequences, and at 15 of these positions there are no differences. The degree of preservation of a nucleotide sequence in DNAs from so varied sources points to its importance.

CONCLUSION

The considerations outlined in this paper provide the pictures of a developmentally complex locus which, as a result of DNA cloning and sequencing, is beginning to show signs of order. The evolution of the globin gene clusters is beginning to be understood. An unexpected class of genes has been demonstrated, the pseudogenes, which may be important in modulating the expression of these globin gene clusters. Finally, the location within the clusters of several repeats which may prove to be origins of DNA replication is now known.

SUMMARY

Recent cloning and sequencing experiments have revealed a great deal of information about mammalian globin genes. DNA corresponding to the whole of the α and β globin clusters of humans has been cloned, the nucleotide sequences of all the genes specifying β-type globin chains have been determined, and a detailed evolutionary tree of the β-type globin has been constructed.

We have studied the nature of the events whereby genes in the globin cluster have been duplicated. The length of the duplication leading to the two fetal globin genes, ${}^{G}\gamma$, and ${}^{A}\gamma$, is 5 kilo-base pairs. Clues to the nature of the initial event leading to this duplication are revealed from the DNA sequence data from the ends of the duplicated regions.

A surprising new finding is that a relatively recent intergenic exchange appears to have occurred between the nonallelic duplicated genes on one of the chromosomes of our DNA donor. This exchange has made the nonallelic genes, ${}^{G}\gamma$ and ${}^{A}\gamma$, on one chromosome more like each other than are the two allelic ${}^{A}\gamma$ genes on the two chromosomes.

Pseudogenes occur in the globin clusters. These are regions of DNA obviously related to the productive globin genes, as judged by their DNA sequences, but clearly unable to lead to the production of functional globin polypeptides because of various anomalies in their structures. A pseudogene has been found between the prenatal and adult globin genes in all species so far studied. Pseudogenes could be "diverting" genes able to modulate the expression of neighboring productive genes by diverting the transcriptional apparatus into a degradative pathway.

Six regions of DNA occur in the human β globin cluster which are related to each other and to the "Alu-family" of repeated DNA. The Alu-repeats may prove to be potential origins of DNA replication. If so, we may soon have some knowledge of the relationships between active and inactive origins of replication vis-a-vis regions of DNA with differential expression during development.

ACKNOWLEDGMENTS

This is paper number 2438 from the Laboratory of Genetics at the University of Wisconsin. This work was supported by NIH grants GM20069 and AM20120.

REFERENCES

1. Efstratiadis A, Posakony JW, Maniatis T, Lawn RM, O'Connell C, Spritz RA, DeRiel JK, Forget B, Weissman SM, Slightom JL, Blechl AE, Smithies O, Baralle FE, Shoulders CC, Proudfoot NJ: The structure and evolution of the human β-globin gene family. Cell, in press, 1980.
2. Dayhoff MO (ed): "Atlas of Protein Sequence and Structure," Vol 5, Chapter 3, Washington, DC: National Biomedial Research Foundation, 1972, pp 17–30.
3. Leder A, Miller HI, Hamer DH, Seidman JG, Norman B, Sullivan M, Leder P: Comparison of Cloned mouse α- and β-globin genes: Conservation of intervening sequence locations and extragenic homology. Proc Natl Acad Sci USA 75:6187–6191, 1978.
4. Nishioka Y, Leder P: The complete sequence of a chromosomal mouse α-globin gene reveals elements conserved throughout vertebrate evolution. Cell 18:875–882, 1979.
5. Lauer J, Shen CKJ, Maniatis T: The chromosomal arrangement of human α-like globin genes: Sequence homology and α-globin gene deletions. Cell 20:119–130, 1980.
6. Smithies O: Chromosomal rearrangements and protein structure. Cold Spring Harbor Symp Quant Biol XXIX:309–319, 1964.
7. Slightom JL, Blechl AE, Smithies O: Human fetal $^{G}\gamma$ and $^{A}\gamma$ globin genes: Complete nucleotide sequences suggest that DNA can be exchanged between these duplicated genes. Cell 21:627–638, 1980.
8. Radding CM: Genetic recombinations: Strand transfer and mismatch repair. Ann Rev Biochem 47:847–880, 1978.
9. Blattner FR, Flechl AE, Denniston-Thompson K, Faber HE, Richards JE, Slightom JL, Tucker PW, Smithies O: Cloning human fetal γ globin and mouse α-type globin DNA: Preparation and screening of shotgun collections. Science 202:1279–1284, 1978.
10. Smithies O, Blechl AE, Denniston-Thompson K, Newell N, Richards JE, Slightom JL, Tucker PW, Blattner FR: Cloning human fetal γ globin and mouse α-type globin DNA: Characterization and partial sequencing. Science 202:1284–1289, 1978.
11. Vanin EF, Goldberg GI, Tucker PW, Smithies: A mouse alpha globin-related pseudogene ($\psi\alpha$30.5) lacking intervening sequences. Nature, in press, 1980.
12. Nishioka Y, Leder A, Leder P: Unusual α-globin-like gene that has cleanly lost both globin intervening sequences. Proc Natl Acad Sci USA 77:2806–2809, 1980.
13. Hamer DH, Leder P: SV40 recombinants carrying a functional RNA splice junction and polyadenylation site from the chromosomal mouse β^{maj} gene. Cell 17:737–747, 1979.
14. Lacy E, Hardison RC, Quon D, Maniatis T: The linkage arrangement of four rabbit β-like globin genes. Cell 18:1273–1283, 1979.
15. Hardison RC, Butler ET, Lacy E, Maniatis T, Rosenthal N, Efstratiadis A: The linkage arrangement of four rabbit β-like globin genes. Cell 18:1285–1297, 1979.
16. Fritsch EF, Lawn RM, Maniatis T: Molecular cloning and characterization of the human β-like globin gene cluster. Cell 19:959–972, 1980.
17. Jahn CL, Hutchison CA III, Phillips SJ, Weaver S, Haigwood NL, Voliva CF, Edgell MH: DNA sequence organization of the β-globin complex in the BALB/c mouse. Cell, submitted for publication, 1980.
18. Weintraub H, Groudine M; Chromosomal subunits in active genes have an altered conformation. Science 193:848–856, 1976.
19. Young NS, Benz EJ Jr, Kantor JA, Kretschmer P, Nienhuis AW: Hemoglobin switching in sheep: Only the γ gene is in the active conformation in fetal liver but all the β and γ genes are in the active conformation in bone marrow. Proc Natl Acad Sci USA 75:5884–5888, 1976.
20. Stalder J, Groudine M, Dodgson JB, Engel JK, Weintraub H: Hb switching in chickens. Cell 19: 973–980, 1980.
21. Nevins JR, Darnell JE Jr: Steps in the processing of Ad2 mRNA: Poly(A)$^{+}$ nuclear sequences are conserved and poly(A) addition precedes splicing. Cell 15:1477–1493, 1978.

22. Early P, Rogers J, Davis M, Calame K, Bond M, Wall R, Hood L: Two mRNAs can be produced from a single immunoglobulin μ gene by alternative RNA processing pathways (II). Cell, in press, 1980.
23. Duncan C, Biro PA, Choudary PV, Elder JT, Wang RRC, Forget BG, DeRiel J, Weissman SM: RNA polymerase III transcriptional units are interspersed among human non-α-globin genes. Proc Natl Acad Sci USA 76:5095–5099, 1979.
24. Jelinek WR, Toomey TP, Leinwand L, Duncan CH, Biro PA, Choudary PV, Weissman SM, Rubin CM, Houck CM, Deininger PL, Schmid CW: Ubiquitous, interspersed repeated sequences in mammalian genomes. Proc Natl Acad Sci USA 77:1398–1402, 1980.

Expression of Globin Genes

Organization and Expression of Globin Genes, pages 119–126

Expression of Globin Genes In Vivo and In Vitro

R.A. Flavell, H. Bud, H. Bullman, H. Dahl, E. De Boer, T. De Lange, J. Groffen, F. Grosveld, G. Grosveld, D. Kioussis, N. Moschonas, and C. Shewmaker

Laboratory of Gene Structure and Expression, National Institute for Medical Research, Mill Hill, London, England

INTRODUCTION

For a number of years the hemoglobin genes have provided a successful model for the study of gene expression in higher eukaryotes. The availability of recombinant DNA technology has permitted the elucidation of the structure and the linkage arrangement of numerous globin genes from diverse species. As a prerequisite for studies on gene expression, we have studied the structure and linkage arrangement of the human and rabbit β-related globin genes. The structural data have been considered in other recent symposium articles [1, 2] and will be reviewed by others in this volume. To determine how globin genes are expressed, and how this expression is regulated, we have analysed the in vivo transcription products of the rabbit β-globin gene in rabbit bone marrow. Transcription of the normal and mutated rabbit β-globin gene has been studied in cell-free extracts. Furthermore, we have characterized a gene affected by a β^0 thalassemic mutation by DNA sequencing and by study of its in vitro transcription.

TRANSCRIPTION OF THE RABBIT β-GLOBIN GENE IN VIVO

The rabbit β-globin gene consists of three noncontiguous blocks of mRNA-coding sequences separated by two introns of 126 and 573 base pairs. S_1 mapping experiments using the procedure of Berk and Sharp [3] have shown that the structure of in vivo RNAs is as shown in Figure 1. The largest globin pre-mRNA detectable is a colinear transcript of the gene, including the introns and exons, but lacking sequences derived from the extragenic DNA. The 5′ and 3′ termini of the colinear transcript have been mapped using 5′ and 3′ terminally labelled probes in S_1 nuclease experiments; the products were sized on a thin polyacrylamide gel. Precise size markers were provided by running the G + A sequencing reaction products of the fragment in a parallel lane. In these experiments, the size of the S_1 nuclease digestion products for both the 5′ and 3′ ends fits with the size expected if the 5′ and 3′ termini of the globin mRNA are coincident with those of the colinear pre-mRNA. We conclude that within experimental error, the 5′ and 3′ ends of

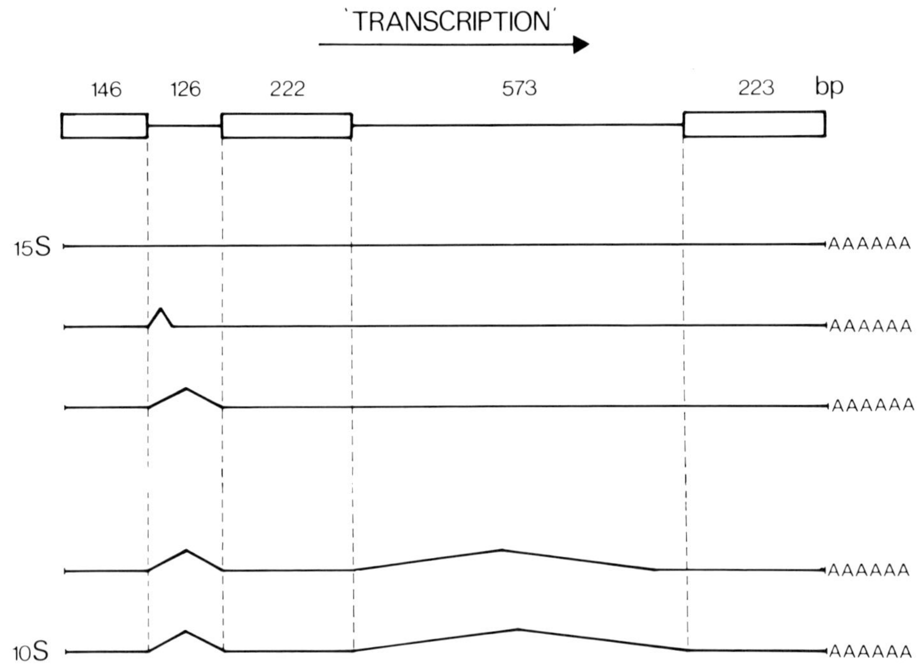

Fig. 1. Structure of β-globin pre-mRNAs in rabbit bone marrow RNA. The β-globin gene is depicted as three coding blocks separated by two intervening sequences. The horizontal lines indicate the RNA present in a given species and _/_ indicates the region removed by splicing. The 5′ splice junction of the intermediate, drawn here at the intron-exon boundary, has not been experimentally verified.

β-globin mRNA and pre-mRNA are co-terminal. A similar conclusion has been drawn for the 5′ end of mouse β-globin pre-mRNA [4].

In addition to the colinear pre-RNA, spliced β-globin pre-mRNA intermediates can be detected by the same method (Fig. 1). Splices are formed within both the small and large introns giving intermediate species. Also, molecules are present where the entire small intron has apparently been spliced out, but where the entire large intron is still present. Inspection of Figure 1 shows that the RNAs can be arranged in a scheme where sequential splicing of first the small intron and then the large intron, each in two steps, could occur. While this is the most plausible explanation, two critical points should be noted. First, it is not proven that the intermediate spliced molecules are indeed processed to give mature β-globin mRNA. These molecules may also be the result of splicing errors in vivo and may only be destined to be degraded. Second, other partially spliced molecules may be present, but undetected in our experiments. Thus, the pathway, suggested by Figure 1 may only be the way most, but not all, β-globin mRNA is produced.

TRANSCRIPTION OF THE RABBIT β-GLOBIN GENE IN VITRO

Weil et al [5] and Manley et al [6] have recently shown that the Adeno virus 2 (Ad2) late genes can be transcribed in vitro in cell-free extracts to give specific products whose 5′ end is capped and coincident in position with the 5′ end of the major late Ad2 RNA. These data suggest strongly that RNA synthesis initiates at

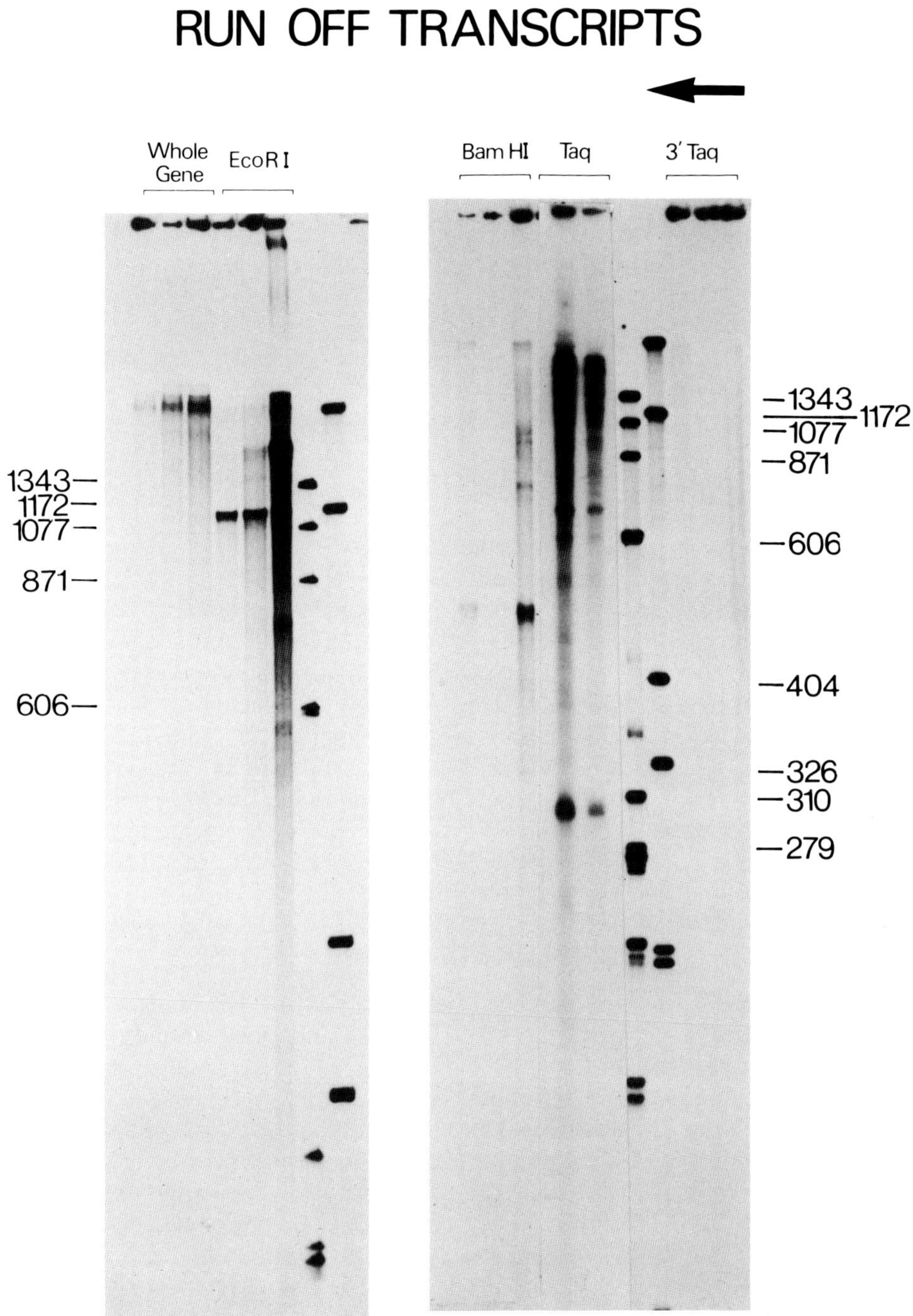

Fig. 2. Transcription of the rabbit β-globin gene in vitro. Restriction fragments of the rabbit β-globin gene truncated at the Taq I (+ 308), Bam HI (+ 475), or Eco RI (+ 1120) were incubated with a Hela cell extract prepared as in reference 6 in a volume of 25 μl with α-^{32}P GTP (16 Ci/mmol) as labelled substrate. The RNA was recovered by proteinase K digestion, phenol extraction, and ethanol precipitation. The RNA and 5′ end labelled DNA markers (ØX DNA × Taq I and ØX DNA × Hae III) were heat denatured in 98% Formamide, 10 mM Hepes pH 6.5 0.1% SDS containing 0.02% each of Bromophenol blue and Xylene cyanol FF, and analysed on a 0.3-mm-thick 5% polyacrylamide-urea gel.

this site in vitro although rapid processing followed by capping of the transcript remains to be excluded. We have used the Hela cell extract of Manley et al [6] to characterize the in vitro transcripts of the rabbit β-globin gene and to define the DNA sequences required to give specific RNA synthesis in vitro. By analogy with bacterial systems we shall refer to these DNA sequences as the rabbit β-globin promotor.

The rabbit β-globin gene is transcribed efficiently in this system. Since the β-globin gene is presumably not expressed in Hela cells we conclude that positive erythroid-cell-specific factors are not required for the transcription of the β-globin gene in this type of extract. If the rabbit β-globin DNA is cut at the intragenic Taq I site (at 308 nucleotides in from the cap site; + 308) and then used as a template in this system, labelled RNAs of 308 and 310 nucleotides can be detected among the in vitro products, together with larger RNAs that derive from nonspecific initiations further upstream. Similarly, if the DNA is cut at the Bam HI site (+475) major RNA products of 471, 475, and 480 are seen; DNA cut at the Eco RI site (+ 1120) gives two products of about 1,160 nucleotides (Fig. 2).

Clearly, specific RNAs are synthesized with 5′ ends at the same site as in vivo globin pre-mRNA. We have also mapped the 5′ ends of the in vitro product by S_1 mapping experiments using as a probe a 5′-labelled Bst NI fragment containing the 5′ end of the β-globin gene. By this criterion, the 5′ end of the in vitro transcript also maps at the same position as that of in vivo β-globin mRNA (not shown). The heterogeneity of the run off transcripts is at least partially the result of heterogeneity of the 3′ ends. The RNA polymerase apparently stutters when it reaches the 3′ end of the DNA molecule; some polymerases terminate at the end of the double-stranded region, while others copy the four-nucleotide sticky end (GATC in the case of Bam HI) on the noncoding strand. This gives two transcripts which differ in length by four necleotides. Removal of the sticky ends by S_1 nuclease, or by DNA repair using DNA polymerase I, essentially eliminates this heterogeneity.

DNA SEQUENCES REQUIRED FOR THE TRANSCRIPTION OF THE RABBIT β-GLOBIN GENE: THE RABBIT β-GLOBIN GENE PROMOTOR

Sequence comparison of a number of eukaryotic genes has pointed to at least three regions that have been conserved in evolution: the "Hogness" box [7] CATAAAAG in the case of the rabbit β-globin gene, located about 30 nucleotides upstream from the cap site; the capping box Py CATTC Pu [8] located just before the cap site; and the so-called CAAT box located about 80 nucleotides upstream from the cap site [9]. To determine if any of these sequences are required for transcription in vitro we have constructed a series of deletion mutants that lack DNA sequences in the 5′ extragenic region of the rabbit β-globin gene. The map-coordinates of some of these deletion mutants are described in Table I, together with their ability to direct transcription. It can be seen that removal of sequences upstream of −100 (counting from the cap site) has no pronounced effect on the transcription efficiency, whereas deletion of DNA between −100 and −12 eliminates transcription entirely. At least part of the rabbit β-globin gene promotor must therefore lie between −100 and −12. We have constructed a large number of deletion mutants in the region close to the Hogness box. A mutant lacking the sequences −16 to −8 is transcribed, albeit poorly, in this system. A mutant lacking −30 to −4, and therefore lacking the Hogness box, is not transcribed at all in this

TABLE I. Deletion Mutant Map Coordinates[1]

DNA sequences deleted	Approximate transcription efficiency
None (wild type)	+++
Upstream of −100	+++
−11 to −7	+
−16 to −8	+
−11 to +14	++
−9 to +164	++
−75 to −10	−
−30 to −4	−
−54 to −4	−

[1]The transcription efficiencies have been estimated from the intensities of the run off products on the polyacrylamide gels. This should only be considered a rough approximation.

system. Since the only conserved region between −17 and −30 is the Hogness box (−31 to −24), we consider it likely that the Hogness box is required for transcription in vitro.

A second point came out of the analysis of mutants lacking DNA downstream from the Hogness box. These deletions cause the 5′ end of the transcripts to be displaced by approximately the same number of nucleotides as the size of the deletion (data to be described in detail elsewhere). We believe therefore that the polymerase is (ultimately) bound to the Hogness box and then measures about 30 nucleotides downstream in the same way that is believed to occur for Escherichia coli polymerase.

TRANSCRIPTION OF A β^0 THALASSEMIA GENE

In β^0 thalassemia no β-globin chains are produced. In some cases, β-globin mRNA can be detected, in others not. cDNA titration experiments [10] and Southern blotting experiments [11] have shown that the β-globin gene is present in β^0 thalassemia, although in a few exceptional cases the 3′ exon of the β-globin gene has been deleted [11, 12]. To determine whether minor differences, either small deletions or point mutations, can account for the failure of β-globin gene expression, we have cloned the β-globin gene from two β^0 thalassemia patients. Detailed restriction analysis shows that in both cases the general structure of these genes is perfectly normal, ie both introns and all three exons are present at the same position as in the normal gene.

We have tested the transcriptional activity of β^0 thalassemic genes from both patients in the Hela cell extract described above [6]. Figure 3 shows that the β^0 thalassemic gene from one of these patients [13] can be transcribed normally in this system. The same was found for the gene from the second patient (not shown). The molecular basis of β^0 thalassemia in these cases remains undetermined.

DNA SEQUENCES FAR FROM GLOBIN GENES ARE PROBABLY REQUIRED FOR EXPRESSION IN VIVO

The human globin genes have been useful in defining DNA sequences required for the expression of eukaryotic genes in vivo. Of particular interest is the DNA sequence requirement for the effective performance of the $\gamma \rightarrow \delta + \beta$ globin switch

Fig. 3. Transcription of the normal human β-globin gene and the β-globin gene from a $\delta\beta^0/\beta^0$ thalassemia patient [13]. The 1.8 kb Bam HI fragment of the human β-globin gene was transcribed in vitro as described in Figure 2. The lower intensity of the normal β-globin gene fragment is not significant. In mixing experiments both genes are transcribed at the same efficiency.

around birth in man. In $\delta\beta^0$ thalassemia and in Hereditary Persistence of Fetal Hemoglobin (HPFH) this switch apparently fails either partially or completely, respectively [see eg, reference 15].

Since these diseases appear to be the result of deletion of DNA in the globin-gene region (but see below), we believe that the analysis of a large number of deletion mutants which exhibit this type of phenotype, and the comparison of these with deletion mutants where the switch is normal, should point to DNA sequences in this region that are essential for the switch. A large number of such deletions have been mapped with respect to the human β-globin gene cluster using Southern blotting by ourselves and Maniatis, Forget, and their respective colleagues, and these results are shown in Figure 4.

These results point to two regions which may be important for the switch. First, removal of the 3′ extragenic regions of the β-globin gene in $\delta\beta^0$ thalassemia appears to cause an enhancement of the γ-gene expression in adults when compared with Hb Lepore patients who retain this DNA region [16]. Second, removal of the sequences 5′ to the δ-globin gene appears to cause still more γ-globin gene expression in adults when two types of ${}^{G}\gamma{}^{A}\gamma$ HPFH are compared with $\delta\beta^0$ thalassemia [17–20] and when the putative deletion in Hb Kenya, which has not yet been mapped at the DNA level, is considered. The working hypothesis [16, 17, 19] is that these deletions remove DNA sequences that are required to organize the chromatin into an active $\delta + \beta$-globin gene domain. Deletion of these sequences inactivates the adult β-globin gene domain and therefore forces expression of the alternative, but in adults less favoured, foetal domain. In support of this idea, deletion of sequences far upstream of the β-globin gene apparently inactivates the β-globin gene in γ-β thalassemia [21].

A note of caution is necessary at this point. It is not established that the phenotype observed in any of these diseases is the result of the deletion observed. It is also possible that the phenotype is the result of some other mutation, perhaps even a point mutation. Indeed, the Greek Aγ-HPFH does not exhibit a detectable deletion in the β-globin gene cluster [19, 20].

In order to test these ideas, it is necessary to set up assay systems for developmental switching. Here the following protocol can be envisaged: first, the isolation of the entire $\gamma\delta\beta$-globin locus as a large recombinant DNA molecule; second, its introduction into committed erythroid cells (either foetal or adult) under conditions that allow developmental stage specific expression in vivo; third, manipulation of the recombinant DNA in vitro to delete regions suspected of being required

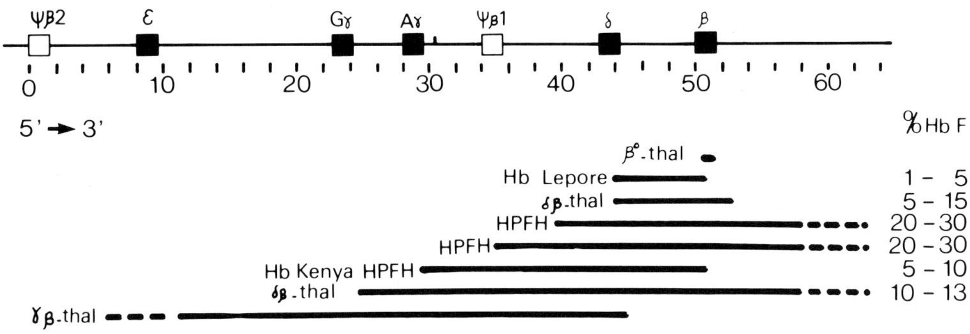

Fig. 4. The structure of the $\epsilon\gamma\delta\beta$ globin region in normal DNA and in a number of hereditary diseases. The DNA deleted in the hereditary diseases indicated is depicted by the horizontal line. A dashed line indicates that the precise location of the end point of the deletion in that area is not known. The deletion in Hb Kenya has not been established at the DNA level.

for the switching process, reintroduction of this DNA in the cells, followed by reassaying the developmental expression of the DNA. As a first step along these lines, we have applied the cosmid technology of Collins and Hohn [22] and Kourilsky and his colleagues [23] to the cloning of the human globin genes. We have at present isolated several overlapping cosmid clones. One of these may be a suitable candidate for the first step above since it contains the $^{G}\gamma$, $^{A}\gamma$, δ, and β-globin genes together with the intergenic DNA regions. We are currently introducing this DNA into erythroid cells to test these ideas.

ACKNOWLEDGMENTS

Work from this laboratory was supported by the British Medical Research Council. We are grateful to Cora O'Carroll for preparing the manuscript.

REFERENCES

1. Flavell RA, Grosveld GC, Grosveld FG, Bernards R, Kooter JM, de Boer E, Little PFR: The structure and expression of normal and abnormal globin genes. In Russell TR, Brew K, Faber H, Schultz J: "From Gene to Protein: Information Transfer in Normal and Abnormal Cells," Miami Winter Symp. Vol 16, New York: Academic Press, 1979, 149–164.
2. Flavell RA, Bernards R, Grosveld GC, Hoeijmakers-van-Dommelen HAM, Kooter JM, de Boer E, Little PFR: The structure and expression of globin genes in rabbit and man. In Axel R, Maniatis T, Fox CF: "Eucaryotic Gene Regulation," ICN-UCLA Symposia on Molecular and Cellular Biology, Vol XIV, New York: Academic Press, 1979, pp 335–354.
3. Berk AJ, Sharp PA: Sizing and mapping of early adenovirus mRNAs by gel electrophoresis of S_1 endonuclease-digested hybrids. Cell 12: 721–732, 1977.
4. Weaver RF, Weissmann C: Mapping of RNA by a modification of the Berk-Sharp procedure; The 5′ termini of 15S β-globin mRNA precursor and mature 10S β-globin mRNA have identical map coordinates. Nucleic Acids Res 6: 1175–1193, 1979.
5. Weil PA, Luse DS, Segall J, Roeder RB: Selective and associate initiation of transcription at the Ad2 major late promotor in a soluble system dependant on purified RNA polymerase II and DNA. Cell 18: 1979.
6. Manley JL, Fire A, Cano A, Sharp PA, Gefter ML: DNA-dependent transcription of adenovirus genes in a soluble whole-cell extract. Proc Natl Acad Sci USA, in press.
7. Proudfoot NJ: Eukaryotic Promotors? Nature 279: 376, 1979.

8. Busslinger M, Portmann R, Irminger JC, Birnstiel ML: Ubiquitous and gene-specific regulatory 5′ sequences in a sea urchin histone DNA clone coding for histone protein varients. Nucleic Acids Res 8: 957–977, 1980.
9. Efstratiadis A, Posakony JW, Maniatis T, Lawn RM, O'Connell C, Spritz RA, DeRiel JK, Forget B, Weissman SM, Slightom JL, Blechl AE, Smithies O, Baralle FE, Shoulders CC, Proudfoot NJ: The structure and evolution of the human β-globin gene family. Cell, in press.
10. Tolstoshev P, Mitchell J, Lanyon G, Williamson R, Ottolenghi S, Comi P, Giglioni B, Masera G, Modell B, Weatherall DJ, Clegg JB: Presence of gene for β globin in homozygous β^0 thalassaemia. Nature 259: 95–98, 1976.
11. Flavell RA, Bernards R, Kooter JM, de Boer E, Little PFR, Annison G, Williamson R: The structure of the human β-globin gene in β-thalassaemia. Nucleic Acids Res 6: 2749–2760, 1979.
12. Orkin SH, Old JM, Weatherall DJ, Nathan DG: Partial deletion of β-globin gene in certain patients with β^0 thalassaemia. Proc Natl Acad Sci USA 76: 2400–2404, 1979.
13. Ottolenghi S, Lanyon WG, Williamson R, Weatherall DJ, Clegg JB, Pitcher CS: Human globin gene analysis for a patient with $\beta^0/\delta\beta$-thalassemia. Proc Nat Acad Sci USA 72: 2294–2299, 1975.
14. Chang JC, Kan YW: β^0 thalassemia, a nonsense mutation in man. Proc Natl Acad Sci USA 76: 2886–2889, 1979.
15. Weatherall DJ, Clegg JB: Recent developments in the molecular genetics of human hemoglobin. Cell 16: 467–479, 1979.
16. Bernards R, Kooter JM, Flavell RA: Physical mapping of the globin gene deletion in $(\delta\beta)^0$ thalassaemia. Gene 6: 265–280, 1979.
17. Fritsch EF, Lawn RM, Maniatis T: Characterization of deletions which affect the expression of fetal globin genes in man. Nature 279: 598–603, 1979.
18. Tuan D, Biro PA, deRiel JK, Lazarus H, Forget BG: Restriction endonuclease mapping of the human γ-globin gene loci. Nucleic Acids Res, 6: 2519–2544, 1979.
19. Bernards R, Flavell RA: Physical mapping of the globin gene deletion in hereditary persistence of foetal haemoglobin (HPFH). Nucleic Acids Res 8: 1521–1534, 1980.
20. Tuan D, Murnane MJ, de Riel JK, Forget BG: Heterogeneity in the molecular basis of hereditary persistence of foetal haemoglobin. Nature 285: 335–337, 1980.
21. Van der Ploeg LHT, Konings A, Oort M, Roos D, Bernini L, Flavell RA: γ-β thalassaemia: Deletion of the γ- and β-genes influences β-globin gene expression in man. Nature 283: 637–642, 1980.
22. Collins J, Hohn B: Cosmids: A type of plasmid gene-cloning vector that is packageable in vitro in bacteriophage λ heads. Proc Natl Acad Sci USA 75: 4242–4246, 1978.
23. Royal A, Garapin A, Cami B, Perrin F, Mandel JL, Le Meur M, Bregegegre F, Gannon F, Lepennec JP, Chambon P, Kourilsky P: The ovalbumin gene region: Common features in the organization of three genes expressed in chicken oviduct under hormonal control. Nature 279: 125–132, 1979.

Accurate Transcription Initiation on Purified β-Globin DNA Fragments in a Reconstituted Cell-Free System

Donal S. Luse, Takashi Matsui, Joel R. Haynes, Eric A. Schon, Paul Rosteck, Jr., Jerry B. Lingrel, and Robert G. Roeder

Departments of Biological Chemistry and Genetics (D.S.L., T.M., R.G.R.), Division of Biology and Biomedical Sciences, Washington University School of Medicine, St. Louis and Department of Biological Chemistry (J.R.H., E.A.S., P.R., J.B.L.), University of Cincinnati College of Medicine, Cincinnati

INTRODUCTION

The regulation of eukaryotic gene expression is presently not well understood. Elucidation of the process of gene regulation must include a detailed description of the mechanisms controlling the initiation of RNA synthesis. The ability to study transcription mechanisms has been greatly facilitated by the development of soluble systems which mediate accurate transcription by class III [1–4] and by class II [5, 6] RNA polymerases. With such systems it should be possible to directly assess the involvement in gene expression of DNA promotor sequences, putative regulatory molecules, and the packaging of the template DNA into nucleosomes.

In our own studies of class II genes, we employed purified RNA polymerase II, a purified DNA fragment(s) and a soluble extract (S100) containing other essential components. The model gene system which we chose for our initial in vitro transcription studies was the major late promotor of adenovirus 2 (Ad2). This promotor had two important advantages for such a study: the 5′ end of the in vivo transcript had been precisely mapped [7], and it had been determined that the late promotor was an extremely strong promotor in vivo [8]. We were able to demonstrate selective and accurate initiation of transcription at the Ad2 late promotor when DNA fragments bearing that promotor were incubated with human RNA polymerase II and human S100 extracts [5]. Partial sequence analysis of the transcripts generated showed that initiation occurred at the same base used in vivo [5].

To ascertain the general applicability of this system for studies of gene regulation, including its reconstitution in vitro, it was necessary to analyze chromosomal genes with well-characterized patterns of expression. We therefore chose to study the in vitro transcription of genes coding for the β-chains of hemoglobin. We first examined transcription of the mouse β-globin major gene in the S100 system; in this case the location of the 5′ end of the in vivo transcript had been precisely mapped

[9]. We found that discrete transcripts were generated in the human RNA polymerase II/human S100 system from truncated DNA templates bearing the mouse globin promotor; a detailed analysis of these transcripts (which will be outlined below) revealed that they were initiated at the same base used in vivo [10].

We report here the results of two studies which extend our initial investigations of in vitro transcription of class II genes. The first set of experiments involves transcription of β-like globin genes from the goat. These DNAs have been isolated and extensively characterized by Lingrel and coworkers [11, 12, and Lingrel et al, this volume]. We demonstrate that genes coding for adult, juvenile, and fetal β-chains are transcribed accurately and to the same extent in the human S100 system. We also examined the transcription of two β-like genes which do not code, on the basis of sequence analysis [Lingrel et al, this volume], for known β-globin proteins. Only one of these genes has a functional promotor in the in vitro system. Examination of putative promotor sequences has allowed us to begin to identify regions of the DNA important in promotor function in these genes.

The second study involves the characterization of the factors involved in accurate in vitro initiation by RNA polymerase II [13]. Chromatography of S100 extracts has thus far yielded four fractions, each of which is essential for reconstitution of active and accurate in vitro RNA synthesis. The most purified components will mediate accurate transcription of all of the promotors which were accurately recognized in the original S100 extract, suggesting that the fractions contain general factors for transcription of class II genes.

GENERAL METHODOLOGY: TRANSCRIPTION OF MAMMALIAN β-GLOBIN GENES

Our primary method for demonstrating the accuracy of in vitro initiation by RNA polymerase II requires the use of DNA fragments cut by restriction enzymes at known distances downstream of the expected initiation site. These DNAs are incubated in the presence of S100 extract and purified RNA polymerase II; the resulting RNAs are purified and electrophoresed on a polyacrylamide gel (details of RNA synthesis, purification, and analysis are given in [5] and [10]). Accurate initiation will result in the production of discrete "run-off" RNAs whose sizes can be measured relative to markers of known length. The generation of a series of such run-off products in the size range of 200–500 bases allows the localization of the in vitro initiation site within plus or minus 10 bases. Exact localization of the 5′ end of the in vitro transcript can then be achieved by partial sequence analysis of one of the run-off RNAs.

The application of these techniques is illustrated by our analysis of the in vitro transcription of the mouse β-globin major gene [10]. An 800 base pair (bp) mouse DNA fragment bearing the in vivo initiation site for the β-globin major gene is diagrammed in Figure 1; the locations of four restriction sites, each unique within this DNA sequence, are indicated. When this mouse DNA fragment is cleaved in turn with each of the four restriction enzymes listed in Figure 1 and the resulting truncated DNAs are used as templates in the human RNA polymerase II/human S100 system, four discrete RNA products are generated (lanes 4–7, Fig. 2). Comparison with the RNA standards (lanes 1–3 and 8–10, Fig. 2) gives estimated lengths

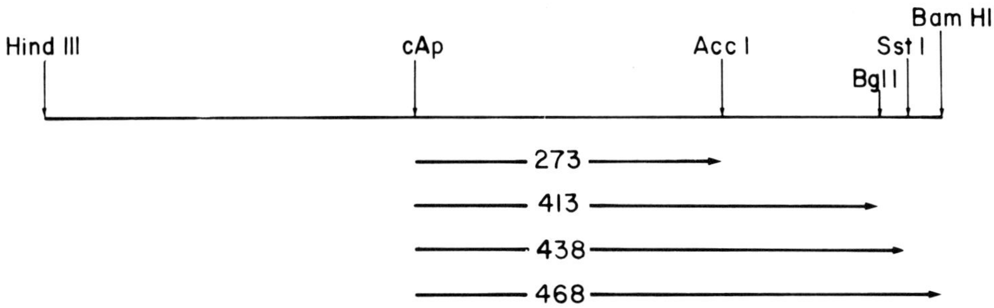

Fig. 1. Map of restriction sites and cap site within mouse β-globin DNA fragment. The 800 base pair mouse sequence diagrammed above was cloned, by insertion between the unique Hind III and Bam Hl sites of pBR322, to give the recombinant plasmid pMGS 1. "cAp" indicates the location of the 5′ end of the mature capped globin mRNA; this is presumed to be the initiation site of the primary transcript. All of the restriction sites mapped are unique within the mouse DNA sequence. The numbered arrows indicate the distances, in bases, from the cap site to the end of the coding strand for each restriction cut. The above data are taken from reference 9; the figure itself is from reference 10.

for the run-off RNAs as follows: Bam Hl cleaved, 468 ± 3 bases; Sst I cleaved, 439 ± 7 bases; Bgl I cleaved, 414 ± 7 bases; Acc I cleaved, 271 ± 5 bases. These values are in very good agreement with those predicted for accurate initiation (Fig. 1).

Exact localization of the 5′ end of the in vitro transcript was achieved by partial sequence analysis of the 468 base run-off RNA (lane 4, Fig. 2). As shown in Figure 3, the DNA sequence of the mouse gene predicted the existence of four large ($\geq$ 12 bases) oligonucleotides in a T_1 RNAase digest of the run-off RNA from the Bam Hl-cleaved template. The sequence also predicted that the 5′ terminal T_1 oligonucleotide would be a heptamer (Fig. 3). On the basis of our results with in vitro transcription of the Ad 2 late promotor, we expected the 5′ terminus of the mouse in vitro transcript to be capped.

When preparative amounts of the run-off RNA synthesized from Bam Hl-cleaved mouse globin DNA are digested with RNAase T_1 and subjected to fingerprint analysis, the pattern shown in Figure 4 is obtained. The digest contains four large oligonucleotides (labeled 1–4 in Fig. 4), as predicted. Nearest neighbor analysis of these oligonucleotides, labeled with either α ^{32}P-GTP or α ^{32}P-ATP, confirmed that they are the large oligonucleotides predicted from the DNA sequence (Fig. 3). Secondary digestion with RNAase T_2 of all oligonucleotides in the fingerprint larger than about 5 bases revealed only one oligonucleotide (labeled 17 in Fig. 4) with a cap structure. Thus, spot 17 must represent the 5′ end of the in vitro transcript. The mobilities of this oligonucleotide and its nearest neighbor analysis (when labeled with α ^{32}P-GTP or α ^{32}P-ATP) are consistent with its being the heptamer predicted as the 5′ terminal T_1 oligonucleotide from an in vivo transcript. Further, digestion of α ^{32}P-ATP-labeled oligonucleotide 17 with nuclease Pl revealed an A residue at its 5′ end and one internal A (our unpublished observations). These data indicate that the in vivo and in vitro transcripts of the mouse β-globin gene have the same initiation site [10].

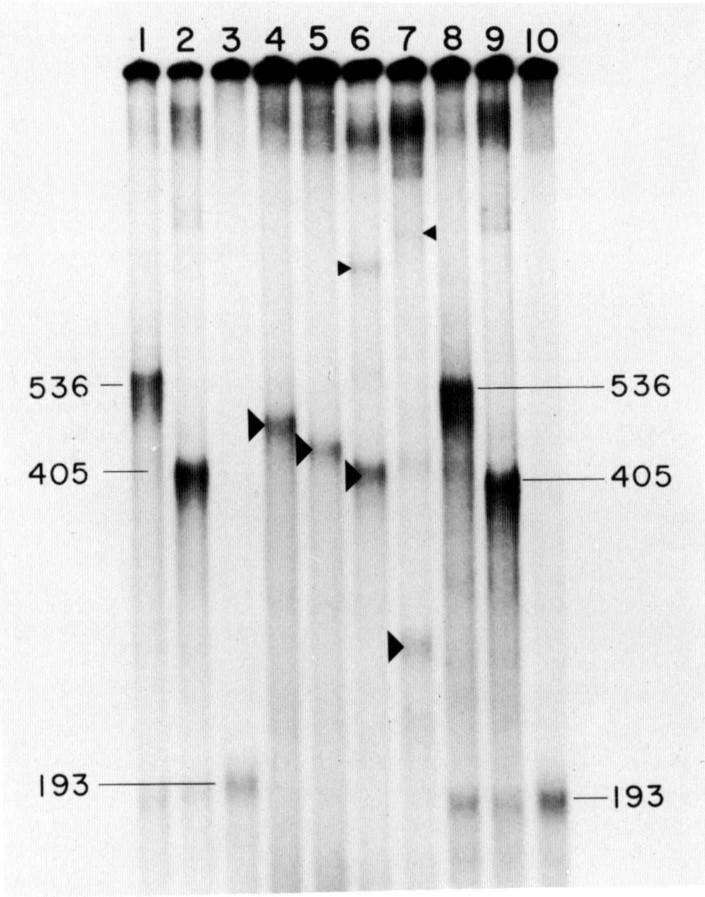

Fig. 2. Electrophoretic analysis of glyoxal-denatured RNAs synthesized in vitro from truncated mouse globin templates. The autoradiogram shows a polyacrylamide gel analysis of glyoxal-denatured RNAs synthesized in vitro by human RNA polymerase II in the presence of human S100. In lanes 4–7, the templates were the mouse β-globin containing plasmid pMGS 1 (see Fig. 1) cleaved with the following restriction enzymes: lane 4, Bam HI; lane 5, Sst I; lane 6, Bgl I; lane 7, Acc I. Transcripts in lanes 4–7 whose production is sensitive to 0.5 μg/ml α-amanitin are labeled with large or small arrowheads; the transcripts in lanes 6 and 7 labeled with small arrowheads are synthesized from vector (pBR322) sequences. Transcripts in lanes 1–3 and 8–10 are run-off RNAs synthesized in vitro from the Ad2 late promotor; the lengths of these RNAs are indicated in the margin of the gel. This figure is taken from reference 10.

The successful demonstration of accurate in vitro transcription of the mouse globin gene encouraged us to extend our analysis to the family of β-like globin genes of the goat. A detailed description of the isolation and characterization of these genes can be found elsewhere [11, 12 and Lingrel et al, these proceedings]. Five genes have thus far been tested in the in vitro system: β_A, the adult β-globin; β_C, the juvenile β-globin; γ, the fetal β-globin, and two other β-homologous sequences referred to as β_X and β_Y. Sequence analysis shows that β_Y could code for a protein

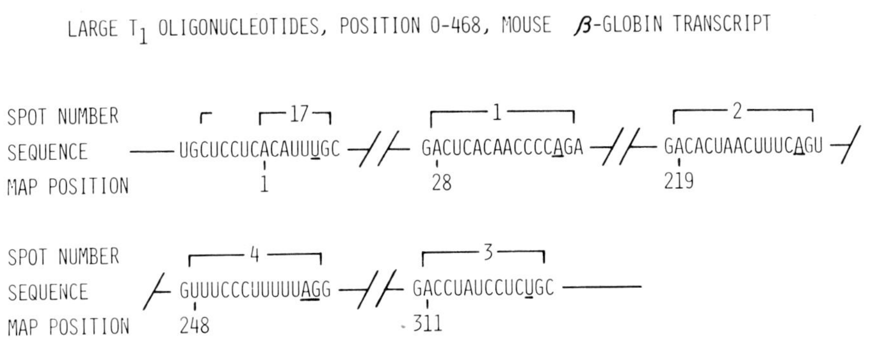

Fig. 3. Selected RNAase T_1 oligonucleotides in a mouse β-globin run-off RNA initiated at the in vivo cap site. The partial RNA sequence is shown for an RNA (complementary to the in vivo coding strand) extending from a position just 5′ distal to the in vivo cap site to a position 468 bases 3′ distal to that site. Five RNAase T_1 oligonucleotides are indicated. These are the four large (≥ 12 bases) oligonucleotides and the 5′ proximal oligonucleotide expected from a transcript initiated at the in vivo cap site and extending downstream to the point of Bam Hl cleavage (see Fig. 1). The number above each oligonucleotide indicates the spot in Figure 4 to which that oligonucleotide corresponds. The figure is taken from reference 10; the source of the sequence data is reference 9.

about 70% homologous to the goat ε chain, while β_X probably could not code for a protein recognizable as a globin chain due to numerous mutations (Lingrel et al, these proceedings).

The strategy for preliminary studies of in vitro transcription of these genes in the human S100 system involved, as above, the synthesis of run-off RNAs from DNA templates cut at restriction sites downstream of the (presumed) initiation site. For the studies reported here, all of the cleaved goat genes except β_X were expected to yield correctly initiated run-off RNAs of about 470 bases; correctly initiated β_X run-offs would be about 405 bases long. When each of the goat DNA fragments is incubated in the human RNA polymerase II/human S100 system, run-off RNAs such as those in Figure 5 are produced. Ad2 run-off RNAs of 536 (lanes 1 and 9) and 405 bases (lane 8), and a mouse globin run-off RNA of 468 bases (lane 2), are included for size comparison. As Figure 5 demonstrates, β_A, β_C, and γ are all transcribed in the S100 system (lanes 3, 4, and 5); the amount of run-off RNA for each template is approximately the same. β_Y is also transcribed in this system (lane 6), but β_X gives no run-off transcript (lane 7). The transcripts for β_A, β_C, γ, and β_Y all comigrate with the 468 base mouse globin run-off; thus, within the level of resolution of such an experiment, the goat transcripts are correctly initiated.

FRACTIONATION OF RNA POLYMERASE II FACTORS FROM S100 EXTRACTS

We have begun to identify and purify the components in human S100 extracts which mediate accurate initiation by RNA polymerase II. Fractions which have been chromatographically separated were assayed, singly or in combination, by

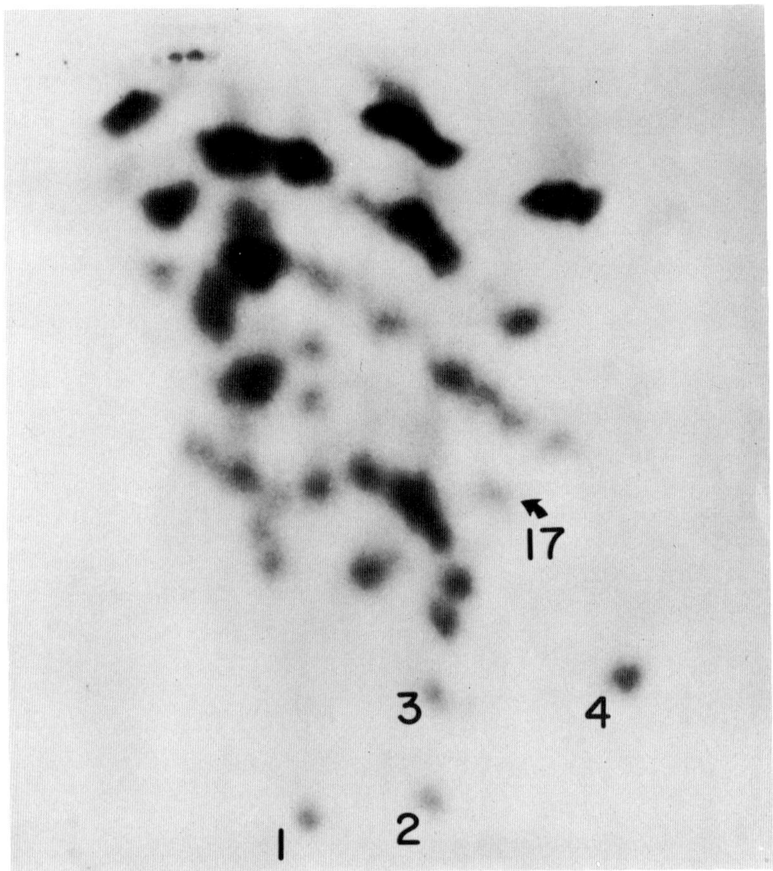

Fig. 4. Fingerprint analysis of mouse β-globin run-off RNA synthesized in vitro. The 468 base mouse globin run-off RNA (lane 4, Fig. 2) was synthesized in preparative quantities, purified, digested with RNAase T_1, and analyzed by ionophoresis in the first dimension (left to right) and homochromatography in the second dimension (bottom to top). The oligonucleotides labeled 1–4 correspond to the four large T_1 oligonucleotides expected in the 468 base run-off RNA on the basis of the DNA template sequence; oligonucleotide 17 is the putative capped 5′ end of the transcript (see text and Fig. 3). This figure is taken from reference 10.

incubation with purified RNA polymerase II and an Ad2 DNA fragment bearing the viral late promotor, which yields a 536 base run-off RNA upon accurate initiation of RNA synthesis. Details of the purification and characterization of these factors will be given elsewhere [13]. In brief, an initial chromatographic fractionation of KB cell S100 yielded three fractions, each of which was required to reconstitute specific transcription of the Ad2 late promotor. Upon further purification, two of these fractions continued to chromatograph as single components while the third fraction was separated into two components. Thus, after our most extensive purification we have four fractions, referred to as TF IIA, TF IIB, TF IIC, and TF IID, all of which are required to reconstitute active and accurate transcription by RNA polymerase II at the Ad2 late promotor. At present none of the factors in TF IIA-D have been

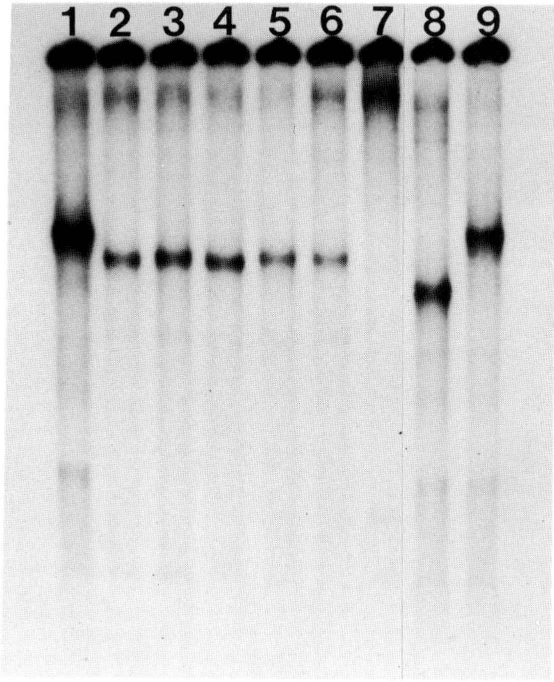

Fig. 5. Electrophoretic analysis of glyoxal-denatured RNAs synthesized in vitro from truncated goat globin templates in the presence of S100. The autoradiogram shows a polyacrylamide gel analysis of glyoxal-denatured RNAs synthesized in vitro by human RNA polymerase II in the presence of human S100. Lanes 1 and 9 contain 536 base Ad2 run-off RNAs; lane 8 contains a 405 base Ad2 run-off RNA. The transcript in lane 2 is the 468 base mouse β-globin run-off RNA which was characterized by fingerprint analysis (see lane 4, Fig. 2, and text). Lanes 3 through 7 contain goat β-globin template cleaved as indicated; the number in parentheses indicates the approximate length in bases of a correctly initiated run-off RNA from that template: lane 3, β_A cleaved with Bam Hl (470); lane 4, β_C cleaved with Bam Hl (470); lane 5, γ cleaved with Bam Hl (470); lane 6, β_Y cleaved with Bam Hl (470); lane 7, β_X cleaved with Sal I (405).

completely purified; thus, a detailed description of their modes of action is not yet possible. The data do suggest, however, that three of the four fractions (TF IIA, B, and D) are absolutely required for initiation, while the fourth (TF IIC) may function by suppressing random transcription (details of the purification and characterization can be found in [13]).

Since these factors (A–D) were isolated based on their ability to reconstitute accurate initiation at the Ad2 major late promotor, it is important to ask whether these same fractions will mediate accurate initiation at the eukaryotic promotors which have been correctly transcribed in the presence of S100. Figure 6A shows the run-off RNAs generated when mouse and goat globin templates, cleaved to generate the same run-off RNAs seen in Figure 5, are transcribed in the presence of the three initial chromatographic fractions, mentioned above, which are needed to reconstitute transcription of the Ad2 late promotor. A comparison with Figure 5

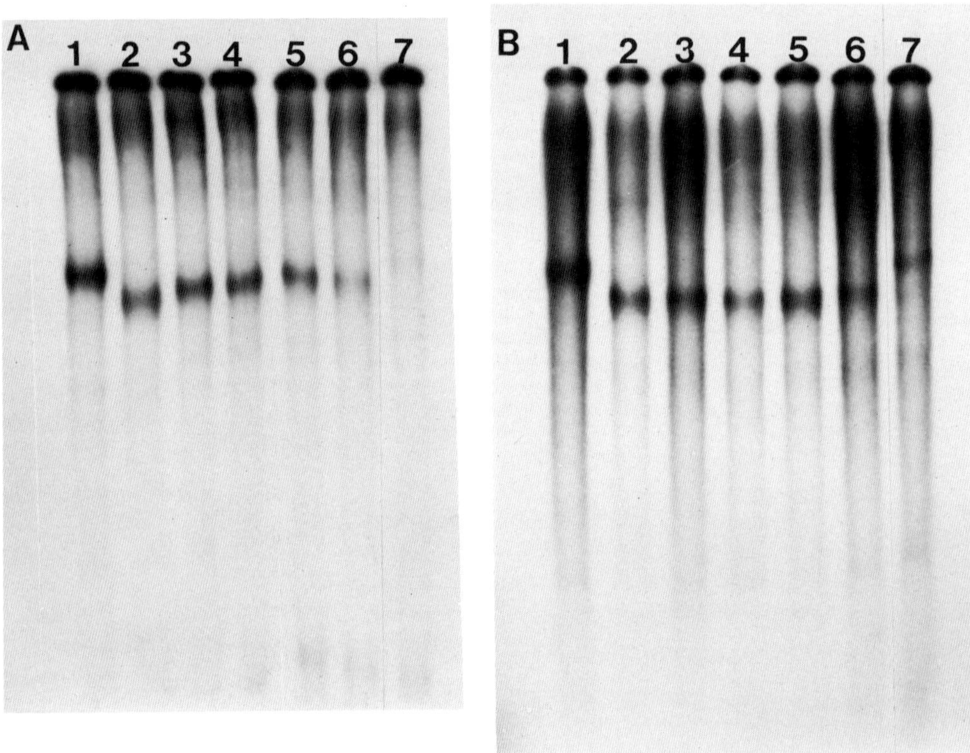

Fig. 6. Electrophoretic analysis of glyoxal-denatured RNAs synthesized in vitro from truncated goat globin templates in the presence of purified transcription factors. The autoradiograms show polyacrylamide gel analyses of glyoxal-denatured RNAs synthesized in vitro by human RNA polymerase II in the presence of different sets of purified transcription factors. In both cases, the templates for each lane, and the expected sizes of correctly initiated run-off RNAs, are: lane 1, Ad2 late promotor (536 bases); lane 2, mouse β-globin major (468 bases); lane 3, goat β_A (470 bases); lane 4, goat β_C (470 bases); lane 5, goat γ (470 bases); lane 6, goat β_Y (470 bases); lane 7, goat β_X (405 bases). (The restriction enzymes used to cleave the goat templates are those listed in the legend to Fig. 5). In panel A, the factors employed were the three factors from the initial chromatographic separation which were needed to reconstitute Ad2 late transcription. In panel B, the factors were the most purified fractions (TF IIA-D), again assayed on the basis of their ability to reconstitute Ad2 late transcription (see text).

shows that all of the transcripts generated in the presence of S100 are also generated by the reconstituted components; further, the relative intensities of the transcripts are approximately the same in both cases. Transcripts generated from the globin templates using the most purified fractions (TF IIA-D) are displayed in Figure 6B. Once again, the same transcripts (in roughly the same amounts) are generated from the globin genes in the presence of TF IIA-D as were generated from these genes in the presence of S100. It thus seems probable that the transcription factors necessary for accurate transcription initiation at the Ad2 late promotor are general class II transcription factors. This topic will be more fully discussed below.

DISCUSSION

We have demonstrated that the human RNA polymerase II/human S100 system will selectively and accurately initiate RNA synthesis at a broad range of class II promotors. In the case of Ad2 major late and mouse β-globin major, initiation in vitro occurs at the same base used in vivo. The goat globin transcripts have not yet been as rigorously characterized (by 5′ end analysis), but the production of discrete run-off RNAs of the expected size from truncated goat globin templates argues for exact initiation at these promotors as well. Fractionation of S100 extracts reveals at least four components involved in specific transcription by RNA polymerase II. These four fractions produce the same transcripts from adenovirus and globin templates as did the original S100.

Several characteristics of our in vitro transcription system are clear from the data presently accumulated. First, the protein components involved (both RNA polymerases and factors) are not species-specific. Thus, a variety of vertebrate RNA polymerases II will transcribe the Ad2 late promotor accurately in the presence of human S100 [5] and factors in the human S100 extract will mediate accurate transcription by human RNA polymerase II on mouse [10] and goat (see above) globin genes. Second, and perhaps more importantly, the tissue from which the S100 extract is made need not be expressing the gene to be transcribed in order to obtain an accurate transcript of that gene in purified form. Thus, it was shown earlier that S100 extracts from uninfected KB cells will direct accurate initiation at the Ad2 late promotor [5]. The most striking example of this lack of tissue specificity is seen with the transcription of the goat β_A, β_C, and γ genes, which are expressed sequentially in development (Fig. 5). Here we observe equal levels of transcription of these three genes in vitro using an S100 extract from cells which are not expressing globin genes at all. These results strongly suggest that the factors in S100 extracts which are involved in specific initiation by RNA polymerase II are general transcription factors. This hypothesis is supported by the fact that initiation factors which have been extensively purified on the basis of their ability to direct accurate initiation at the Ad2 late promotor retain the ability to direct accurate initiation at the mouse and goat globin promotors as well (Fig. 6A, B).

It is important to ask whether the observed generality of transcription factors for RNA polymerase II is correlated with the presence of certain conserved DNA sequences at class II promotors. It has been observed that almost all promotors recognized by RNA polymerase II have an A-T rich sequence (TATAAA, or some variation thereof) about 25 bases upstream of the initiation site [see, for example, 14]. Several observations made with the in vitro transcription system support the idea that this sequence may be involved in the initiation of transcription. The only goat globin gene which failed to yield an in vitro transcript was β_X; examination of the DNA sequence near the putative initiation site shows that β_X has lost the TATAAA sequence [see Fig. 6 of Lingrel et al, these proceedings]. When the β_A gene is mutated in vitro by cleavage at a restriction site 9 bases upstream of the initiation site and ligation to the pBR322 carrier plasmid, run-off RNA is no longer produced in the in vitro transcription system [J.R. Haynes, D.S. Luse, and J. Lingrel, unpublished results]. This deletion eliminates almost all upstream goat DNA, not only the TATAAA sequence; thus the lack of transcription of this deleted gene does not necessarily result from the loss of the TATAAA sequence. Nevertheless, the result is consistent with the importance of this sequence in directing accurate initiation. The result also demonstrates that sequences contained within the transcribed region of the gene are not in themselves sufficient to direct the initiation of transcription, as has been recently demonstrated to be the case with 5S RNA genes [15, 16]. Further in vitro deletion and rearrangement experiments are currently in progress, in order to more precisely map DNA sequences involved in the initiation process.

It is somewhat disappointing that no evidence of gene control is found in transcription using the S100 extracts. However, the seemingly universal applicability of the S100 factors in directing specific initiation at class II promotors may in itself be informative. One simple model of gene regulation consistent with our observations would invoke the presence of promotors of nearly equal strength for all genes, with differential expression mediated by selective binding of control proteins and/or selective changes in the local nucleoprotein structure. The removal of such transcriptional constraints in our current in vitro system by the use of purified DNA templates would then account for the observed lack of regulated gene expression. Such ideas can clearly be tested by using purified initiation factors to transcribe templates in their native chromatin configuration, with or without putative regulatory components (such as proteins which bind selectively to the gene in question). We anticipate that such experiments will ultimately lead to a detailed understanding of the mechanisms of eukaryotic gene control.

SUMMARY

We have recently demonstrated that extracts of human cells (S100s) will mediate the selective and accurate initiation of transcription by human RNA polymerase II on purified DNA fragments containing the adenovirus 2 major late promotor [see 5]. We report here that this soluble system will also accurately transcribe purified β-globin genes from the mouse and the goat. Transcription of the mouse β-globin major gene has been most extensively analyzed; partial sequence analysis demonstrates that initiation of RNA synthesis on this gene occurs at the same base in vivo and in vitro. The adult, juvenile, and fetal β-globin genes of the goat are all transcribed accurately and to the same extent in our in vitro system. We have also transcribed two β-homologous genes of the goat, βx and βy, whose existence was not expected on the basis of known goat β-globin proteins [see Lingrel et al, this volume]. Only one of these two genes has a functional promotor in the in vitro system. Chromatography of the S100 extracts has thus far revealed four fractions which are necessary to reconstitute accurate transcription of the adenovirus late gene. These purified components will also mediate the accurate transcription of the β-globin genes.

REFERENCES

1. Wu GJ: Adenovirus DNA directed transcription of 5.5S RNA in vitro. Proc Nat Acad Sci USA 75:2175–2179, 1978.
2. Birkenmeier EH, Brown DD, Jordan E: A nuclear extract of Xenopus laevis oocytes that accurately transcribes 5S RNA genes. Cell 15:1077–1086, 1978.
3. Ng SY, Parker CS, Roeder RG: Transcription of cloned Xenopus 5S RNA genes by X. laevis RNA polymerase III in reconstituted systems. Proc Nat Acad Sci USA 76:136–140, 1979.

4. Weil PA, Segall J, Harris B, Ng SY, Roeder RG: Faithful transcription of eukaryotic genes by RNA polymerase III in systems reconstituted with purified DNA templates. J Biol Chem 254:6163–6173, 1979.
5. Weil PA, Luse DS, Segall J, Roeder RG: Selective and accurate initiation of transcription at the Ad2 major late promotor in a soluble system dependent on purified RNA polymerase II and DNA. Cell 18:469–484, 1979.
6. Manley JL, Fire A, Cano A, Sharp PA, Gefter ML: DNA-dependent transcription of adenovirus genes in a soluble whole-cell extract. Proc Nat Acad Sci USA 77:3855–3859, 1980.
7. Ziff EB, Evans RM: Coincidence of the promotor and capped 5′ terminus of RNA from the adenovirus 2 major late transcription unit. Cell 15:1463–1475, 1978.
8. Flint J: The topography and transcription of the adenovirus genome. Cell 10:153–166, 1977.
9. Konkel DA, Tilghman SM, Leder P: The sequence of the chromosomal mouse β-globin major gene: Homologies in capping, splicing and poly(A) sites. Cell 15:1125–1132, 1978.
10. Luse DS, Roeder RG: Accurate transcription initiation on a purified mouse β-globin DNA fragment in a cell-free system. Cell 20:691–699, 1980.
11. Haynes JR, Rosteck P, Schon EA, Gallagher PM, Burks DJ, Smith K, Lingrel JB: The isolation of β_A, β_C and γ globin genes and a presumptive embryonic globin gene from a goat DNA recombinant library. J Biol Chem 255:6355–6367, 1980.
12. Haynes JR, Rosteck P, Lingrel JB: Unusual sequence homology at the 5′ ends of the developmentally regulated β_A, β_C and γ globin genes of the goat. Proc Nat Acad Sci USA 77:in press, 1980.
13. Matsui T, Segall J, Weil PA, Roeder RG: Multiple factors required for accurate initiation of transcription by RNA polymerase II. J Biol Chem 255:11992–11996, 1980
14. Baker CC, Herisse J, Courtois G, Galibert F, Ziff E: Messenger RNA for the Ad2 DNA binding protein: DNA sequences encoding the first leader and heterogeneity at the mRNA 5′ end. Cell 18:569–580, 1980.
15. Sakonju S, Bogenhagen DF, Brown DD: A control region in the center of the 5S RNA gene directs specific initiation of transcription: I. The 5′ border of the region. Cell 19:13–25, 1980.
16. Bogenhagen DF, Sakonju S, Brown DD: A control region in the center of the 5S RNA gene directs specific initiation of transcription: II. The 3′ border of the region. Cell 19:27–35, 1980.

Organization and Expression of Globin Genes, pages 139–159

Selective Transcription of Mouse Globin Genes In Vitro: Characterization of the Region Recognized by RNA Polymerase II

Carol A. Talkington and Philip Leder

Laboratory of Molecular Genetics, National Institute of Child Health and Human Development, National Institutes of Health, Bethesda, Maryland

INTRODUCTION

The globin genes are of particular interest because they represent a gene family whose expression is both temporally and coordinately regulated throughout the lifetime of the mouse. Although the adult α and β globin genes reside on different chromosomes [1,2], their expression is coordinated such that equal amounts of each polypeptide are produced. Within the β-globin gene cluster, at least four nonallelic genes are active and their products have been identified. Two are expressed only in the fetal mouse (y and z) and two only in the adult (βmaj and βmin). The adult β genes appear to be regulated with respect to one another (80% of adult mouse hemoglobin contains the βmaj globin polypeptide) [2–4]. The α-globin gene cluster is equally complex. An embryonic globin polypeptide (designated x) has been identified that resembles the α-globin chain [2], and there are at least two nonallelic adult α-globin genes (α_1 and α_2) that are coordinately expressed in both fetal and adult Balb/c mice [5]. Thus, activation of adult α-globin gene expression precedes that of the adult β-globin gene [6]. In addition, cloning of the α and β-globin genes has revealed several α and β-globinlike genes whose significance is unknown (A. Leder, unpublished; [7]; and unpublished results; [8]).

Very little is understood about the mechanisms that control coordinate and temporal expression of globin or other eukaryotic genes. However, in addition to representing a well-defined developmental system, the mouse globin genes offer an advantage for studying the regulation of gene expression since the nucleotide sequences of βmaj [9], βmin [10] and α_1 [11] have been determined and compared; also several mutant globinlike genes are available for study ([7] and unpublished).

The α and β-globin genes share short segments of sequence homology in the region most likely to contain sequences involved with the initiation of transcription, namely in the 5′ portion. There is also limited homology between the genes at and around the capping site (the capping site is the nucleotide corresponding to the 5′

end of the nascent RNA to which a cap structure is attached); most striking is the shared hexanucleotide TATTA$^{G}{}_{A}$ whose 5′ end is located exactly 30 nucleotides preceding the capping site (ie, the putative initiation site) of each of these genes. This DNA segment closely resembles the "Hogness box" sequence found at the same location in many eukaryotic genes transcribed by RNA polymerase II [12]; however, its function is not yet understood.

Although nucleotide sequence data is very important in elucidating gene organization, at present it alone cannot explain the regulation of globin gene expression. However, the recent development of cell-free extracts which accurately initiate transcription of purified eukaryotic DNA templates in vitro [13–16] provides a first step in this direction. These extracts offer an excellent tool for the identification of the DNA sequences selectively recognized by the RNA polymerase and for the isolation of factors that direct the RNA polymerase to the initiation site. In order to apply this methodology, we have simplified the preparation of cell-free extract based on that described by Weil et al [14, 15] and applied it to study the transcription of mouse globin genes. We have tested the ability of this cell-free system (which requires no exogenous RNA polymerase II) to accurately initiate transcription of the α_1 globin gene, of two α-globinlike genes, of the βmaj globin gene, and of a truncated gene. Using the results of these initial studies we have identified the specific region(s) of α-globin DNA that directs initiation of transcription by RNA polymerase II in our cell-free extract.

METHODS

Plasmid DNA Templates

The $\alpha_1 p_1$•pBR clone was constructed by Y. Nishioka as described [17]. All cesium chloride-purified globin gene-containing plasmid DNAs (grown and purified as in [11]) were restricted utilizing the assay conditions provided with the enzymes, except that we substituted 1 mM dithiotreitol for B-mercaptoethanol. Restricted DNAs were extracted with phenol and chloroform, precipitated with ethanol, and resuspended in distilled H_2O.

RNA Polymerase II Purification

We partially purified RNA polymerase II from frozen TEPC mouse tumor cells in a manner similar to that described [18]. (All procedures were carried out at 4°C. The centrifugations referred to were done at 10,000 × g in an SS34 rotor for 20–30 minutes each.) Briefly, 20 gm of mouse tumor cells were thawed, minced, homogenized, and the cells washed twice with phosphate-saline buffer, pH 7.2. The cells were resuspended in 25 ml TGMED (50 mM Tris-HCl, pH 7.9/10% glycerol/5 mM $MgCl_2$/0.1 mM EDTA/0.5 mM dithiothreitol) and brought to 320 mM ammonium sulfate. The cells were sonicated and the mixture centrifuged. The resulting supernatant was diluted to 100 mM ammonium sulfate, centrifuged as above, and the supernatant diluted to 75 mM ammonium sulfate. The RNA polymerase was precipitated from the supernatant by the addition of 5% polymin P (20 μl/ml supernatant). After centrifugation, the pellet was resuspended in 250 mM ammonium sulfate-TGED (TGED is 50 mM Tris-HCl, pH 7.9/10% glycerol/0.1 mM EDTA/0.5 mM DTT) and centrifuged. The polymerase in the resulting supernatant was subsequently precipitated with solid ammonium sulfate, and that pellet, resuspended in TGED, was submitted to DEAE cellulose chromatography (DE52 from Whatman, Clifton, NJ), and RNA polymerase II eluted from the column at approximately 200 mM ammonium sulfate. The resultant RNA polymerase II, at a concentration of approximately 20,000 units/ml, contained less than 5% RNA polymerases I and III. (One unit of activity catalyzes the incorporation of 1 pmole of UMP into RNA in 20 minutes at 37°C under optimal conditions, using calf thymus DNA as template.) RNA polymerase II was stored in aliquots at –80°C.

Preparation of Cell-Free Extract

Crude S-100 extracts from HeLa cells were prepared in a manner similar to that described by Weil et al [14], with a significant modification, as noted. HeLa cells were grown to a density of 0.5–1.0 × 10^6 cells/ml (cells were a gift from S.P. Lai) and harvested. The cells were washed two times with phosphate-saline buffer, pH 7.2, and the packed-cell-volume recorded after the second centrifugation (for 1 liter of cells, this volume was about 3 ml). The cells were then washed one time with hypotonic buffer (10 mM Hepes, pH 7.9/10 mM KCl/1.5 mM $MgCl_2$/0.5 mM DTT). The pellet was resuspended in two packed-cell-volumes of hypotonic buffer and placed on ice for 15 minutes. After swelling, the cells (kept at 4°C) were lysed by Dounce homogenization; ten strokes of a "B" pestle were sufficient to break open the cells, without significantly damaging the nuclei. Next, 0.11 volume of buffer (300 mM Hepes, pH 7.9/1.4 M KCl/30 mM $MgCl_2$) was added to the cell lysate. Subsequently, the extract was quick-frozen in liquid nitrogen and stored in a –20°C freezer overnight; this latter step was our addition to the protocol of Weil et al [14] and apparently resulted in a large accumulation of RNA polymerase II in the soluble fraction of the extract. The lysate was thawed slowly and centrifuged at 100,000 × g for 60 minutes. The supernatant (S-100) was dialyzed at least 6 hours against S-100 dialysis buffer (20 mM Hepes, pH 7.9/20% glycerol/0.2 mM EDTA/0.5 mM DTT/ 100 mM KCl). The dialysate was collected and centrifuged at 10,000 × g for 10 minutes. The resultant supernatant was aliquoted and stored at –80°C for use in the transcription experiments described below. This supernatant, namely an RNA polymerase II-enriched S-100 extract, contained approximately 10^8 HeLa cell equivalents/ml, 1,000 units of RNA polymerase II/ml, and 500 units of RNA polymerases I and III/ml (as measured by a standard assay utilizing calf thymus DNA as template, see previous section). Approximately 2 × 10^6 cell equivalents were used per transcription reaction.

In Vitro Transcription and Analysis of RNA Products

The in vitro transcription reactions (50 μl) consisted of approximately 2 μg (unless otherwise noted) of restricted plasmid DNA (the specific templates utilized are described in the text), 7.5 mM $MgCl_2$, 10 mM Hepes, pH 7.9, 10% glycerol, 0.1 mM EDTA, 0.25 mM dithiothreitol, 50 mM KCl, 600 μM each of adenosine, cytidine, and uridine triphosphates, 60 μM guanosine [α^{32}P] triphosphate (specific activity 4.5 Ci/mmol—purchased from ICN, Irvine, CA), and approximately 2 × 10^6 cell equivalents of HeLa cell S-100 extract in the presence or absence of 2 μg/ml α-amanitin (Calbiochem, La Jolla, CA); 20–100 units of mouse tumor RNA polymerase II was added as indicated. The reactions were incubated at 30°C for 30 or 60 minutes and terminated by the addition of 150 μl of 66 mM sodium acetate, pH

5.5, 0.3% sodium dodecyl sulfate, and 100 μg tRNA. RNA synthesis for S1 nuclease mapping experiments was accomplished in the same manner except 100 μM unlabeled guanosine triphosphate was substituted for the [^{32}P]GTP and the reaction was treated with 75 μg/ml DNAse I (Worthington, Wilmington, DE) at 30°C for 5 minutes prior to the addition of the sodium acetate solution. The resultant transcripts were extracted twice with redistilled phenol and chloroform (volume:volume), extracted once with chloroform, and precipitated with ethanol. The dried pellet was resuspended in 10 μl of 75% deionized formamide/0.05% xylene cyanol, bromophenol blue, heated at 70°C for 2 minutes, and submitted to electrophoresis through a 4% polyacrylamide-7 M urea gel (in 0.05 M tris-borate-EDTA buffer, pH 8.3) [19]. The separated RNA species were visualized by autoradiography (on Kodak XR-5 film with high intensity screens); exposure times are indicated in the figure legends. In order to determine the sizes of the RNA species, the migration distances of the RNA polymerase II-dependent transcripts through the polyacrylamide gel were compared with those of denatured DNA fragments whose sizes in base pairs (bp) were known. The molecular weights of the end-to-end transcripts generated from a HincII-restricted template of α_1•pBR were determined by glyoxalating the RNA species and electrophoresing them through a 1.4% agarose gel in Tris-phosphate buffer as described [20]. These RNAs migrated coincident with the denatured DNA restriction fragments.

S1 Nuclease Mapping

The hybridization reactions consisted of RNA (either RNA synthesized in vitro from a HincII-restricted α_1•pBR template, 2 ng of which was the 445-base RNA, or 50 ng of adult Balb/c mouse globin RNA) and 10–15 ng of an in vivo [^{32}P] uniformly labeled alpha globin gene-containing DNA fragment (the 2,100 bp Eco RI and 580 bp HpaII DNA fragments, each at a specific activity of 2×10^5 cpm/μg, were gifts from D. Hamer) resuspended in 30 μl of 80% formamide/0.4 M NaCl/1 mM EDTA/40 mM Pipes, pH 6.5. The samples were heated at 70°C for 10 minutes and then incubated overnight at 57°C to allow for the formation of RNA•DNA hybrids. Four hundred microliters of 250 mM NaCl/1 mM $ZnSO_4$/5% glycerol/30 mM NaAc, pH 4.6/25 μg/ml heat-denatured salmon sperm DNA plus 4,000 units of S1 nuclease (Miles, Elkhart, IN) were added to the hybridization reactions and incubation was continued at 45°C for 30 minutes. The S1 nuclease digestion was stopped upon the addition of ethanol; the nucleic acid was precipitated and resuspended in 15 μl of 5 mM Tris-HCl, pH 7.4/65% deionized formamide/10 mM NaOH/1 mM EDTA/0.02% xylene cyanol, bromophenol blue. Samples were heated at 90°C for 2 minutes and loaded immediately onto an 8% polyacrylamide-7 M urea gel. After electrophoresis, the [^{32}P] DNA fragments protected from S1 nuclease digestion by the in vivo or in vitro RNAs were visualized by autoradiography. The autoradiograph shown in Figure 4 was of a gel exposed to Kodak XR-5 film with high intensity screens for 4 days.

RESULTS

In Vitro Transcription of α_1 and βMaj Genes: Characterization of the Simplified Transcription System

Analysis of RNA products synthesized by monkey kidney cells infected with SV40 recombinants containing the adult mouse α_1 globin gene has suggested that

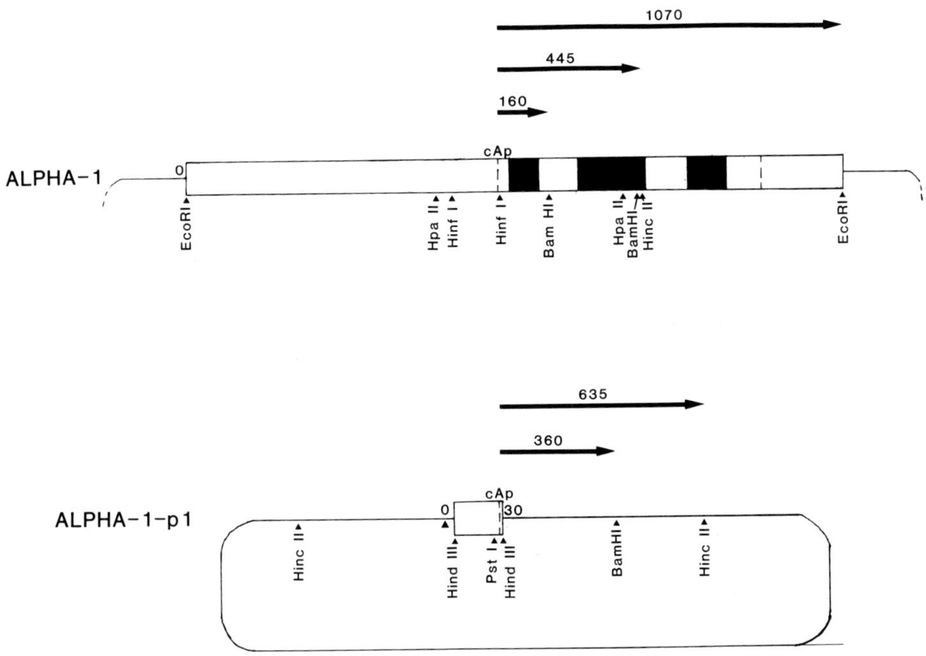

Fig. 1. Schematic maps of plasmid DNAs α_1•pBR and α_1p_1•pBR. Restriction endonuclease sites referred to in the text are denoted by the small vertical arrowheads. The cAp-labeled and unlabeled vertical dotted lines identify the locations on the DNA of the capping and polyA addition sites, respectively, of the α-globin mRNA. The black boxes indicate the location and sizes of the three α-globin gene coding regions, separated by the two intervening sequences. The RNA polymerase II-dependent transcripts generated from each of the restricted α_1•pBR or α_1p_1•pBR templates described in the text are represented by the thick lines; the arrowheads refer to the direction of transcription, which terminated at the end of the DNA fragment cleaved by the respective restriction endonuclease. The size, in bases, of each of the transcripts is shown above the arrows. Each of the α-globin gene-containing DNA fragments is oriented such that transcription proceeds from left to right. Each fragment's orientation with respect to the pBR322 DNA (denoted by the single lines) is indicated by the 0 at the EcoRI restriction site; that is, if the nucleotides of pBR322 are numbered as described (35) then in α_1•pBR the vector nucleotide numbers increase in a counterclockwise direction starting at 0 (the EcoRI site on the left), while in α_1p_1•pBR the vector nucleotide numbers increase in a clockwise direction. The number 30 on α_1p_1•pBR refers to the orientation of the vector with respect to the α gene containing DNA (ie, α_1p_1) inserted at the HindIII site of pBR322.

the α_1-containing fragment utilized to construct the recombinant possesses a functional promoter that is accurately recognized in vivo [21]. Thus, we have selected this 2.1 kilobase (kb) genomic fragment (ligated into pBR322 [M. Kaehler, unpublished]) to test its ability to be transcribed in vitro by RNA polymerase II in a crude cell-free extract. In order to assay for accurate initiation of α_1 transcription, we have measured the sizes of RNA species synthesized in vitro from templates of endonuclease-restricted α_1•pBR. A map displaying the relevant restriction sites and the predicted sizes of run-off transcripts of α_1•pBR is shown in Figure 1.

When HincII-restricted α_1•pBR DNA was transcribed in the presence of a HeLa cell S-100 extract supplemented with mouse tumor RNA polymerase II, RNA species of 445, 1,100, 2,100, and 3,300 bases were synthesized (Fig. 2, lane b). However, only synthesis of the 445-base RNA was completely inhibited by low

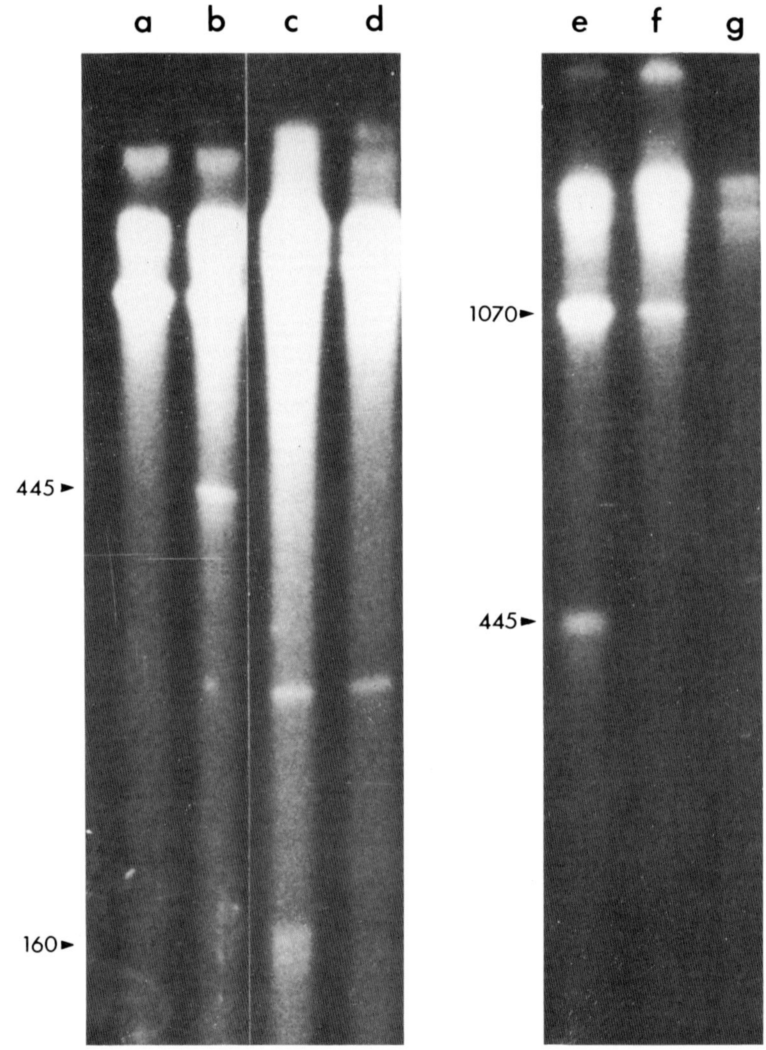

Fig. 2. Autoradiograph of RNAs synthesized in vitro from restricted α_1•pBR templates. [^{32}P]-RNAs were synthesized at 30°C for 1 hour, purified, and electrophoresed through a 4% polyacrylamide-7 M urea gel as described in the Experimental Procedures. The transcription reactions analyzed in lanes a–g each contained approximately 2×10^6 cell equivalents of our HeLa cell S-100 extract, 60 units of mouse tumor RNA polymerase II, and approximately 2 μg of the following templates and 2 μg/ml α-amanitin where indicated: lane a, HincII-restricted α_1•pBR plus α-amanitin; lane b, HincII-restricted α_1•pBR; lane c, BamHI-restricted α_1•pBR; lane d, BamHI-restricted α_1•pBR plus α-amanitin; lane e, HincII-restricted α_1•pBR; lane f, EcoRI-restricted α_1•pBR; lane g, EcoRI-restricted α_1•pBR plus α-amanitin. Note that the gel containing lanes a–d was run for a shorter period of time than the gel containing lanes e–g. The transcripts were visualized by exposing the gels to Kodak XR-5 film with intensifying screens for 10 hours. The sizes, in bases, of the RNA polymerase II-dependent transcripts indicated to the left of lanes a and e were determined by comparing the mobilities of BstNI- or HaeIII-restricted pBR 322 DNA (35) (the fragments were denatured prior to electrophoresis) to those of the [^{32}P]-transcripts.

concentrations of α-amanitin (Fig. 2, lane a), indicating that this transcript was synthesized by RNA polymerase II. The larger RNAs, whose synthesis was insensitive to α-amanitin, corresponded in size to the α_1•pBR HincII restriction fragments, and probably resulted from end-to-end transcription by the RNA polymerases in the S-100 extract. (Eukaryotic RNA polymerase tends to initiate transcription at ends and nicks; [22].) Synthesis of the 445-base transcript was dependent upon factors in the S-100 extract, since transcription in the presence of RNA polymerase II alone exhibited no specificity [not shown].

In order to simplify the preparation of cell-free extracts and to eliminate the requirement of exogenous RNA polymerase II, we prepared the HeLa cell S-100 extract in a manner slightly different from that described by Weil et al [14]. The modification, ie, freezing of the extract between the homogenization and centrifugation steps (see Experimental Procedures) resulted in an S-100 preparation highly enriched with RNA polymerase II. Hence, incubation of HincII-restricted α_1•pBR with the S-100 extract alone resulted in the same proportion of the RNA polymerase II-specific 445-base transcript as did incubation of the same template with S-100 supplemented by exogenous mouse tumor RNA polymerase II (Fig. 3, lane c, as compared to lanes d and e). Therefore, all experiments described below were carried out in the presence of our HeLa cell S-100 extract without added RNA polymerase II.

Further characterization of the transcription system demonstrated the importance of the ratio of template to S-100 (see Fig. 3). At low ratios of DNA to S-100, no specific transcript (ie, the 445-base RNA) was synthesized (Fig. 3, lane a), whereas at high DNA to S-100 concentrations, the large amount of nonspecific transcription obscured the specific 445-base RNA (Fig. 3, lane h).

To ascertain whether the 445-base RNA did indeed represent the properly initiated run-off transcript of α_1, we compared its size to that of RNAs synthesized from BamHI- or EcoRI-restricted α_1•pBR templates. When BamHI-restricted α_1•pBR DNA was incubated with the HeLa S-100 extract, an α-amanitin-sensitive 160-base transcript was synthesized (Fig. 2, lanes c and d). In contrast, an RNA polymerase II-dependent transcript of 1,070 bases was synthesized from EcoRI-restricted α_1•pBR DNA (Fig. 2, lanes f, g) (lane e depicts RNA synthesized from the HincII-restricted α_1•pBR template for size comparison). With this template, transcription apparently was initiated accurately at the capping site and extended rightward through the entire length of the α-globin gene and beyond to the EcoRI site; thus the HeLa cell S-100 extract apparently lacked the factors required to terminate transcription properly. Also, under these conditions, no splicing of the two intervening sequences of the α-globin precursor RNA was detected.

We have also tested the ability of our HeLa cell S-100 extract to accurately initiate transcription of the adult mouse βmaj globin gene, utilizing as template a 7-kb EcoRI fragment containing the β-globin gene ligated into pMB9 [9] (see Fig. 4). When this plasmid was restricted with HindIII and the digest incubated with the HeLa cell S-100 extract, a 690-base α-amanitin-sensitive RNA was synthesized (Fig. 5, lanes a and b). In contrast, transcription of a BamHI-restricted plasmid template in the presence of the S-100 extract yielded an α-amanitin-sensitive RNA 465 nucleotides in length (Fig. 5, lanes c and d). (Transcription of both templates also generated large, α-amanitin-resistant RNAs corresponding to end-to-end transcripts of the DNA restriction fragments.) Mapping of the 690- and 465-base RNAs onto β-globin DNA indicated that, for each RNA, synthesis was initiated at or very near the βmaj globin gene capping site and extended downstream through the gene to the end of the respective restriction fragment. Thus, the HeLa cell S-100 extract was able to accurately initiate transcription not only of α but also of β adult mouse globin genes in vitro.

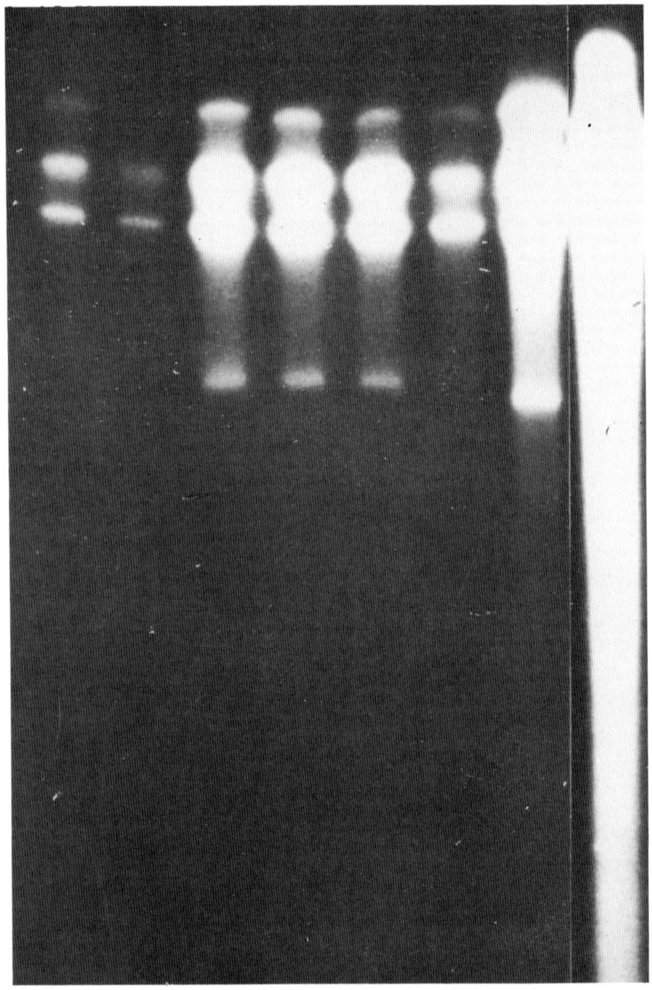

Fig. 3. Effects of exogenous RNA polymerase II and DNA concentrations on transcription of the HincII-restricted α_1•pBR plasmid template. The autoradiograph portrays the [^{32}P] RNAs synthesized in transcription reactions containing 2 × 10^6 cell equivalents of the HeLa cell S-100 extract and the following amounts of HincII-restricted α_1•pBR template, mouse tumor RNA polymerase II, and α-amanitin: lane a, 0.5 μg DNA; lane b, 0.5 μg DNA, 2 μg/ml α-amanitin; lane c, 1.5 μg DNA; lane d, 1.5 μg DNA, 20 units RNA polymerase II; lane e, 1.5 μg DNA, 60 units RNA polymerase II; lane f, 1.5 μg DNA, 20 units RNA polymerase II, 2 μg/ml α-amanitin; lane g, 3 μg DNA; lane h, 6 μg DNA. RNAs were synthesized at 30°C for 1 hour, purified, and analyzed as described in Figure 1. After electrophoresis, the gel was exposed to film with intensifying screens for 4 hours.

S1 Nuclease Mapping of the α_1 Globin Gene Transcript

To verify the initiation point and orientation of the α-globin gene transcript, we have located its coding region on α-globin genomic DNA by the S1 nuclease mapping technique developed by Berk and Sharp [23]. Specifically, the RNA synthesized from a HincII-restricted α_1•pBR template was hybridized to a uniformly labeled [^{32}P]-DNA probe containing α-globin sequences under conditions

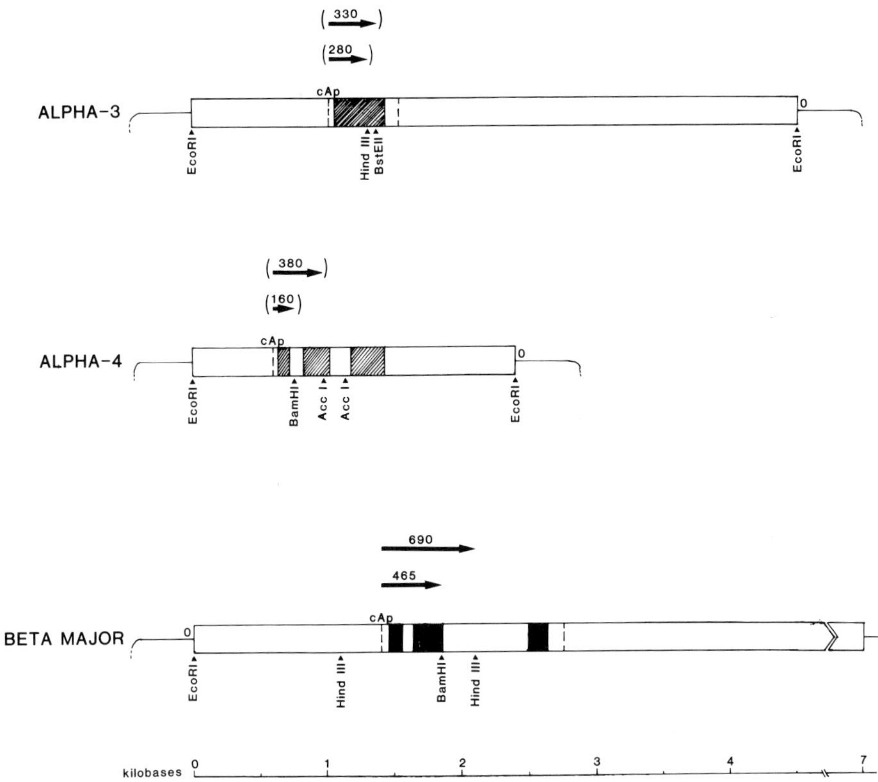

Fig. 4. Schematic maps of plasmid DNAs α_3•pBR, α_4•pBR, and βmaj•pMB9. As in Figure 1, only the restriction sites described in the text are indicated, and the gene-containing fragments are oriented such that transcription would proceed from left to right. Since α_3 and α_4 are apparently not expressed, the hatched boxes indicate the locations and sizes of the α-globin-like coding regions as determined by nucleotide sequence comparison to α_1; α_3 has no intervening sequences, but in α_4 the coding-like regions are separated by intervening sequences. The putative capping sites (identified by aligning α_1, α_3, and α_4 globin sequences) are identified by the vertical dotted lines labeled cAp. Only α_3 has sequences resembling the polyA addition site, denoted by the vertical dotted line to the right of the coding region. The predicted size and orientation of the RNA polymerase II transcripts that would be generated from the truncated templates of α_3•pBR or α_4•pBR described in the text are denoted by the thick arrows, with sizes, in bases, above the arrows. Since these transcripts are not made in vitro by the HeLa cell extract, they are surrounded by parentheses. The βmaj•pMB9 gene, in contrast, has well-defined coding regions, capping and polyA addition sites, and is utilized as a template in the HeLa cell S-100 system; each of these features is denoted as in Figure 1.

that permitted the formation of RNA-DNA but not DNA-DNA duplexes (Fig. 6, lane a). The resultant hybrids were treated with S1 nuclease, denatured, and analyzed by polyacrylamide-urea gel electrophoresis and autoradiography. When the [^{32}P]-DNA probe was the entire 2.1-kb EcoRI α-globin gene-containing fragment, a 445-base DNA segment was protected from S1 digestion by the RNA polymerase II-dependent transcript (Fig. 6, lane c; denoted by upper arrow). This segment was not protected from S1 digestion by RNA synthesized in the presence of low α-amanitin (Fig. 6, lane b). The larger protected fragments of approximately 630, 1,450, and 2,100 bases (Fig. 6, lanes b and c) resulted from S1-resistant duplexes formed between the DNA probe and nonspecific end-to-end transcripts.

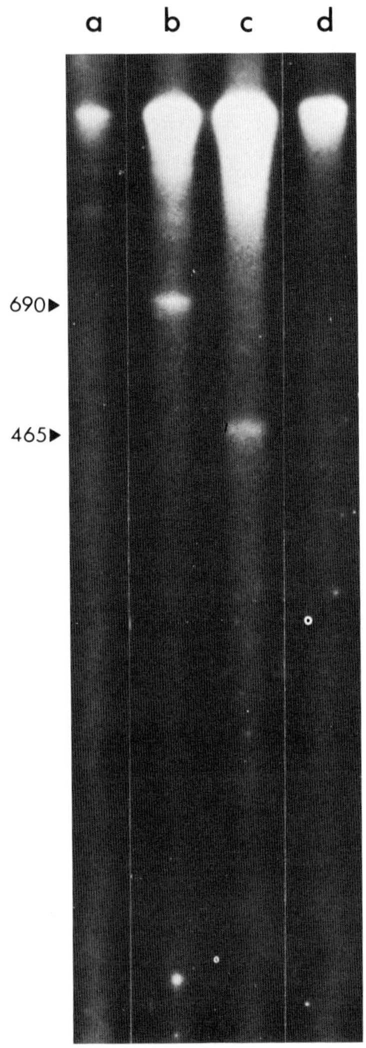

Fig. 5. Autoradiograph of RNAs synthesized in vitro from restricted templates of βmaj•pMB9 DNA. [^{32}P]-RNAs were synthesized at 30°C for 30 minutes, purified, and electrophoresed through a 4% polyacrylamide-7 M urea gel as described in the Experimental Procedures. The transcription reactions each contained approximately 2×10^6 cell equivalents of HeLa cell S-100 extract, 2 μg/ml of α-amanitin were indicated, and approximately 2 μg of the following templates: lane a, HindIII-restricted βmaj•pMB9 plus α-amanitin; lane b, HindIII-restricted βmaj•pMB9; lane c, BamHI-restricted βmaj•pMB9; and d, BamHI-restricted βmaj•pBM9 plus α-amanitin. The transcripts were visualized by exposing the gels to film with intensifying screens for 8 hours. The sizes, in bases, of the RNA polymerase II-dependent transcripts are indicated to the left of lane a.

Fig. 6. S1 nuclease mapping both of RNA synthesized in vitro from the HincII-restricted α_1•pBR template and of mouse α-globin mRNA. Details of the transcription, hybridization, and S1 nuclease digestion reactions, as well as of polyacrylamide gel electrophoresis of the samples, are presented in the Methods. The hybridization reactions each contained either the 2100-bp EcoRI or the 580-bp HpaII α-globin gene-containing DNA fragments (see Fig. 1) that had been uniformly labeled in vivo with [^{32}P] (4,000 cpm/reaction) and the following unlabeled RNA species: lane a, 2,100-bp EcoRI DNA without RNA; lane b, 2,100-bp EcoRI DNA plus RNA synthesized from the HincII-restricted α_1•pBR

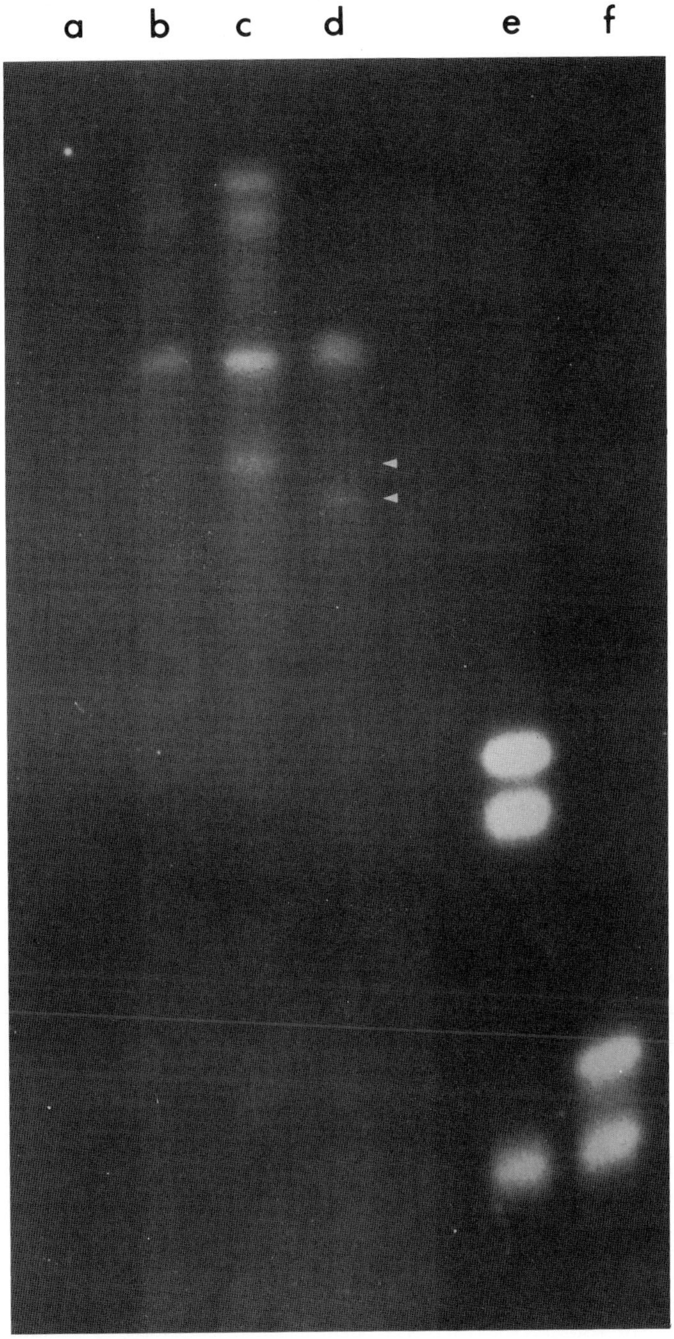

template in the presence of 2 μg/ml α-amanitin; lane c, 2,100-bp EcoRI DNA plus RNA synthesized in vitro from the HincII-restricted α_1•pBR template; lane d, 580-bp HpaII DNA plus RNA synthesized in vitro as for lane c; lane e, 2,100-bp EcoRI DNA plus globin mRNA isolated from the mouse; and lane f, 580 bp-HpaII DNA plus globin mRNA isolated from the mouse. The [^{32}P]-DNA fragments protected from S1 nuclease digestion by the transcripts were visualized by autoradiotraphy of the gel; exposure time on film with intensifying screens was 4 days. The positions of the 445 and 385 base DNAs are denoted by the white arrows to the right of lane d.

We also utilized a second DNA probe, namely the 580-nucleotide HpaII restriction fragment whose endpoints are shown in Figure 1. Since the HpaII site is located 60 nucleotides to the left of the HincII site (as oriented in Fig. 1), we predicted that when this probe was hybridized to the 445-base RNA, the size of the protected DNA segment would be smaller; in fact its size should represent the distance between the initiation site of transcription and the HpaII restriction site in the α-globin gene (see Fig. 1). Figure 6, lane d, shows that a 385-base DNA segment (denoted by lower arrow) was protected by the RNA polymerase II-dependent transcript (the larger DNA was protected from digestion by an end-to-end transcript). Thus, the initiation site of the 445-base transcript synthesized from the HincII-restricted α_1•pBR template is 385 bases to the left of the HpaII site in the α-globin gene. That location coincides with the capping site and the site predicted for initiation of transcription of the α-globin gene by the studies cited above. The S1 mapping analysis also indicated that even though the 445-base RNA spans the first intervening sequence of the α-globin gene, the S-100 extract did not splice out the RNA corresponding to the first intervening sequence.

As a control to these experiments, we also hybridized adult mouse α-globin mRNA to each of the two [^{32}P]-DNA probes. When the probe was the 2.1-kb EcoRI restriction fragment, three segments of 125, 200, and 230 bases were protected from S1 nuclease action by the mature mRNA (Fig. 6, lane e). These DNAs corresponded in size to the 5′ untranslated plus first coding segment, to the second coding segment, and to the third coding segment plus 3′ untranslated region, respectively, of the α-globin gene. When the probe was the 580-bp HpaII DNA fragment, the protected DNAs were 125 bases (the 5′ untranslated and first coding segment) and 140 bases (the second coding segment extending to the HpaII site) in size (see Fig. 6, lane f).

Identification of the Transcription Initiation Region Recognized by RNA Polymerase II

Having localized the initiation site of α-globin gene transcription, we sought to identify the stretch of nucleotides recognized by RNA polymerase II. To do this, the HindIII-linked 148-bp HinfI fragment of α_1 containing the putative capping site and its flanking sequences was ligated into pBR322 [Y. Nishioka, 17] (a map of $\alpha_1 p_1$•pBR is shown in Fig. 1). We then asked whether that fragment could be selectively recognized by the RNA polymerase. From the nucleotide sequence, we knew that $\alpha_1 p_1$ extended 141 bases upstream from (ie, to the left of) and seven nucleotides downstream from (ie, to the right of) the capping site of the α-globin gene. In order to determine whether RNA polymerase II could accurately initiate transcription at the capping site on $\alpha_1 p_1$, we assayed the ability of the polymerase to transcribe various restricted templates of $\alpha_1 p_1$•pBR. When $\alpha_1 p_1$•pBR was restricted with HincII and subsequently transcribed by the HeLa S-100 cell extract, a 635-base α-amanitin-sensitive RNA was synthesized (Fig. 7, lanes a and b). On the other hand, when a BamHI-restricted $\alpha_1 p_1$•pBR template was utilized, a 360-base α-amanitin-sensitive RNA was synthesized (Fig. 7, lanes c and d). Mapping of the 635-base and 360-base RNAs onto the plasmid in relation to the HincII and BamHI restriction sites indicated that each transcript was initiated at or very near the capping site of the α-globin gene and was elongated through the short stretch of α-globin sequences and distal pBR322 sequences to the end of the respective

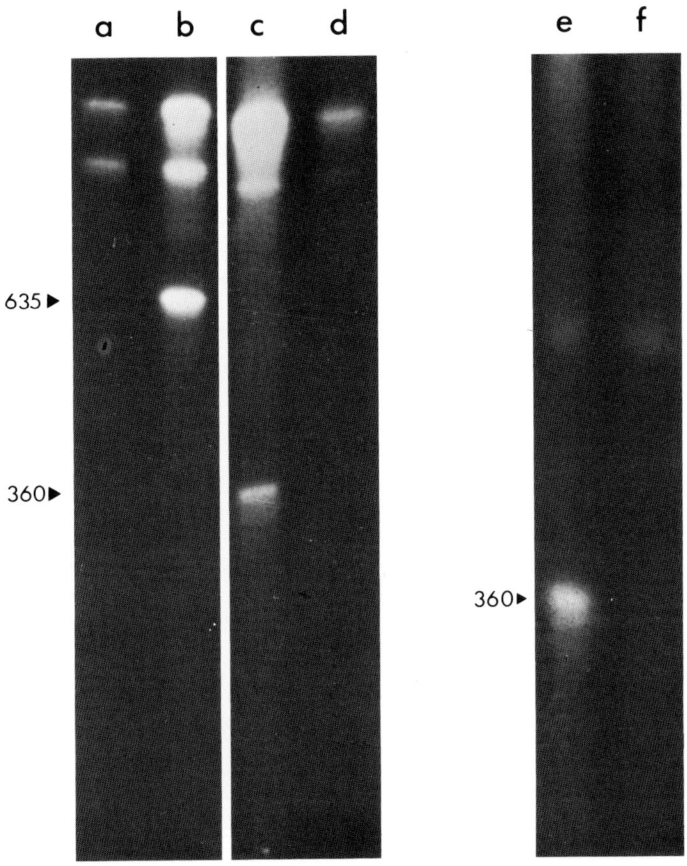

Fig. 7. Autoradiograph of RNAs synthesized in vitro from restricted α_1p_1•pBR templates. [^{32}P]-RNAs were synthesized at 30°C for 30 minutes, purified, and electrophoresed through a 4% polyacrylamide-7 M urea gel as described in the Experimental Procedures. The transcription reactions each contained approximately 2×10^6 cell equivalents of our HeLa cell S-100 extract, 2 μg/ml of α-amanitin where indicated, and approximately 2 μg of the following templates: lane a, HincII-restricted α_1p_1•pBR plus α-amanitin; lane b, HincII-restricted α_1p_1•pBR; lane c, BamHI-restricted α_1p_1•pBR; lane d, BamHI-restricted α_1p_1•pBR plus α-amanitin; lane e, BamHI-ClaI-restricted α_1p_1•pBR; lane f, BamHI-ClaI-restricted α_1p_1•pBR plus α-amanitin. The transcripts were visualized by exposing the gels to film with intensifying screens for 5 hours. The sizes, in bases, of the RNA polymerase II-dependent transcripts are indicated to the left of lanes a and e. Note that the gel containing lanes e and f was electrophoresed for a long time and that the top of this gel, with the 4,000 b end-to-end transcripts is not shown here.

restriction fragment. Therefore, the 148-nucleotide long HinfI fragment α_1p_1 from the α-globin gene apparently contained sufficient information to allow the RNA polymerase to initiate transcription properly.

This result implies that the RNA polymerase did not discriminate between specific nucleotide sequences located either upstream from the HinfI site at position −141 or downstream from the HinfI site located at position +7 when initiating transcription at position +1 (the capping/initiation site). However, still unresolved was whether any DNA preceding position −141 is required for RNA polymerase II binding and initiation. To answer this, we digested α_1p_1•pBR DNA with both

BamHI and ClaI, the latter of which cleaves pBR322 DNA five nucleotides to the left of the leftmost HindIII site (as oriented in Fig. 1). This digestion yielded two fragments, 511 and 4,010 nucleotides. The smaller fragment contained 151 nucleotides preceding and 360 nucleotides following the putative α-globin gene capping site. When BamHI-ClaI-restricted $\alpha_1 p_1$•pBR DNA was transcribed by the HeLa cell S-100 extract, an RNA polymerase II-dependent RNA of 360 bases was indeed synthesized (Fig. 7, lanes e and f). Thus RNA polymerase II in the presence of the S-100 does not require more than 151 nucleotides preceding the capping site (141 of which were mouse globin gene DNA) in order to recognize and initiate transcription accurately.

Transcription of α-Globin-like Sequences in Naturally Occurring Mutant Genes

We mentioned above the complexity of the mouse α and β-globin loci; eg, each cluster includes several globin-like genes containing a variety of mutations that prevent them from encoding normal globin polypeptides. Several EcoRI DNA fragments containing mouse α-globin-like sequences have been cloned and characterized [24; A. Leder, unpublished]. The nucleotide sequences of two of these, α_3 and α_4, have been determined [7; and unpublished results] and compared to that of α_1. Even though these two genes apparently are not expressed in the mouse and do not appear to be transcribed in globin-producing cells, they share a great deal of sequence homology with α_1, with some interesting exceptions: α_3 lacks both intervening sequences and α_4 contains mutations analogous to those found in certain human hemoglobinopathies. However, since α_3, α_4, and α_1 also share a relatively high degree of homology 5′ to the putative capping site, we investigated the transcriptional properties of these genes in vitro, hoping to distinguish which sequences were important for promoter recognition.

Schematic maps of α_3•pBR and α_4•pBR indicating the predicted size of run-off transcripts from the restricted templates analyzed are shown in Figure 4. Neither HindIII- nor BstEII-restricted α_3•pBR DNA exhibited accurate initiation of transcription by RNA polymerase II (Fig. 8, lanes a–c; BstEII results, not shown). Similarly, there was no RNA polymerase II-dependent specific transcription of α_4•pBR templates restricted by AccI or BamHI (Fig. 8, lanes d–g, h–k). For each plasmid, a variety of DNA: S-100 ratios were tried but in all cases only end-to-end α-amanitin resistant transcription was observed. That there was not an inhibitor of specific transcription in the plasmid preparation was demonstrated by the synthesis of 635-base $\alpha_1 p_1$ RNA in a reaction containing both HincII-restricted $\alpha_1 p_1$•pBR and AccI-restricted α_4•pBR templates (not shown).

DISCUSSION

Specific Transcription of Mouse Globin Genes In Vitro

The recent development of a cell-free extract that can direct RNA polymerase II to properly initiate transcription of DNA templates in vitro [15, 16] has provided an excellent opportunity to study eukaryotic gene expression. We have modified the procedure originally described by Weil et al [14] for the preparation of HeLa cell S-100 extracts so as to overcome the need to add RNA polymerase II to this system.

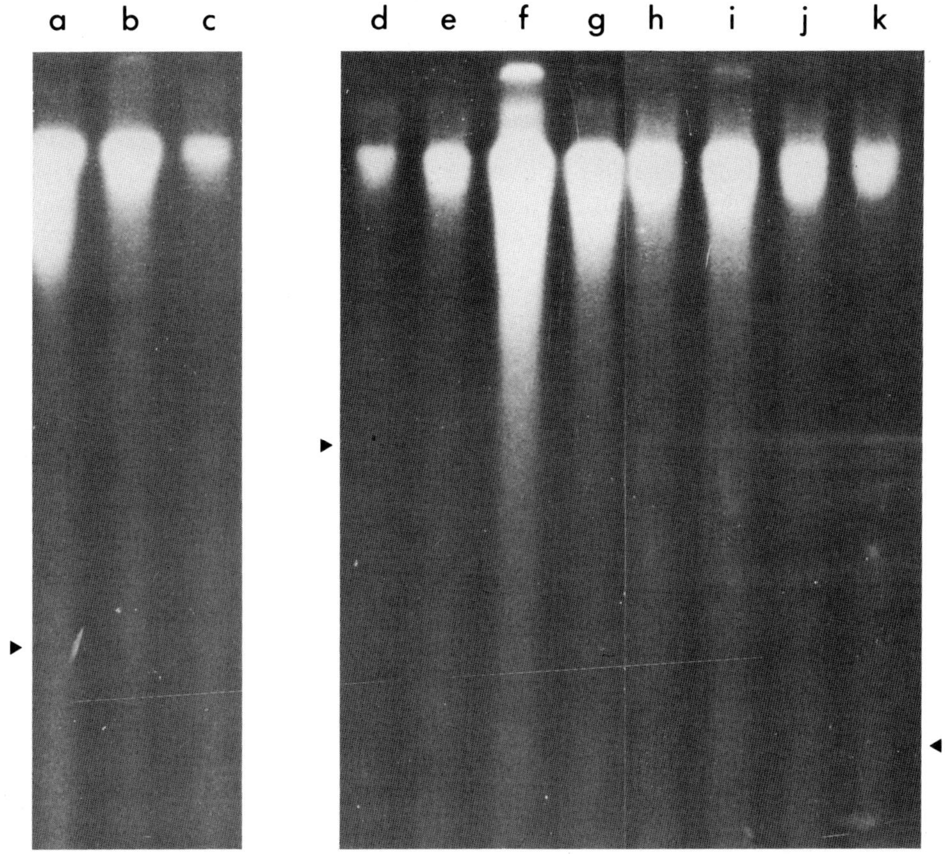

Fig. 8. Autoradiograph of RNAs synthesized in vitro from restricted templates of naturally occurring α-globin gene mutants. [^{32}P] RNAs were synthesized at 30°C for 1 hour, purified, and electrophoresed through a 4% polyacrylamide-7 M urea gel as described in the Experimental Procedures. The transcription reactions each contained approximately 2×10^6 cell equivalents of HeLa cell S-100 extract, 2 μg/ml α-amanitin where indicated, and the following templates: lane a, 5 μg of HindIII-restricted α_3•pBR; lane b, 2 μg of HindIII-restricted α_3•pBR; lane c, 2 μg of HindIII-restricted α_3•pBR plus α-amanitin; lane d, 2 μg of AccI-restricted α_4•pBR plus α-amanitin; lane e, 2 μg of AccI-restricted α_4•pBR; lane f, 5 μg of AccI-restricted α_4•pBR; lane g, 5 μg of AccI-restricted α_4•pBR plus α-amanitin; lane h, 5 μg of BamHI-restricted α_4•pBR plus α-amanitin; lane i, 5 μg of BamHI-restricted α_4•pBR; lane j, 2 μg of BamHI-restricted α_4•pBR; and lane k, 2 μg of BamHI-restricted α_4•pBR plus α-amanitin. The transcripts were visualized by exposing the gels to film with intensifying screens for 6 hours. The arrows to the left of lane a, to the left of lane d, and to the right of lane k indicate the predicted locations of RNA polymerase II transcripts generated from HindIII-restricted α_3•pBR, AccI-restricted α_4•pBR, and BamHI-restricted α_4•pBR templates, respectively.

By freezing the homogenized cells at –80°C prior to centrifugation at 100,000 × g, a large amount of RNA polymerase II apparently leaks out of the cell nuclei, thus simplifying the preparation of the system considerably.

Using the α_1•pBR plasmid as template, we have found that such a HeLa cell S-100 extract could accurately initiate transcription of the mouse α-globin gene in vitro. This conclusion was based on two forms of analysis. First, we measured the sizes of α-amanitin-sensitive "run-off" transcripts synthesized from plasmid templates that had been restricted at either the HincII or BamHI sites within the α-globin gene. The resultant 445- and 160-base RNAs generated from the respective templates indicated that transcription was initiated at or near the α-globin gene capping site. Second, we mapped the 445-base RNA to the α-globin gene utilizing the transcript's ability to hybridize to the DNA segment that encoded it and thus protect that DNA from S1 nuclease digestion. Comparison of that mapping data to S1 mapping analysis of mouse globin mRNA indicated that the initiation site of the 445-base in vitro-synthesized RNA matched that of mouse α-globin mRNA. Therefore, transcription of the α-globin gene was initiated in vitro at or very near the capping site and extended downstream through α-globin sequences to the HincII restriction site. Similarly, transcription of the mouse β-maj globin gene is initiated accurately by RNA polymerase II in the HeLa S-100 extract. It will be of interest to determine whether the same factors in the HeLa cell S-100 extract direct the polymerase to recognize both α and β-globin gene promoters.

In contrast, when partially purified RNA polymerase II alone was incubated with an α-globin gene-containing template, no selective initiation of transcription was detected. Thus, there were factors in the HeLa cell S-100 extract that suppressed random RNA synthesis and allowed selective recognition of the α-globin gene promoter by RNA polymerase II. Since these factors are part of a crude extract, it is not now possible to determine whether they interact with the template, with the RNA polymerase, or with both.

It is, however, important to note that globin gene-containing DNA templates can be specifically transcribed by HeLa cell extracts even though, in vivo, globin is only produced in reticulocytes. Obviously, the purified DNA templates used in these experiments lack the nucleoproteins of chromatin, some of which may function as negative transcriptional control elements to effect the developmental and organ specificity seen in vivo. It must also be kept in mind, that although in vitro transcription of the α-globin gene is accurate in the presence of HeLa cell extracts, it is very inefficient; specific transcription apparently reaches its maximal amount after 30 minutes in this system, and we estimate that there are only 0.02 or fewer RNA molecules synthesized per DNA template. It is not known whether this low level of transcription reflects an inappropriate physical state of the template or a relatively low level of RNA polymerases that are capable of initiating specifically. Perhaps HeLa cells lack or contain only a very limited quantity of a factor, or factors, present in hemoglobin-producing cells that greatly stimulate transcription of globin genes. Thus, while this system may reflect the ability of polymerase II to specifically recognize given promoter sequences, some of the modulating variables utilized in vivo are doubtless missing from it. The S-100 extract also apparently lacks at least some of the factors required for proper termination and processing of the RNA transcript, a fact that is not surprising since these processes occur in the nucleus, and the S-100 is essentially a cytoplasmic extract "contaminated" by some nuclear proteins (eg, RNA polymerase II, RNA polymerase III).

Identification of Promoter Region

Since a DNA segment (ligated into pBR322) extending 141 nucleotides preceding and seven nucleotides following the α-globin gene capping site possessed sufficient information to direct the RNA polymerase to initiate transcription accurately at the capping site, we conclude that a maximum of 148 nucleotides is required for promoter recognition. Included within this DNA segment are the two stretches of homology shared by many eukaryotic RNA polymerase II-controlled genes and implicated as control regions (particularly for the histone and chick conalbumin genes)—namely, the TATAAG "Hogness box"-like sequence and the region immediately surrounding the capping site [12, 25, 26].

Since our 148-nucleotide α-globin DNA promoter segment contains only seven nucleotides downstream from the capping site, the promoter region recognized by RNA polymerase II is in a location that distinguishes it from that of the RNA polymerase III in the 5S ribosomal genes [27–29]. The latter control region is contained within a 30- to 40-nucleotide stretch of DNA positioned in the middle of the 5S ribosomal gene, beginning about 50 nucleotides 3′ to the initiation site. In contrast, the major segment (if not all) of the RNA polymerase II promoter recognition region is located immediately 5′ to the initiation site, at a position similar to that of promoters for bacterial and phage genes [30, 31].

Use of Naturally Occurring α-Globin Gene Mutants: Alterations in the TATAAG Sequence and Capping Site

As we have indicated above, evolutionary processes have generated an array of mutant globin genes or pseudogenes some (perhaps all) of which are positioned near normally active adult and embryonic globin sequences [7, 8, 32, 33]. We have characterized two of the mouse α-like globin genes in considerable detail [7; and unpublished] and found that, in addition to containing base alterations and deletions in critical coding and noncoding regions, these pseudogenes fail to form stable hybrid structures with mRNAs present in either adult or fetal red cells. It is therefore possible that they are transcriptionally inactive; clearly they do not direct the synthesis of significant amounts of stable mRNAs ($<5\%$ of the normal globin gene mRNA level). These genes, because they contain base alterations in regions implicated in promoter as well as other critical functions, provide a useful repertoire of naturally occurring mutants that can be tested as in vitro assays for promoter function, transcript splicing, termination, and other regulatory phenomena become available. They are of particular interest in terms of promoter function because they contain specific base alterations in the "Hogness box" and capping site, regions closely conserved between the adult α_1 and βmaj globin genes.

When the two pseudogenes were treated over a wide range of DNA concentrations and under conditions in which the α_1 and β-globin promoters were fully active, both genes failed to show any RNA polymerase II promoter specific activity (Fig. 8). Naturally, lack of in vitro activity may be unrelated to the activity of these genes in vivo, but the interesting point is that the pseudogenes behave quite differently from the normal genes under comparable conditions and appear to reflect their in vivo behavior.

At present we cannot distinguish between the possibility that this lack of promoter activity reflects a deficiency in the HeLa S-100 system (for example, the lack of a specific factor) or a specific sequence deficiency in the promoter (RNA polymerase II-binding) sites. It is nevertheless useful to compare the sequences of the normal α- and β-globin genes with those of the α-globin pseudogenes in the 150-base region that we have established to contain the promoter sequence (Fig. 9). The region between positions –50 and +10, though highly conserved between the α

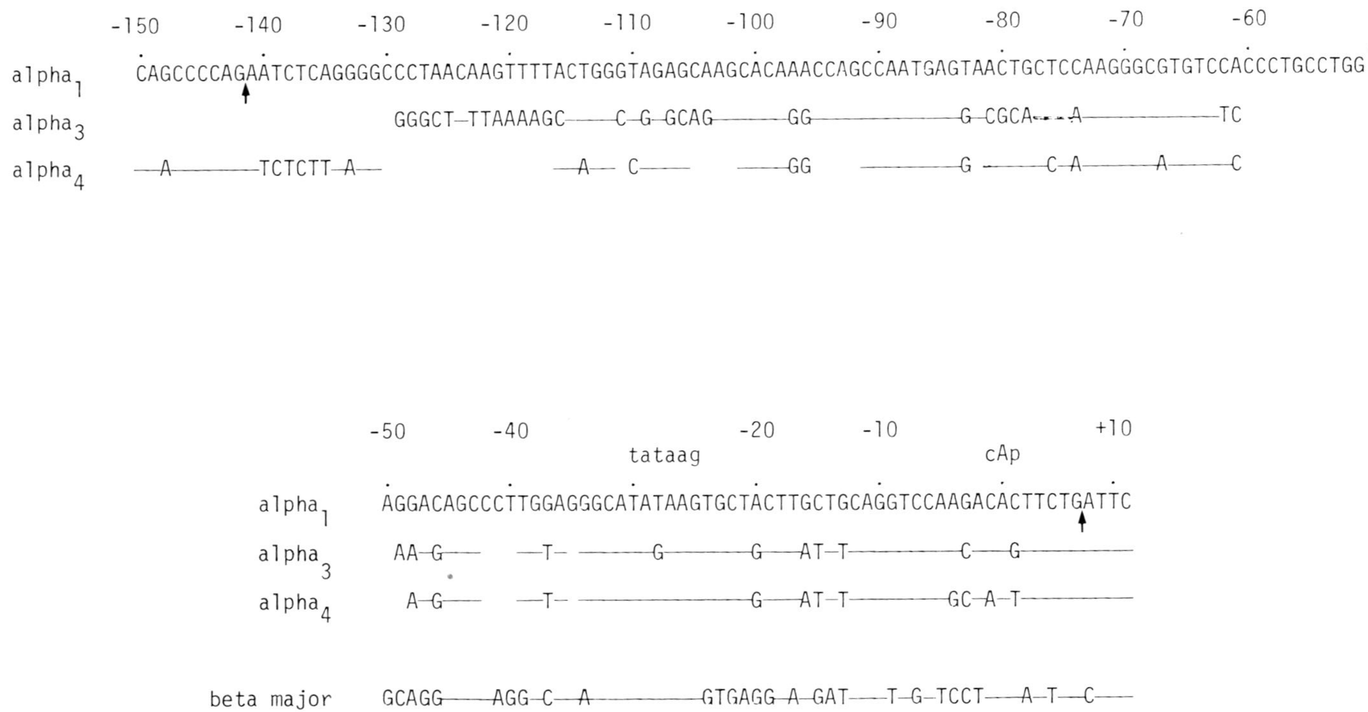

Fig. 9. Nucleotide sequence comparison of the promoter regions of α_1, α_3, α_4, and βmaj. Since the α_1 globin gene promoter is contained within the stretch of DNA between the HinfI restriction sites (see text) denoted by the arrows, only sequences between −150 and +11 are compared in this figure. (− and + numbers refer to bases upstream and downstream, respectively, from the transcription initiation site located at position +1; only the nontemplate strand is presented here.) The α_1 and βmaj globin DNA sequences are aligned by their respective capping sites (at +1). α_3 and α_4 sequences are aligned to maximize homology with α_1, thereby resulting in gaps in the α_3 and α_4 sequences with respect to that of α_1. The nucleotides shared between α_3, α_4, or βmaj and α_1 are indicated with horizontal lines. Those nucleotides that differ are denoted by their respective letter. Only those nucleotides between positions −50 and +11 are depicted for the βmaj globin gene because the α and β-globin genes diverge greatly upstream from position −50. (Note that the nucleotide sequence of α_3 has been determined upstream only to position −129 [7].

and α-like genes, contains distinct discrepancies, the most interesting of which fall within the region conserved between active α and β-globin gene sequences. In particular, three of these sites lie within either the "Hogness box" or the capping site. The T at position –28 is one of the most highly conserved bases within the "Hogness box" and has been altered to a G (ie, TAGAAG) in the α_3 sequence, while the highly conserved capping sequence (pyrimidine-A-pyrimidine; usually CAT) that spans positions –1 to +2 has been altered at position +2 (G) in α_3 and at position –1 (A) in α_4. These base changes and the functional activity of the DNA segments we have tested do not yet allow us to define completely the structural basis for the RNA polymerase II recognition, though they do allow us to focus on highly restricted segments of the gene. Thus, these results, the availability of other naturally occurring mutant genes, and the ability both to chemically induce mutagenesis [34] and to shuffle DNA segments so as to form hybrid promoters, suggest that it will be experimentally possible to derive the complete structural definition of a promoter region.

SUMMARY

We have transcribed a variety of α-globin gene-containing templates using a simplified in vitro system and have shown that the 5′ end of each of the RNA polymerase II-specific transcripts corresponds to that of authentic α-globin mRNA. Similarly, transcription of the adult mouse β-globin gene is also initiated accurately in vitro. Using a truncated α-globin gene fragment, we then localized the promoter region of the adult α-globin gene to a 148-bp fragment of α-globin DNA, 141 nucleotides of which precede the initiation site. In contrast, when two naturally occurring α-globin gene mutants were tested as templates, no RNA polymerase II-dependent initiation was detected. A nucleotide sequence comparison of the normal and mutant α-globin genes has allowed us to focus upon nucleotide positions that might be essential for promoter activity.

ACKNOWLEDGMENTS

We would especially like to thank Tony Weil and Bob Roeder for sending us aliquots of KB cell S-100, KB RNA polymerase II, and plasmid pSmaF DNA so that we could set up the in vitro transcription system in our laboratory. Their generosity and cooperation were absolutely invaluable to us in the development of this work. We would also like to thank Sing-Ping Lai for HeLa cells, Yutaka Nishioka for plasmid α_1p_1•pBR, Aya Leder for plasmids α_3•pBR and α_4•pBR, Dean Hamer for the [^{32}P]-α-globin gene-containing DNA probes, and Grant Bitter for many helpful discussions, particularly with regard to RNA polymerase II purification. We are most grateful to Terri Broderick for her expert assistance in the preparation of this manuscript.

REFERENCES

1. Popp RA, St. Amand W: Studies on the mouse hemoglobin locus. I. Identification of hemoglobin types and linkage of hemoglobin with albinism. J Hered 51:141–144, 1960.
2. Russell ES, McFarland, EC: Genetics of mouse hemoglobins. Ann NY Acad Sci: 241:25–38, 1974.
3. Gilman JG: Hemoglobin beta chain structural variation in mice: Evolutionary and functional implications. Science 176:873–874, 1972.

4. Gilman JG, Smithies O: Fetal hemoglobin variants in mice. Science 160:885–886, 1968.
5. Hilse K, Popp RA: Gene duplication as the basis for amino acid ambiguity in the alpha-chain polypeptides of mouse hemoglobins. Proc Natl Acad Sci USA 61:930–936, 1968.
6. Fantoni A, Bank A, Marks PA: Globin composition and synthesis of hemoglobins in developing fetal mice erythroid cells. Science 157:1327–1329, 1967.
7. Nishioka Y, Leder A, Leder P: Unusual α-globin-like gene that has cleanly lost both globin intervening sequences. Proc Natl Acad Sci USA 77:2806–2809, 1980.
8. Leder P, Hansen JN, Konkel D, Leder A, Nishioka Y, Talkington C: Mouse globin system: A functional and evolutionary analysis. Science 209:1336–1342, 1980.
9. Konkel DA, Tilghman SM, Leder P: The sequence of the chromosomal mouse β-globin major gene: Homologies in capping, splicing and poly(A) sites. Cell 15:1125–1132, 1978.
10. Konkel DA, Maizel JV Jr, Leder P: The evolution and sequence comparison of two recently divergent mouse chromosomal beta globin genes. Cell 18:865–873, 1979.
11. Nishioka Y, Leder P: The complete sequence of a chromosomal mouse alpha globin gene reveals elements conserved throughout vertebrate evolution. Cell 18:875–882, 1979.
12. Gannon F, O'Hare K, Perrin F, LePennec JP, Benoist C, Cochet M, Breathnach R, Royal A, Garapin A, Cami B, Chambon P: Organization and sequences at the 5′ end of a cloned complete ovalbumin gene. Nature 278:428–434, 1979.
13. Wu GJ: Adenovirus DNA directed transcription of 5.5S RNA in vitro. Proc Natl Acad Sci USA 75: 2175–2179, 1978.
14. Weil PA, Segall J, Harris B, Ng S-Y, Roeder RG: Faithful transcription of eukaryotic genes by RNA polymerase III in systems reconstituted with purified DNA templates. J Biol Chem 254: 6163–6173, 1979.
15. Weil PA, Luse DS, Segall J, Roeder RG: Selective and accurate initiation of transcription at the Ad2 major late promoter in a soluble system dependent on purified RNA polymerase II and DNA. Cell 18:469–484, 1979.
16. Manley JL, Fire A, Cano A, Sharp PA, Gefter ML: DNA-dependent transcription of adenovirus genes in a soluble whole cell extract. Proc Natl Acad Sci USA 77:3855–3859, 1980.
17. Talkington C, Nishioka Y, Leder P: In vitro transcription of normal, mutant, and truncated mouse alpha globin genes. Proc Natl Acad Sci USA 77, 1980.
18. Hodo HG, Blatti SP: Purification using polyethylenimine precipitation and low molecular weight subunit analyses of calf thymus and wheat germ DNA-dependent RNA polymerase II. Biochemistry 16:2334–2343, 1977.
19. Peacock AC, Dingman CW: Molecular weight estimation and separation of ribonucleic acid by electrophoresis in agarose-acrylamide composite gels. Biochemistry 7:668–674, 1968.
20. McMaster GK, Carmichael GC: Analysis of single- and double-stranded nucleic acids on polyacrylamide and agarose gels by using glyoxal and acridine orange. Proc Natl Acad Sci USA 74: 4835–4838, 1977.
21. Hamer D, Kaehler M, Leder P: A mouse globin gene promoter is functional in SV40. Cell 21:697–708, 1980.
22. Roeder RG: Eukaryotic nuclear RNA polyermases. In Losick R, Chamberlain M (eds): "RNA Polymerase." New York: Cold Spring Harbor Laboratory, 1976, pp 285–329.
23. Berk AJ, Sharp PA: Sizing and mapping of early adenovirus mRNAs by gel electrophoresis of S1 endonuclease-digested hybrids. Cell 12:721–732, 1977.
24. Leder A, Miller HI, Hamer DH, Seidman JG, Norman B, Sullivan M, Leder P: Comparison of cloned mouse α- and β-globin genes: Conservation of intervening sequence locations and extragenic homology. Proc Natl Acad Sci USA 75:6187–6191, 1978.
25. Grosschedl R, Birnstiel ML: Identification of regulatory sequences in the prelude sequences of an H2A histone gene by the study of specific deletion mutants in vivo. Proc Natl Acad Sci USA 77: 1432–1436, 1980.
26. Wasylyk B, Kedinger C, Corden J, Brison O, Chambon P: Specific in vitro initiation of transcription on conalbumin and ovalbumin genes and comparison with adenovirus-2 early and late genes. Nature 285:367–373, 1980.
27. Sakonju S, Bogenhagen DF, Brown DD: A control region in the center of the 5S RNA gene directs specific initiation of transcription: I. The 5′ border of the region. Cell 19:13–25, 1980.
28. Bogenhagen DF, Sakonju S, Brown DD: A control region in the center of the 5S RNA gene directs specific initiation of transcription: II. The 3′ border of the region. Cell 19:27–35, 1980.

29. Engelke DR, Ng S-Y, Shastry BS, Roeder RG: Specific interaction of a purified transcription factor with an internal control region of 5S RNA genes. Cell 19:717–728, 1980.
30. Rosenberg M, Court D: Regulatory sequences involved in the promotion and termination of RNA transcription. Ann Rev Genet 13:319–353, 1979.
31. Talkington C, Pero J: Distinctive nucleotide sequences of promoters recognized by RNA polymerase containing a phage-coded "σ-like" protein. Proc Natl Acad Sci USA 76:5465–5469, 1979.
32. Fritsch EF, Lawn RM, Maniatis T: Molecular cloning and characterization of the human β-like globin gene cluster. Cell 19:959–972, 1980.
33. Lauer J, Shen C-KJ, Maniatis T: The chromosomal arrangement of human α-like globin genes: Sequence homology and α-globin gene deletions. Cell 20:119–130, 1980.
34. Shortle D, Nathans D: Local mutagenesis: A method for generating viral mutants with base substitutions in preselected regions of the viral genome. Proc Natl Acad Sci USA 75:2170–2174, 1978.
35. Sutcliffe JG: pBR322 restriction map derived from the DNA sequence: Accurate DNA size markers up to 4361 nucleotide pairs long. Nucleic Acids Res 5:2721–2728, 1978.

Regulation of the Level of Chromatin

Organization and Expression of Globin Genes, pages 163–173

The Structure and Expression of Globin Chromatin During Hematopoesis in the Chicken Embryo

Mark Groudine, Mary Peretz, and Harold Weintraub

Department of Genetics, Fred Hutchinson Cancer Research Center, Seattle (M.G., M.P., H.W.) and Department of Radiation Oncology, University of Washington, Seattle (M.G.)

INTRODUCTION

In the developing chicken embryo, Hb first appears in first-generation erythroblasts at approximately 30–35 hours of development. These cells then serve as the progenitors of six subsequent "maturational" generations of erythroblasts, each of which contains characteristic amounts of Hb and displays unique morphologic characteristics. These cells mature rather synchronously and, by day 6 of embryonic development, most have become erythrocytes. Between day 6 and 7, another lineage of hematopoetic cells, the definitive embryonic lineage enters the circulating red cell population, arising from the endothelial lining of the blood vessels. The exact relationship between the precursor cells for the primitive line and those for the definitive lines is unknown; however, a number of papers in these volumes deal directly with this question.

The definitive series enters the circulating erythroid compartment between days 6 and 7 as fairly mature erythroblasts and by days 11–13 matures into inactive erythrocytes. These cells are the major erythroid cells in the circulation until hatching, at which time a second definitive line appears. While minor differences occur, by and large both definitive lines produce similar types of Hb chains [1], and both differ from the "primitive" series in the synthesis of specific β-chains.

In this and former studies, we have focused on the switch in β-chain synthesis between the primitive and first definitive series. Given the similarities in β-chain synthesis between the two definitive lines, we have not distinguished between the two types of definitive cells, and, for convenience, refer to both as "adult" red cells since they both produce "adult" β-chains. In contrast, given the distinctive embryonic β-chains synthesized in the primitive series, we refer to these cells as "embryonic" red cells.

In examining the control of embryonic and adult globin gene expression, we previously reported that steady-state globin mRNA is not detectable in precursor hematocytoblasts present in the area vasculosa before 20–25 hours of development, and that adult-specific globin RNA is not detectable in steady-state cytoplasmic or nuclear RNA from embryonic red blood cells [2]. Similarly, as assayed by solution hybridization, globin chromatin is resistant to DNase I before 25 hours, but becomes sensitive to DNase I after overt erythroid differentiation at 35 hours. In addition, adult and embryonic-specific globin genes are in an "inactive" conformation as measured by DNase I sensitivity and solution hybridization in the heterologous red blood cell nuclei [3].

In the experiments reported here, we use a different assay of DNase I sensitivity, the "blot" hybridization assay, to further probe the structure of globin chromatin in the hematopoetic lineages, and demonstrate that the adult β-globin genes are not only preferentially sensitive to DNase I in adult red cell nuclei, but they are also sensitive to DNase I in embryonic red cell nuclei. In addition, employing adult and embryonic β-globin gene-containing recombinant clones from a library of random chicken DNA fragments [4, 5], and "run-off" endogenous nuclear transcription [6, 7], we demonstrate that little, if any, transcription from the embryonic or adult β-globin genes is detectable in the heterologous red cell nuclei, even under conditions in which RNA processing is most likely minimized. Thus, we suggest that the "blot" DNase sensitivity of the adult genes in embryonic cells represents a chromosomal state present prior to overt transcription of the gene.

SENSITIVITY OF GLOBIN GENES AS MEASURED BY LOW LEVEL DIGESTION OF NUCLEI WITH DNASE I

Nuclei were isolated from "embryonic" 4–5-day chick red blood cells and from "adult" 14-day red cells. The respective nuclear preparations were digested with very low levels of DNase I [8] and the DNA isolated. At the highest level of DNase used in the experiments reported here, the average size of DNA was greater than 10 kilobase pairs (kbp) and in most cases had an average size between 15 and 20 kbp. The purified DNA from these preparations was then restricted with Bam H1, separated by electrophoresis on neutral 1% agarose gels, transferred to nitrocellulose filters [9], and hybridized to a ^{32}P-nick-translated adult globin cDNA clone pHb1001 [10] or to ovalbumin cDNA as a control (Fig. 1A,B). At the level of DNase digestion used, no loss of ovalbumin restriction fragments was observed. As an additional control, nonerythroid MSB cell nuclei were digested using the highest DNase I digestion conditions, and this DNA also showed no preferential sensitivity for either the globin probe or the ovalbumin probe. (The ovalbumin Bam HI pattern is different in MSB cells than in the flock of chickens used for red cell isolation. This probably reflects a genetic polymorphism associated with the ovalbumin gene [11].)

When the DNase-digested DNA blots are hybridized to the adult globin probe pHb1001, there is a rather steady and gradual loss of specific Bam HI restriction bands as the concentration of DNase increases. This result contrasts with the random and variable intensities displayed by the ovalbumin bands (Fig. 1B), a variability that probably arises from difficulties in loading precise quantities of very viscous DNA onto the gel. In adult nuclei, the 4.5 kilobase (kb) adult gene is strikingly more sensitive than the 8.5 kb embryonic gene. The former begins to disappear at about 0.05 μg/ml of DNase, whereas the latter disappears at about 1 μg/ml DNase. These respective values are highly reproducible and are also observed using a λ genome clone containing both an embryonic β-globin gene and an adult β-globin gene, $\lambda C\beta G1$ (β_1, see Fig. 3) as a probe. (For example, Fig. 2 is an Eco RI-Hind III double digest hybridized to $\lambda C\beta G1$, showing that in embryonic

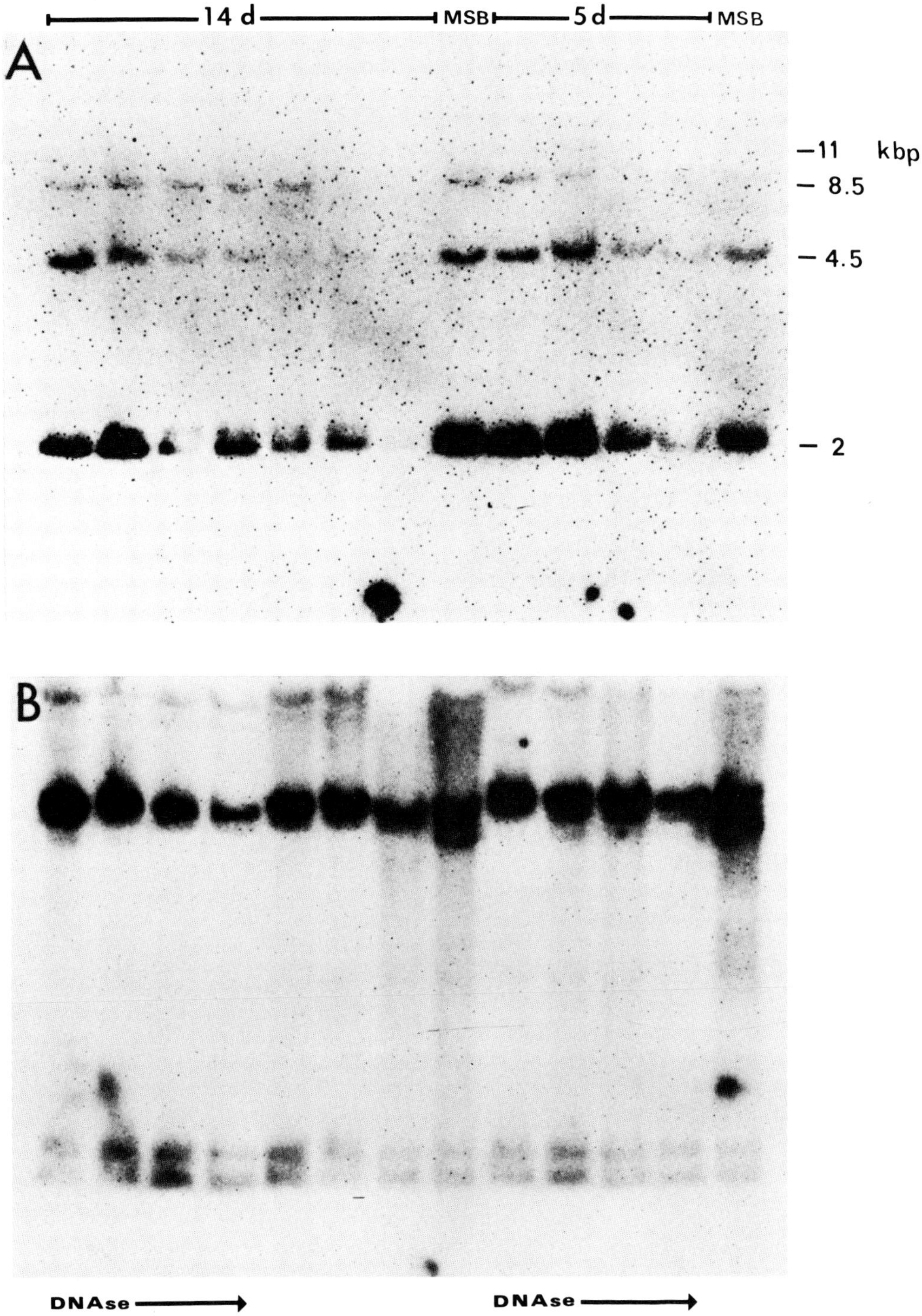

Fig. 1. DNase I sensitivity of the adult and embryonic β-globin genes in 5-day (embryonic) and 14-day (adult) red blood cell nuclei. Control DNA and DNA from 14-day and 5-day nuclei treated with increasing concentrations of very low levels of DNase I were digested with Bam H1, run on 1% agarose gels, blotted, and hybridized to (A) nick-translated pHb 1001 clone DNA and (B) ^{32}P-labeled ovalbumin cDNA. As an additional control, DNA from MSB cell nuclei, incubated with the highest (1.5 μg/ml) DNase I concentration used in the red blood cell nuclear digests, was analyzed.

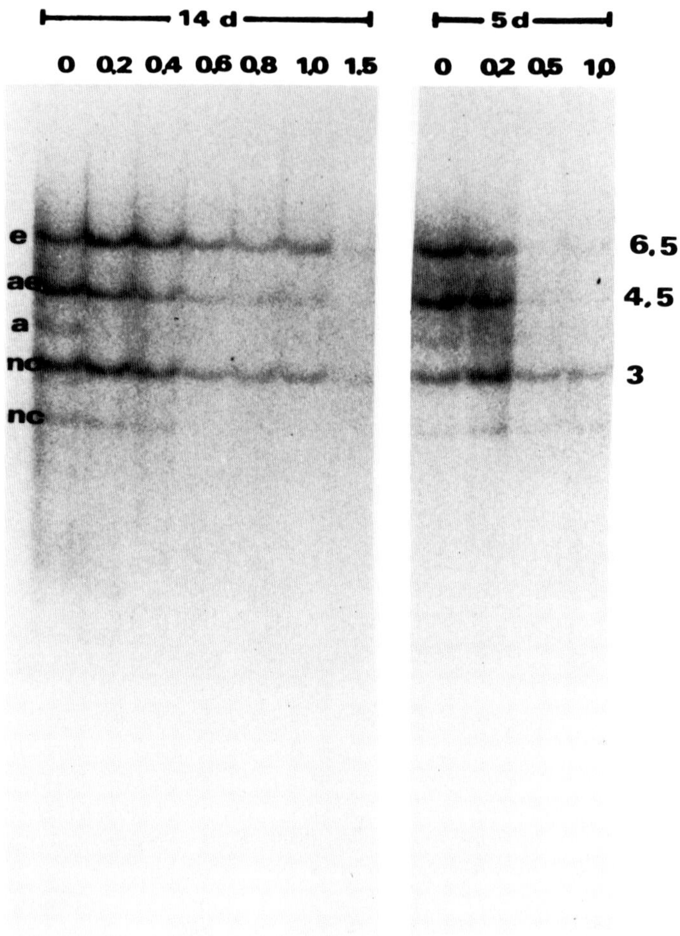

Fig. 2. Kinetics of DNase I digestion as assayed by hybridization to λCβG1. DNase-treated DNA from 14- and 5-day red cells was digested with R1 and Hind III and the corresponding blots were hybridized to λCβG1. (a) represents segments coding exclusively for adult genes; (e) segments coding exclusively for embryonic genes; (ae) segments containing both adult and embryonic genes; (nc) segments containing noncoding fragments. Note that these fragments are digested in the highest digest of adult nuclei. We should also point out that the embryonic gene (e) is larger than predicted from restriction analysis of the corresponding genome clone, λCβG1 [4, 5]. We believe that this result may reflect modified Hind III sites in the nuclear DNA; they are unlikely to represent partially digested fragments because the same digest yields the predicted restriction map when hybridized to different globin gene containing recombinant DNA clones.

cells both adult- and embryonic-specific fragments are both degraded rapidly but that the embryonic-specific fragment becomes much more resistant in adult cells. Similar results are also obtained using a second genomic clone, λCβG2, which carries the 8.5 kb embryonic gene and the 2 kb adult gene [data not shown].) Assuming that the initial rate of DNase cutting is roughly proportional to DNase concentration, it is a reasonable first guess that in adult nuclei the adult gene is about 20 times more sensitive than the embryonic gene, which in turn is more sensitive than the ovalbumin gene. Adjusting for the differences in target size, the

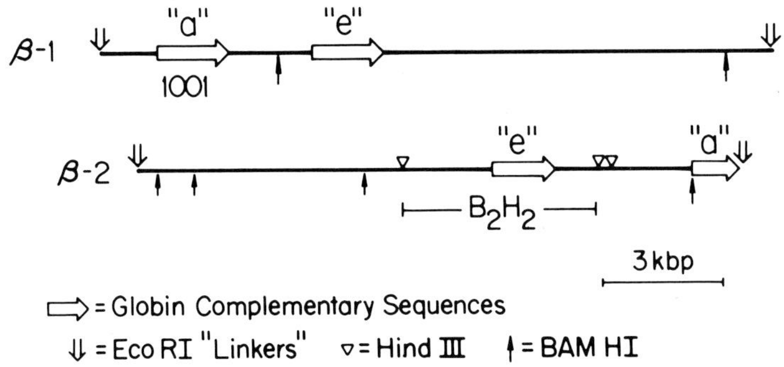

Fig. 3. Restriction maps for two λ Charon 4a recombinant chicken clones carrying globin genes. "a" and "e" denote adult and embryonic β-globin genes extending over the distance indicated by the thick arrows. The direction of transcription (5′ → 3′) was determined by Dodgson et al (1978) and M. Dolan, J.B. Dodgson, and J.D. Engel (unpublished results). The "1001" refers to the cDNA plasmid Hb1001 (Saltser et al, [10]), which is homologous to the adult β_1-globin gene, as indicated. "B_2H_2" is a subclone of λCβG-2, containing the sequences indicated on the diagram of β-2.

adult gene may actually be as much as 50 times more sensitive than the embryonic gene. Clearly, this is a very rough estimate and the true value might be anywhere between 10 and 100.

A different picture emerges in comparing the relative sensitivity of the embryonic and adult genes in embryonic cells. In embryonic nuclei, both adult and embryonic genes begin to disappear at about 0.1 μg/ml of DNase I. Given the larger target size of the embryonic gene, it is possible that this gene may be slightly more sensitive to DNase digestion; however, this slight increase in sensitivity is marginal when considered against the factor of 20 estimated for the relative sensitivity of the adult gene compared with the embryonic gene in adult cells. Finally, the DNA sizes for the respective DNase concentrations are comparable in 5- and 14-day nuclei. Thus it is also clear that at comparable levels of digestion, the embryonic gene is more sensitive in embryonic nuclei than in adult nuclei, in which approximately ten times more DNase is required for complete digestion. In an absolute sense, the embryonic gene therefore becomes more resistant during the transition from primitive to adult red cells.

LINEAGE-DEPENDENT TRANSCRIPTION OF EMBRYONIC AND ADULT β-GLOBIN GENES

In the following experiments, we have employed the two recombinant DNA clones containing chicken β-globin genes (β_1 and β_2) mentioned in the previous section to assay the transcription of these various genes in embryonic and adult red blood cells. Relevant restriction maps of these clones are illustrated in Figure 3. Purified cloned DNAs were digested with various restriction enzymes, separated on 1% agarose gels, blotted to nitrocellulose filters [9], and hybridized to purified ^{32}P-nuclear RNA (nRNA) synthesized (2.5 minute reactions) by endogenous "run-off" transcription [6] from embryonic or adult red cell nuclei.

Figure 4a shows the ethidium bromide staining patterns and resultant hybridization to specific regions of λCβG-1 after digestion of the clone with Bam HI and hybridization with ^{32}P-nRNA from either adult or embryonic red cell nuclei. An-

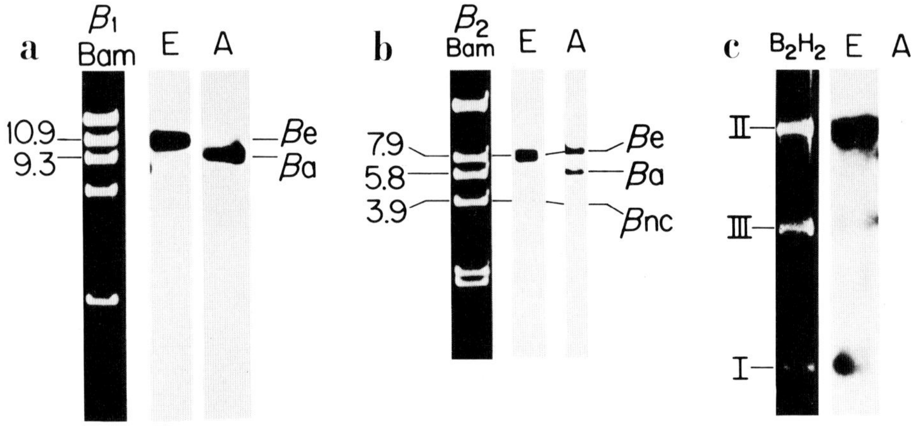

Fig. 4. Hybridization of in vitro synthesized ^{32}P-RNA from nuclei of 5-day embryonic (e) and 12-day adult (a) red blood cells to a) BAM restriction fragments of λCβG-1, b) BAM restriction fragments of λCβG-2 and β, and c) plasmid B_2H_2. The first lane in each series shows the ethidium bromide stain of DNA from the various clones after restriction and electrophoresis in a 1% agarose gel: β_e and β_a indicate those fragments containing embryonic globin sequences and "adult" globin genes, respectively.

nealing of "blots" from Bam-digested β1 with ^{32}P-RNA isolated from nuclei of 5-day embryonic red blood cells results in almost exclusive hybridization to the 10.9 kbp embryonic fragment, although a very small signal (< 1% of that observed from the 10.9 kbp fragment) can be detected from the 9.3 kbp adult β1 fragment. Conversely, hybridization with ^{32}P-nRNA from 12-day "adult" red cell nuclei results in the almost exclusive labeling of the 9.3 kbp adult fragment, although, again, a small amount of hybridization is observed with the 10.9 kbp embryonic fragment with very long exposures. Given the previous reports of the cross-hybridization of the embryonic and adult β-globin genes, we suspect that the small signal observed from the heterologous fragments most likely represents cross-hybridization. Additional evidence for this comes from the observation that low salt washes preferentially reduce hybridization to the heterologous β-globin gene fragments [data not shown].

Additional data supporting the lineage-dependent transcription of the "adult" β-globin gene is derived from experiments in which we hybridized ^{32}P-nRNA from either embryonic or adult red blood cell nuclei to cDNA plasmid Hb1001, which contains the coding sequences for the adult β-1 globin gene (see Fig. 3). Significant hybridization to this cDNA clone is observed only with ^{32}P-nRNA from the 12-day "adult" red cells, and not from ^{32}P-nRNA from embryonic cells (not shown).

We next analyzed transcription of the embryonic and adult globin genes located on λCβG-2 (β2). These sequences lie to the immediate 5′ side of β1. Figure 4b shows the results of hybridizing either adult or embryonic ^{32}P-nRNA to blots of Bam HI-digested β2. As can be seen, when embryonic ^{32}P-nRNA is used, major hybridization is observed to the 7.9 kb embryonic gene-containing fragment, with less, but detectable signal from the 3.9 kbp, noncoding fragment. We believe, however, that the signal obtained from the 3.9 kbp fragment is ambiguous since this fragment is highly repetitive in the chick genome and therefore, the hybridization signal may actually reflect transcription from other regions of the genome rather than transcription from the region in the globin domain. Some evidence for this view comes from the fact that this fragment is not digested by DNase I in red blood cells. Lastly, overexposure of the blot reveals a signal from the 5.8 kbp adult $\beta 2$ gene fragment equal to less than 1% of the signal observed from the embryonic $\beta 2$ fragment. Again, this may reflect cross-hybridization.

When ^{32}P-nRNA from 12-day, adult, red blood cell nuclei is hybridized with the Bam-digested $\beta 2$ blot, almost equal signals are observed from the 5.8 kbp (adult) and 7.9 kbp (embryonic) fragments. In addition, as with embryonic ^{32}P-nRNA, minor signals from the 3.9 kbp, noncoding (and repetitive sequence-containing) fragment is also observed with the adult ^{32}P-nRNA. Given the large size of the embryonic 7.9 kbp $\beta 2$ Bam HI fragment, it is possible that hybridization between this region and adult red blood cell ^{32}P-nRNA could result from transcription from sequences outside of the embryonic gene. To test this possibility, we hybridized ^{32}P-nRNA from embryonic to adult red blood cells to a blot containing a PBR subclone of λCβG-2, β2H2, which contains the embryonic $\beta 2$ gene and its 5′ and 3′ flanking sequences defined by the adjacent Hind III sites. As can be seen in Figure 4c, significant hybridization is observed between β2H2 and ^{32}P-nRNA from 5-day embryonic nuclei, but essentially no hybridization is detected between 12-day adult ^{32}P-nRNA and β2H2 at similar exposure. Thus, we feel these results indicate that hybridization between ^{32}P-nRNA from adult red blood cells and the 7.9 kbp Bam embryonic $\beta 2$ gene-containing fragment is a consequence of transcription from a region outside the embryonic $\beta 2$ coding sequences, possibly just to the left of the Bam site at the 5′ side of the adult gene in $\beta 2$. Also, since these bordering regions contain repetitive DNA sequences, again, it is possible that the transcription signal actually comes from another region of the genome besides the globin domain. DNase I sensitivity experiments support this possibility. Thus, we conclude that as our probes approach the size of the known coding regions for adult and embryonic β-globin genes, our ability to detect lineage specific "run-off" transcription is greatly enhanced. In general, both embryonic genes are transcribed in embryonic cells, while both adult genes are transcribed in adult cells. Heterologous transcription is at least 100-fold lower in the respective red cell lineages and may even be zero if the weak signal observed proves to come from cross-hybridization.

While the experiments described above indicate that transcription of adult and embryonic β-globin genes is restricted to their respective developmental compartments, two alternative possibilities could also explain our results: (1) Both embryonic and adult globin genes are transcribed in both adult and embryonic red cells, but the heterologous transcripts are degraded so rapidly as to avoid detection, even in our 2.5 minute pulses. (2) Polymerases are present on both embryonic and adult globin genes in red cells of both developmental compartments, but are unable to elongate in the heterologous red cells due to "inhibitory" chromosomal proteins. Some evidence against preferential degradation comes from the observation that the adult to embryonic signal (in either adult or embryonic nuclei) is unchanged during either a 2.5 or 15 minute incubation. Nevertheless, we have attempted to test these possibilities by performing the elongation assay in the presence of sarkosyl or heparin. These compounds remove histones and most other chromosomal proteins, but leave initiated RNA polymerases still bound to DNA and capable of elongation, but not new initiation [12–15]; thus, "blocked" RNA polymerases would most likely elongate if histones are usually preventing such elongation in the intact nuclei. In addition, since heparin is a known inhibitor of ribonuclease [16, 17], and since either heparin or sarkosyl would be expected to strip proteins from hnRNP and decrease the specificity of "sequence-specific" RNA nucleases, degradation of specific RNA sequences in the presence of these compounds is less likely.

When either adult or embryonic red blood cell nuclei are incubated in the presence of sarkosyl or heparin, no change in the ratio of transcripts from the adult or embryonic β-globin genes is observed in the presence of these compounds (data not shown). If either sequence-specific posttranscriptional degradation or "blocked" RNA polymerases were major regulatory mechanisms in globin gene switching, we would have expected a significant decrease in the ratio of adult to embryonic or embryonic to adult globin gene transcription in the adult and embryonic red cell nuclei, respectively. Thus, our results indicate that a major regulatory mechanism in governing hemoglobin switching in the chicken embryo resides at the level of transcription of the respective β-globin genes by RNA polymerase.

DISCUSSION

We feel our current findings may have several implications for the regulation of expression of the globin genes developmentally. We previously reported that the switch from primitive to definitive erythropoiesis is accompanied by a decrease in DNase I sensitivity of the embryonic globin genes [3]. Our current demonstration that expression of the embryonic β-globin genes is limited to the primitive red blood cells is consistent with the notion that a specific chromosomal conformation is coupled to transcription of this gene. In contrast, as communicated above, by blot hybridization (which measures the first cut introduced into the gene), both the embryonic and adult β-globin genes are digested at approximately the same rate in primitive erythroid cells. As assayed by solution hybridization (which assays the frequency of introducing many cuts into the gene), however, the embryonic β-globin genes are more sensitive than the adult β-genes in embryonic cells [3]. Thus, the "blot" sensitivity of the adult gene in embryonic cells may reflect a "preactivation" chromosomal state. The "run-off" experiments support the conclusion that the adult genes are in a preactivation structure since they show that transcription of the adult genes is limited to the adult cells, and is barely, if at all, detectable in embryonic cells. Thus, while an altered chromosomal state is present prior to overt expression and may, in fact, be required for subsequent expression, that altered conformation of chromatin is insufficient to permit transcription, per se. In fact, run-off transcription experiments in the presence of sarkosyl and heparin reveal that few, if any, polymerases are bound to the adult gene in the embryonic cells.

We feel the experiments reported here can be considered in terms of a currently popular model for chromosome structure based upon the organization of lampbrush loops and, to some extent, polytene chromosomes (Fig. 5). We propose that the globin gene cluster is organized into a functional chromosomal domain of about 50–100 kb and that such a domain is represented by a chromosomal loop. When the loop is relaxed, a moderately sensitive DNase I domain results. This may be the state of the globin domain, but not of the ovalbumin domain in the red cell stem cell (hematocytoblast). With terminal differentiation to an embryonic red cell, additional elements (ie, HMG 14 and 17, undermethylation, as well as other currently undefined modifications) become associated with the newly activated

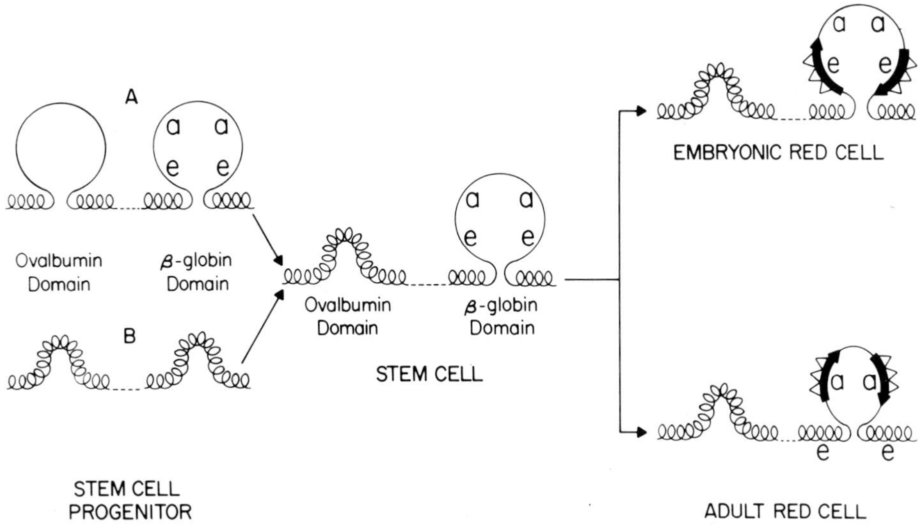

Fig. 5. Interpretation of moderate sensitivity of adult β-globin genes to DNase I in embryonic red cells in terms of "lampbrush" chromosomelike loops or domains. Chromosome domains are either condensed or relaxed. Transcription units become activated in relaxed domains by virtue of their association with HMG 14 and 17, and other modifications (open triangles). The red cell stem cells (hematocytoblasts) are unique in that they are predicted to be defined by a relaxed globin chromatin domain not yet fully activated by HMGs, etc. The hematocytoblast progenitors can have either relaxed (A) or constrained (B) domains; however, whichever proves to be the case, the state of the domains is seen to be the same for ovalbumin and globin.

embryonic gene, resulting in transcription and increased sensitivity to DNase I. Given the orientation of the embryonic (e) and adult (a) β-globin genes in the chicken chromosome (5′ . . . e . . . a . . . a . . . e . . . 3′), overt expression of the embryonic genes requires the "looping out" of the adult β-globin genes, as well. Thus, the "blot" DNase I sensitivity of the adult genes in embryonic cells would be a structural consequence based upon their location in the β-globin domain. As the embryo ages, the hematocytoblasts (or their descendants) switch to adult red cell production, and the adult gene is assembled with the modifications mentioned above. However, since the assembly of active adult β-globin genes does not require the concommitant relaxation of the embryonic β-globin gene regions, the embryonic genes are "reeled in," resulting in a greatly reduced sensitivity of these embryonic genes to DNase I in the adult red cells. (Alternatively, the β-globin domain in the immediate precursor cells could be in a completely repressed state, and the whole domain relaxed only in the transition to embryonic red cell. Transition to the adult red cell would be accompanied by relaxation of only the adult genes and surrounding sequences.) Besides focusing one level of control to the base of the loops, Figure 5 also illustrates two extreme states for the progenitor to the hematocytoblast. Ovalbumin and globin domains are both seen to be "open" initially and then differentially "repressed," or repressed initially and then differentially activated.

In attempting to further analyze the regulation of expression of these genes developmentally, it will be important to extend this type of assay to nonglobin-producing cells and, most important, to precursor erythroblasts present in the area vasculosa of the 20-hour chick blastoderm. Our preliminary results show that not only are the globin genes DNase I resistant in these precursors, but that they are also not being transcribed.

SUMMARY

We have examined the structure in chromatin and transcriptional expression of embryonic and adult β-globin genes during hematopoesis in the developing chicken embryo. Employing the "blot" hybridization assay, in which preferential sensitivity of genes is monitored by the disappearance of specific restriction bands after DNase I digestion in nuclei, we demonstrate that both adult and embryonic β-globin genes are sensitive to DNase I in embryonic red cells; however, in adult erythroid cells, the embryonic β-globin gene becomes resistant to digestion, while the adult gene remains highly sensitive. In attempting to investigate the transcriptional expression of embryonic and adult β-globin gene during hemoglobin switching in the chicken embryo, we have hybridized ^{32}P-nuclear RNA, synthesized by endogenous run-off nuclear transcription from embryonic and adult red cell nuclei, to restricted recombinant chicken DNA clones containing adult and embryonic β-globin genes. Our results indicate that little, if any transcription from the adult or embryonic β-globin genes is detectable in the heterologous red cell nuclei, even under conditions in which RNA processing is most likely minimized. Thus, while the chromosomal structure of both adult and embryonic genes is altered in embryonic red cell nuclei, significant transcription of the adult gene is undetectable in the embryonic red cells. These results suggest that the sensitivity of the adult globin genes in embryonic red cells represent a preactivation state of the chromosome present prior to overt transcription of these genes.

ACKNOWLEDGMENTS

This work was supported by the National Science Foundation and the National Institutes of Health.

REFERENCES

1. Brown J, Ingram V: Structural studies on chick embryonic hemoglobins. J Biol Chem 249: 3960–3972, 1974.
2. Groudine M, Holtzer H, Scherrer K, Therwath A: Lineage-dependent transcription of globin genes. Cell 3:243–247, 1974.
3. Weintraub H, Groudine M: Chromosomal subunits in active genes have an altered conformation. Science 93:848–853, 1976.
4. Dodgson J, Strommer J, Engel JD: Isolation of the chicken β-globin gene and a linked embryonic β-like globin gene from a chicken DNA recombinant library. Cell 17:879–887, 1979.
5. Dodgson J, Strommer J, Engel JD: Organization of globin genes in the chicken chromosome. "ICN-UCLA" Symposium 66:1979.
6. McKnight GS, Palmiter R: Transcriptional regulation of the ovalbumin and conalbumin genes by steroid hormones in chick oviduct. J Biol Chem 254:9050–9058, 1979.
7. Yang W, Binger M, Flint SJ: Transcription of adenoviral genetic information in isolated nuclei. J Biol Chem 255:2097–2108, 1980.
8. Wu C, Bingham P, Livak K, Holmgren R, Elgin SCR: The chromatin structure of specific genes: 1: Evidence for higher order domains of defined DNA sequence. Cell 16:797–806, 1979.
9. Southern EM: Detection of specific sequences among DNA fragments separated by gel electrophoresis. J Mol Biol 98:503–517, 1975.

10. Saltser W, Cummings I, Liv A, Strommer J, Padayatty J, Clarke P: Molecular regulation of hemoglobin switching. In Stamatoyannopoulos G, Neinhuis A (eds): New York: Grune and Stratton, 1979, pp 621–645.
11. Weinstock R, Sweet R, Weise M, Cedar H, Axel R: Intragenic DNA spacer interrupt the ovalbumin gene. Proc Natl Acad Sci USA 75:1299–1303, 1978.
12. Gariglio P, Buss J, Green M: Sarkosyl activation of RNA polymerase activity in mitotic mouse cells. FEBS Lett 44:330–333, 1974.
13. Green M, Buss J, Gariglio P: Activation of nuclear RNA polymerase by sarkosyl. Eur J Biochem 53:217–225, 1975.
14. Gariglio P: Effect of sarkosyl on chromatin and viral-RNA synthesis isolation of SV40 transcription complex. Differentiation 5 (vol 2–3):179–183.
15. Ferencz A, Seifart K: Comparative effect of heparin on RNA synthesis of isolated rat-liver nucleoli and purified RNA polymerase A. Eur J Biochem 53:605–613, 1975.
16. Palmiter R, Christensen A, Schimke R: Organization of polysome from pre-existing ribosomes in chick oviduct by a secondary administration of either estradiol of progesterone. J Biol Chem 245:833–845, 1970.
17. Cox R: Quantitation of elongating form A and B RNA polymerases on chick oviduct nuclei and effects of estradiol. Cell 7:455–465, 1976.

Organization and Expression of Globin Genes, pages 175–190

Changes in Globin Chromatin Structure During Red Cell Differentiation in Chick Embryos

Harold Weintraub, Stuart Weisbrod, Alf Larsen, and Mark Groudine

Department of Genetics, Fred Hutchinson Cancer Research Center, Seattle (H.W., S.W., A.L., M.G.), and Department of Radiation Oncology, University of Washington, Seattle (M.G.)

INTRODUCTION

The previous paper has described changes in chromatin structure that accompany the switch from embryonic to adult β-globin chain synthesis in chick embryos. Measurements of both the number of nascent RNA polymerases present on the respective β-globin genes and also their respective sensitivities to preferential digestion by DNase I indicate that the observed switch is transcriptionally controlled and that this is correlated with changes in the accessibility of the individual genes to DNase I. In addition we showed that regions of chromatin surrounding actively transcribed genes displayed an "intermediate" level of DNase I sensitivity. The intermediate level of sensitivity was dependent on the maintenance of "higher order" chromosome structure, and for convenience, we have defined this structure a "chromosomal domain," and suggest that the relaxation of such a domain is one of the first steps toward commitment to (for example) β-globin synthesis.

In this paper, we deal with the following problems: 1) The interaction of HMG 14 and 17 with active nucleosomes; 2) tissue-specific cutting by DNase I (hypersensitive sites) at different sites in the β-globin domain in adult and embryonic red cells; 3) and the transition between the avian red cell stem cell and the first generation of Hb-producing embryonic red cells.

ASSOCIATION OF HMG 14 AND 17 WITH ACTIVE CHROMATIN

The sensitivity of active genes to DNase I has been shown to depend on the association of two nuclear proteins with these genes [1, 2, 3]. These proteins, HMG 14 and 17 [4], are highly basic and acidic and can be eluted from nucleosomes with 0.35 M NaCl. After elution, the actively transcribed nucleosomes retain the general features of their structure, but lose their sensitivity to preferential digestion; however, when pure HMG 14 or HMG 17 is added back to the depleted nucleosomes, the actively transcribed genes regain their preferential sensitivity to DNase I. An example of such a reconstitution is shown in Figure 1. Here chromatin is depleted of

HMG 14 and 17 using 0.35 M NaCl. The sample is split in two and half is reconstituted with pure HMG 14 and 17. Both samples are then digested to equivalent degrees with increasing concentrations of DNase I. The DNA is then isolated, restricted with the restriction endonuclease, MSP, and then blot hybridized to a nick-translated adult β-globin cDNA clone (pHb 1001). Clearly, the globin gene is much more sensitive to DNase I digestion after reconstitution with HMGs. As controls, no preferential digestion of the inactive ovalbumin gene is observed using the same blots.

Several very general features of HMG binding to actively transcribed nucleosomes have been established: 1) There is no tissue specificity associated with the HMGs. Thus, HMG 14 and 17 from brain nuclei are capable of restoring DNase I

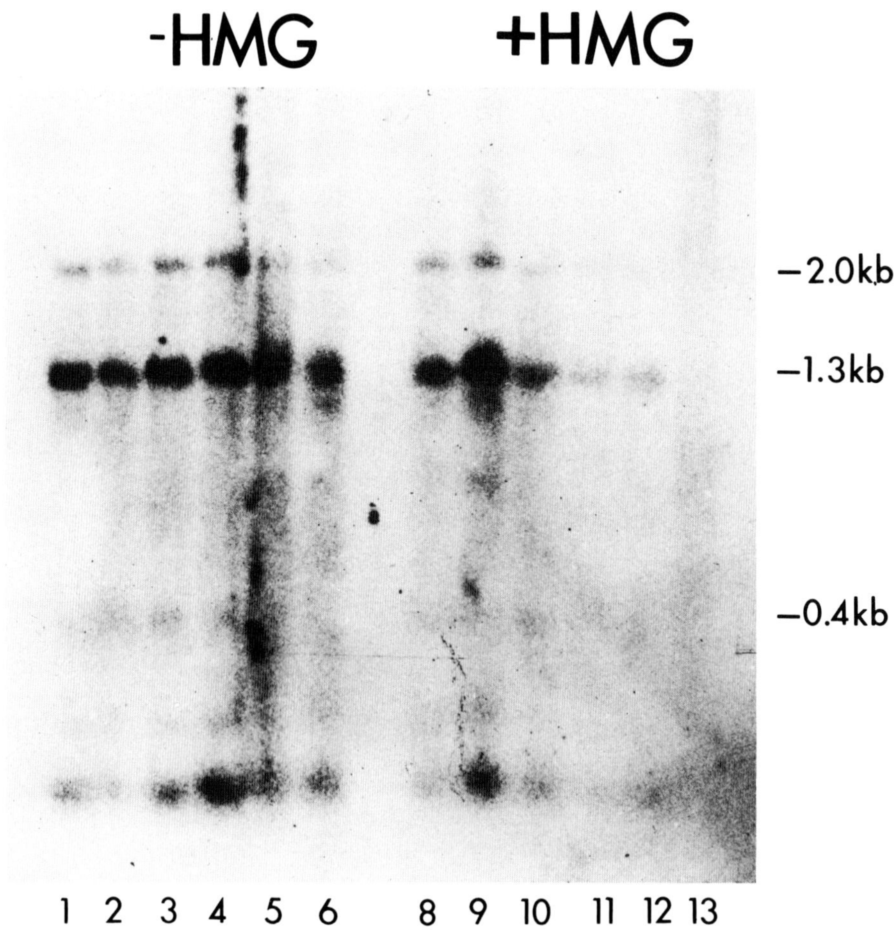

Fig. 1. Reconstitution of DNase I sensitivity of the globin gene with HMG 14 and 17. HMG-depleted and reconstituted chromatin was digested with increasing concentrations of DNase I and the purified DNA was digested with MSP, separated on 1.4% agarose gels, transferred to nitrocellulose paper, and hybridized to a ^{32}P nick-translated β-globin cDNA clone (2×10^8 cpm/μg). Lanes 1–6 represent depleted chromatin digested with 0, 0.01, 0.03, 0.06, 0.1, and 0.3 μg/ml DNase I. Lanes 8–13 represent HMG-reconstituted chromatin digested similarly.

sensitivity when reconstituted with depleted red cell chromatin. 2) The specificity for HMG binding seems to be in the recipient depleted chromatin. Thus, HMG 14 and 17 from red cell chromatin fail to induce DNase I sensitivity when reconstituted with HMG-depleted chromatin from brain. 3) Either HMG 14 or HMG 17 can sensitize most actively transcribed genes to DNase I. 4) Genes transcribed at different rates have about the same affinity for HMG 14 and 17. 5) HMG 14 and 17 bind stoichiometrically to actively transcribed nucleosomes. 6) HMG 14 and 17 can restore DNase I sensitivity to HMG-depleted nucleosome core particles containing 145 base pairs (bp) of DNA. 7) HMGs bind to essentially all of the nucleosomes present in a given transcription unit and consequently, it is observed that essentially all of the nucleosomes in a transcription unit are preferentially sensitive to DNase I. 8) As judged by sensitivity to DNase I, actively transcribed nucleosomes become associated with HMG 14 and 17 within three minutes after these genes are replicated.

One of the most interesting features of HMG activation is the specificity of the binding of these proteins to actively transcribed nucleosomes. When about 1 mole each of pure HMG 14 and 17 was reconstituted onto about 20 moles of HMG-depleted red cell nucleosomes, the DNase I sensitivity of the globin chromatin was restored. The ovalbumin gene used as a control showed no increased sensitivity either before or after reconstitution with HMG 14 and 17. A similar type of result can also be shown by direct measurements of HMG binding. Figure 2 shows a typical result. Labeled HMGs are challenged to bind to a large excess of either pure,

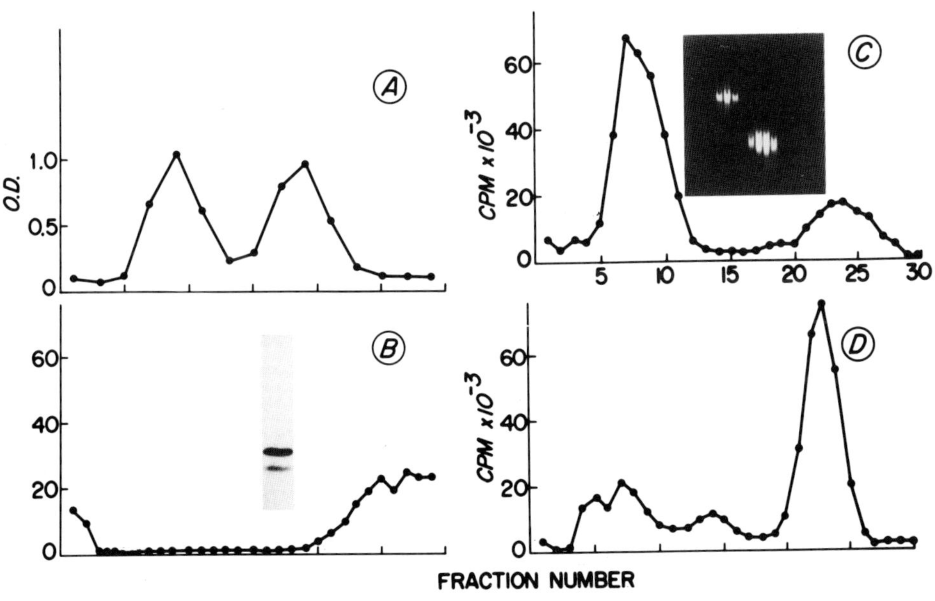

Fig. 2. Preferential binding of ^{3}H-HMG 14 and 17 to HMG-depleted nucleosome monomers. Each panel shows a 5 → 20% sucrose gradient of a mixture of various components as described in the text below. Migration is from right to left. A, Optical density (260 nm) profile of monomers and monomer DNA. B. ^{3}H-succinimidyl propionate-labeled-HMG 14 and 17 alone. Insert is a fluorograph of the ^{3}H-HMG preparation electrophoresed on an 18% SDS-polyacrylamide gel. C. Depleted monomers and monomer DNA were reconstituted with ^{3}H-HMG 14 and 17. Insert is a 4.5% polyacrylamide slab gel profile of each fraction shown in A. D. Undepleted monomers and monomer DNA were reconstituted with ^{3}H-HMG 14 and 17.

200-bp DNA or to nucleosomes. After binding, the mixture is run on sucrose gradients where the 5S DNA is clearly separated from the 11S nucleosomes, and the 2S, unbound HMGs. Figure 2C shows that HMGs prefer depleted nucleosomes to free DNA; however, as shown in Figure 2D, HMGs prefer to bind to pure DNA rather than to undepleted nucleosomes. In separate experiments without addition of the competitor DNA, it is possible to show that the HMGs will bind very readily to even undepleted nucleosomes. Thus, taken as a whole, HMGs prefer to bind to actively transcribed nucleosomes; however, they will also bind to pure DNA and to inactive nucleosomes, probably in that order of preference. It follows that the preferential binding of HMGs to active nucleosomes indicates that there is some feature of these nucleosomes that mark them and allow them to be recognized by HMGs.

In order to gain a better understanding of these additional features of actively transcribed nucleosomes, we have attempted to isolate them using a column composed of HMG 14 and 17 covalently coupled to agarose. As shown in Figure 3, when chromatin is passed over such a column, about 10% of the material binds and is not eluted until the NaCl in the elution buffer is raised to 0.3–0.4 M. The material that binds to the column reruns true, as does the material that does not bind. Most important, the material that binds hybridizes to 40%–50% when driven in a hybridization reaction with an excess of nuclear RNA; the material that does not bind is depleted in sequenceshydridizing to nuclear RNA (Fig. 4). Assuming asymmetric transcription, it is possible to estimate that the column yields a population of nucleosomes that is between 80% and 100% pure active nucleosomes.

To assay the binding of specific nucleosomes to the HMG column, DNA from red cell nucleosomes from the bound and unbound fraction was spotted onto nitrocellulose and hybridized to a variety of probes. Figure 5 shows that when adult red cells are used, the adult β-globin cDNA clone (pHb 1001) hybridized to DNA in the bound fraction; a flanking region (FL) of this gene that is not transcribed, but displays an intermediate level of DNase I sensitivity, is present in the unbound fraction, as is the ovalbumin gene; and a probe that hybridizes predominantly to an embryonic β-globin gene is also found mainly in the unbound fraction. The latter

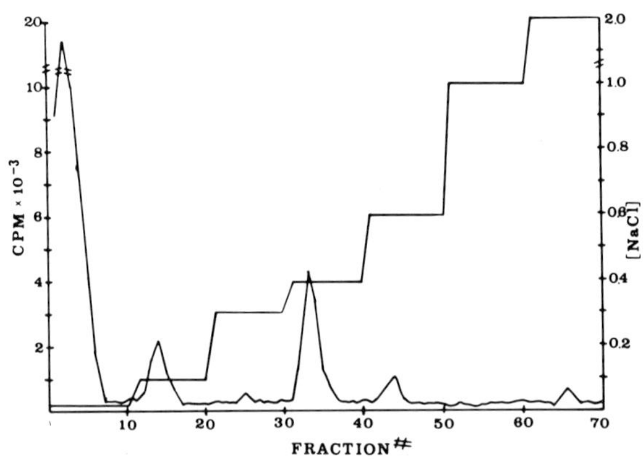

Fig. 3. Binding of chromatin to HMG 14–17 agarose. Chromatin was dialyzed onto the column resin[2] and eluted stepwise with increasing concentrations of NaCl.

probe also hybridized significantly to DNA in the bound fraction and we think that this is probably due to either cross-hybridization with adult sequences, or more likely, to repetitive DNA sequence elements that have been previously localized to regions contained in this probe.

To obtain a more detailed description of the fine structure of active chromatin, we decided to concentrate on the chicken α-globin genes. Restriction maps for this cluster are shown in Figure 6B. The clones and the restriction maps were kindly provided by Dr. J.D. Engel. In particular, we wished to know whether the "spacer" DNA between the two active α-genes, α^A and α^D, was DNase I sensitive and also whether the associated nucleosomes bound HMG 14 and 17. Figure 6A (left) shows that after digestion with Bam HI and Eco RI, and hybridization to αB1.5 (a subclone of α2 in pBR322 that has the 1.5 kilobase (kb) Bam fragment containing both the coding region of α^D and part of the spacer region between α^A and α^D), both α-related fragments are extremely sensitive to DNase I, and most important, they both display approximately equivalent levels of sensitivity. In the same blot, a relatively resistant, cross-hybridizing fragment is also observed at about 4 kb. This is probably the inactive U-gene present in α5 since a pure U-gene probe gives exactly the same DNase I pattern. The U-gene may code for the π-embryonic (α-like) globin [5]. This gene is clearly relatively resistant to DNase I and is known not to be expressed in the 14-day definitive adult red cell line used here. It corresponds to a region right at the 5′ side of the active α^A-α^D complex.

To look at the sensitivity of the region at the 3′ side of α^A, the same blot was hybridized to αB1.2 (a subclone of α2 containing the 1.2 kb Bam fragment located at the 3′ side of α^A). Figure 6A (right) shows that this fragment is as sensitive to DNase I as the coding region and the spacer region. Previously, we showed that the 7.1 and

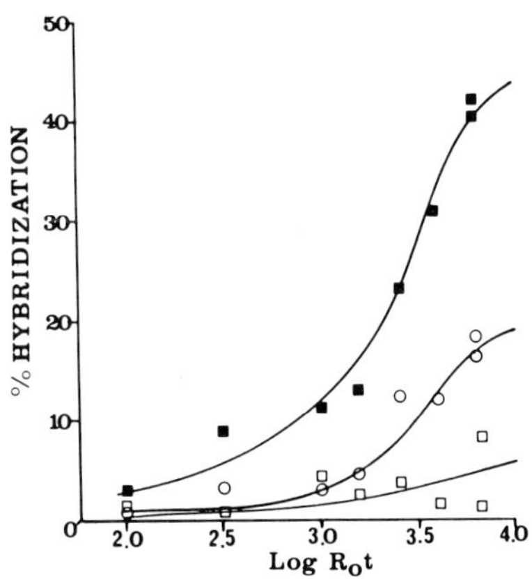

Fig. 4. Sequence analysis of bound and unbound fractions. HMG-agarose affinity chromatography was performed on sheared, ^{3}H-Tdr-depleted chromatin from MSB cells. Hybridization of the purified DNA from each fraction was to an excess of hnRNA. Each reaction contained hnRNA at 20 mgm/ml. ^{3}H-DNA was at 100,000 cpm per reaction (20 μl). ^{3}H-specific activity was about 10^5 cpm/ugm. Parallel hybridizations in the absence of hnRNA were also performed and the hybridization due to DNA alone subtracted. This hybridization ranged between 5% and 10%. (■) Fractions 32–26 (Fig. 1a), the "unbound" chromatin, (□) fractions 1–6 (Fig. 1a), the "bound" chromatin, (○) total ^{3}H-MSB DNA.

3 kb Bam fragments more distal to α^A and α^D at their 3′ side are relatively resistant to DNase [6]. In separate experiments (not shown), we have also shown that the 0.8 kb Bam HI coding fragment located at right at the 5′ side of α^A is DNase I sensitive. Thus, taken as a whole, the DNase I sensitivity seems to begin very close to the coding region in α^A and extends to about 1 kb beyond the 3′ side of α^D, including the spacer region between α^A and α^D.

Using the same probes, we also examined DNA from the bound and unbound fractions from the HMG column for their specific α-related sequences. Figure 6B shows that nucleosomes containing sequences corresponding to the spacer region, to the coding region, and to the 3′ side of the α-genes bind to the HMG column, but a region 4 kb beyond the coding region (on the 3′ side) does not. Thus, in general, DNase I-sensitive regions correspond to regions that bind to HMG 14 and 17. Moreover, flanking sequences that are located near the DNA coding for the primary stable transcript are also DNase I-sensitive and also bind HMG 14 and 17. Regions of intermediate sensitivity (eg, the 7.1 kilobase pair (kbp) Bam HI fragment 4 kbp beyond the 3′ side of α^A) do not bind to the HMG 14 and 17 column.

The previous results have shown that after depletion and reconstitution, HMG 14 and 17 prefer to bind to actively transcribed nucleosomes. The question arises

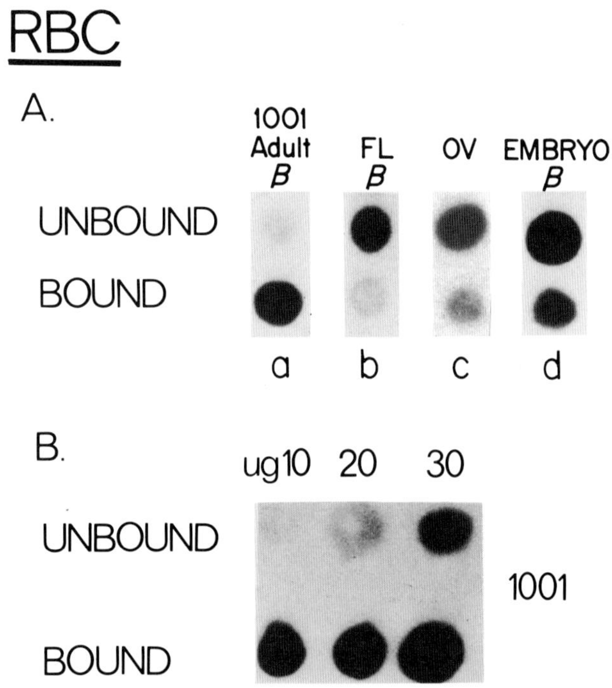

Fig. 5. Dot hybridization analysis of DNA from 14-day RBC bound and unbound monomers after HMG-agarose chromatography. A. Hybridization of 5 μg each of unbound monomer DNA and bound monomer DNA to: A. the adult β-globin cDNA clone (pHb 1001), B. a 9 kb Bam fragment flanking (FL) the coding sequence for the adult β-globin gene in β-1, and C. ovalbumin cDNA. D. a 2 kb Sac I embryonic globin (EM) fragment from $\lambda C\beta G2$. The DNA was covalently linked to DBM paper. B. Hybridization of pHb 1001 to increasing concentration of bound and unbound monomer DNA.

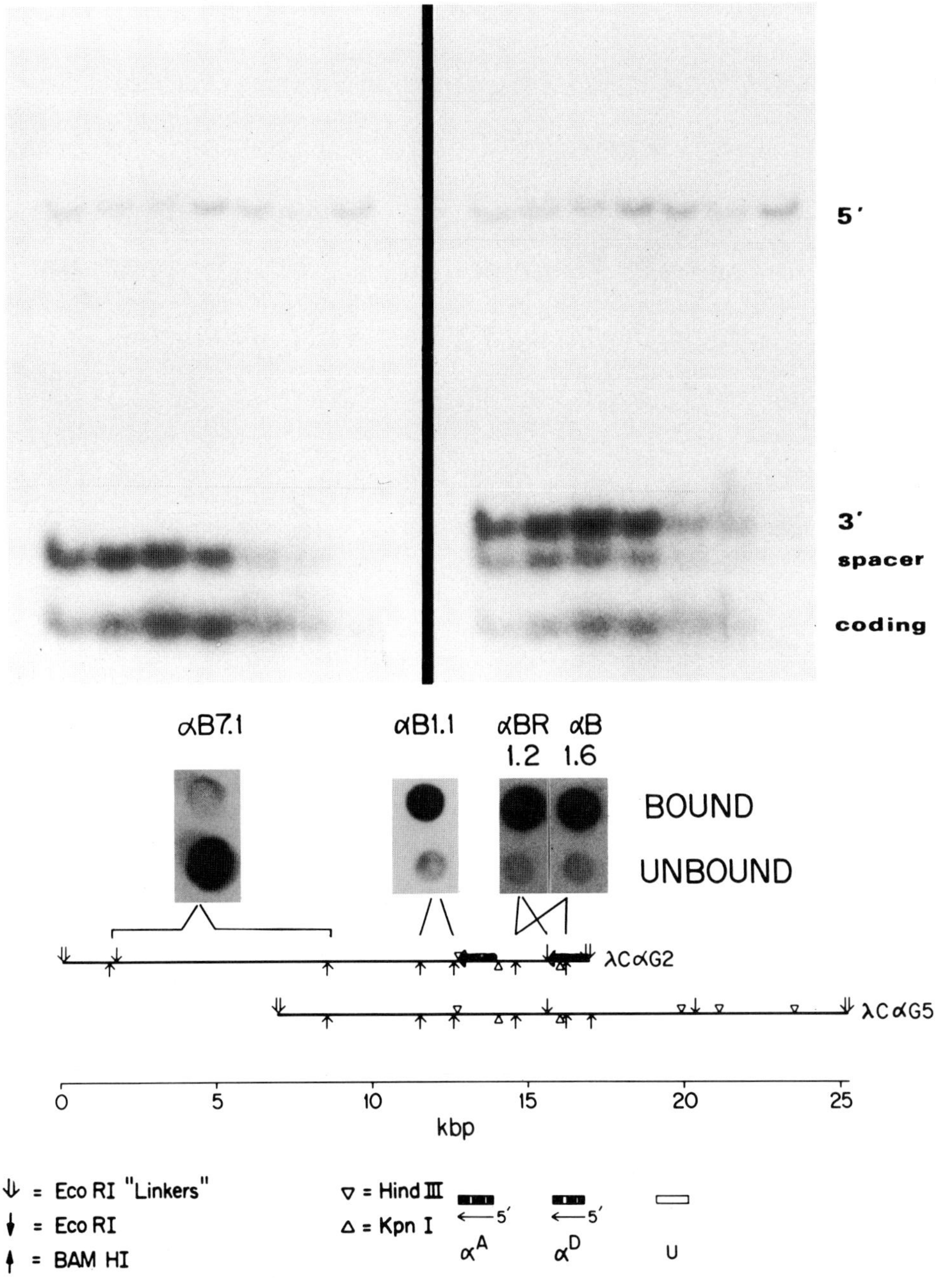

Fig. 6. DNase I sensitivity and HMG binding of chromatin associated with the α Genes. A. Left: Nuclei from 14-day, adult definitive red cells were digested with increasing concentrations of DNase I (left to right) and the purified DNA restricted with Bam and R1, blotted, and hybridized to αB1.6 (a subclone of $\alpha2$ in PBR322 containing part of the 1.6 kb α^D gene and the spacer region between α^A and α^D (see restriction map below). Right: The blot in A was rehybridized to αB1.2 (see restriction map), a subclone of $\alpha2$ containing the 3′ region adjacent to α^A. The DNase I-resistant fragment probably represents cross-hybridization to the gene indicated as "U." This may correspond to the embryonic "π-gene" (α-like). As indicated, this fragment would map to the 5′ side of α^D. The λCαG2 and λCαG5 were isolated and mapped by Dr. J.D. Engel and J. Dodgson. B. Dot blots of bound and unbound erythrocyte nucleosome fractions were hybridized to the labeled probes as indicated in the restriction map below.

whether the direct association of these proteins with active nucleosomes can be demonstrated. In the trout testes system, such an association has been shown by Levy and Dixon [7]. Similar results have also been obtained for the avian globin genes [8]. Figure 7 shows an electrophoretic separation of monomer, dimer, trimer, etc, nucleosomes on an acrylamide gel. Notice that the monomer region is distinctly heterogeneous. By running these particles on a second-dimension protein gel, it is clear from Figure 8B that the monomer heterogeneity is associated with the presence of HMG 14 and 17 which comigrate with the slowest migrating monomers. In this gel system, histones H1 and H5 bind weakly to the nucleosomes and are removed during the electrophoresis (Fig. 8A). In order to study the sequences present in these monomers, the DNA from five fractions across the nucleosome monomer peak was eluted, purified, and run on an agarose gel. The DNA was then transferred to nitrocellulose and hybridized to either a globin probe or an ovalbumin probe (Fig. 9). The data show that the nucleosomes that migrate slowest are enriched in globin sequences, while the nucleosomes that migrate fastest are depleted in globin sequences. Thus, as isolated directly from nuclei, a fraction of nucleosomes that migrate slowest in acrylamide gel electrophoresis contains HMG 14 and 17 and is enriched in globin sequences, but depleted in ovalbumin sequences.

TISSUE-SPECIFIC DNA CLEAVAGES IN THE GLOBIN CHROMATIN DOMAIN INTRODUCED BY DNASE I

In assaying the sensitivity of the respective α and β genes to DNase I using the blot hybridization assay (see previous article), we noticed that as a consequence of the DNase treatment, specific discrete new bands appeared after hybridization of the blots to the appropriate probes [9]. These bands contain one end that represents a specific site where DNase I cuts the nuclear chromatin and a second end defined by a restriction endonuclease cleavage. This follows since the discrete new bands do not appear if the purified DNA is not cut with a restriction enzyme. For the most part, these specific cutting sites are not observed in nonerythroid cells, and surprisingly, the specific cutting sites in the β-globin domain are different, depending on whether the digestion occurs in embryonic or adult nuclei. Figure 10 shows the hybridization of an embryonic β-globin probe to DNase-treated nuclear DNA from embryonic or from adult red cell nuclei. In embryonic cells, a discrete new band appears at 2 kb, while in adult cells, this hypersensitive region disappears and

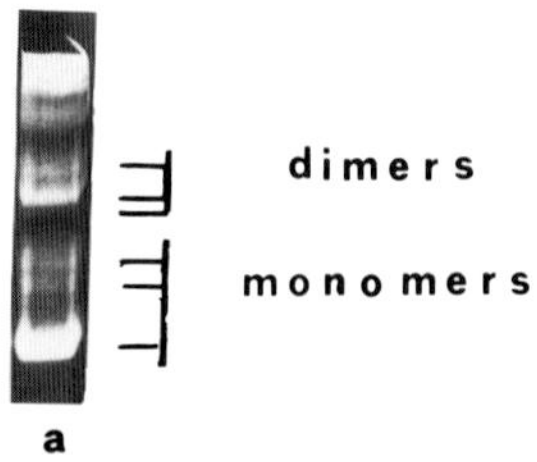

Fig. 7. Electrophoretic pattern of nucleosomes. Nuclease-treated chromatin, prepared by micrococcal nuclease digestion of chick erythrocyte nuclei as described in the Methods and resuspended in 5mM Na phosphate, pH 6.8, was diluted with 1 vol of sample buffer and loaded onto a 5% polyacrylamide slab gel.

a different region of hypersensitivity appears, yielding a subfragment of about 4 kb as compared to the parental fragment of 4.5 kb. Figure 11 is a summary of the DNase I hypersensitive regions in the α- and β-gene clusters that have been observed to date. While there is a definite tendency for these structures to be localized near the 5′ side of several genes, some genes do not contain this type of modification. In addition, hypersensitive sites are also observed in regions that are thought not to be involved in transcription. The possible significance of these hypersensitive regions and their possible role in Hb switching will be discussed elsewhere [H. Weintraub, manuscript in preparation].

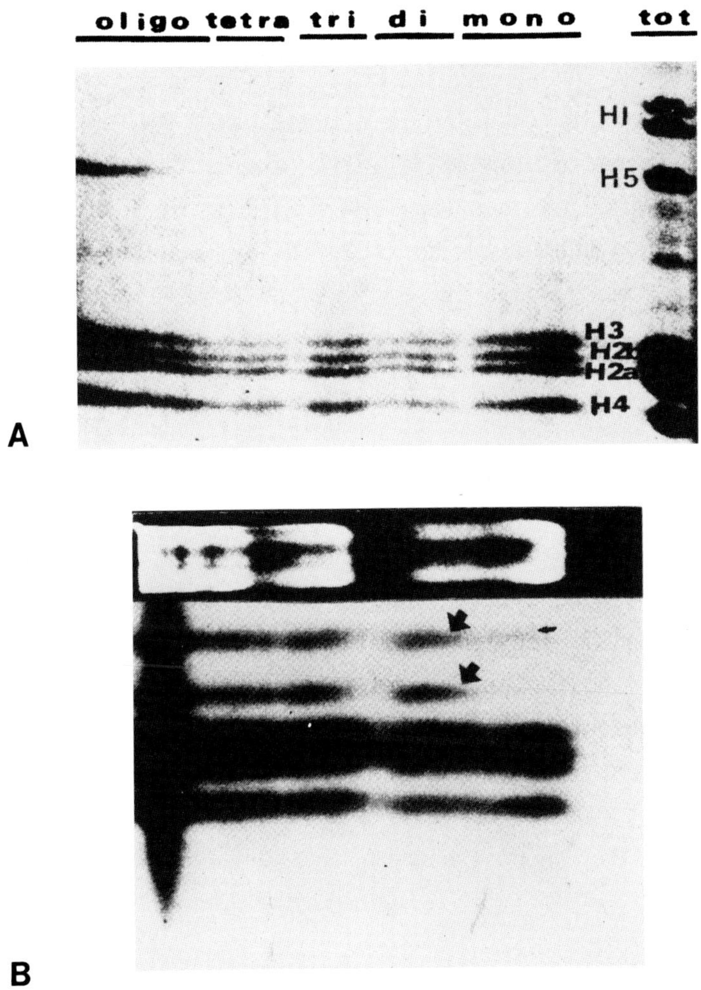

Fig. 8. Second-dimension analysis of the histones associated with electrophoretically fractionated nucleosomes. A. Nuclease-treated chromatin was fractionated on a 5% polyacrylamide gel, in standard conditions. The entire slot of the gel corresponding to the sample was then loaded onto a SDS polyacrylamide gel, in order to analyze the protein content of each nucleosome fraction. A total nucleosome sample was also loaded (slot on the right). B. ^{3}H-lysine-labeled nucleosomes were fractionated as in A in two dimensions. The thin small arrow indicates A24; the thick arrows, HMG 14 and 17.

MOLECULAR EVENTS ASSOCIATED WITH THE TRANSITION BETWEEN PRECURSOR RED CELL AND ERYTHROBLAST

In the chick embryo, red cells first appear at about 35 hours of development. There are about 2×10^6 cells in the whole embryo at this time, and of these, some 5×10^5 are red cell precursors [10]. These cells are localized in a specific region of the developing embryo called the area vasculosa. It is possible to estimate that at least half of the cells in this region must be red cell precursors in order to account for the number of red cells that are produced during the first few days of development [10]. To some extent this estimate can be confirmed directly by measuring the percentage of total erythroid colonies obtained from this region when individual cells are plated out in tissue culture. Our preliminary results suggest that while the plating efficiency is very low (about 2%–5%), the total percentage of erythroid colonies is very high (80%–90%). Since we have no reason to believe that the culture conditions select against nonerythroid colonies, it is a reasonable first guess that most of the cells in this region of the embryo are in fact erythropoietic, as also indicated by the previous numerological estimate [10].

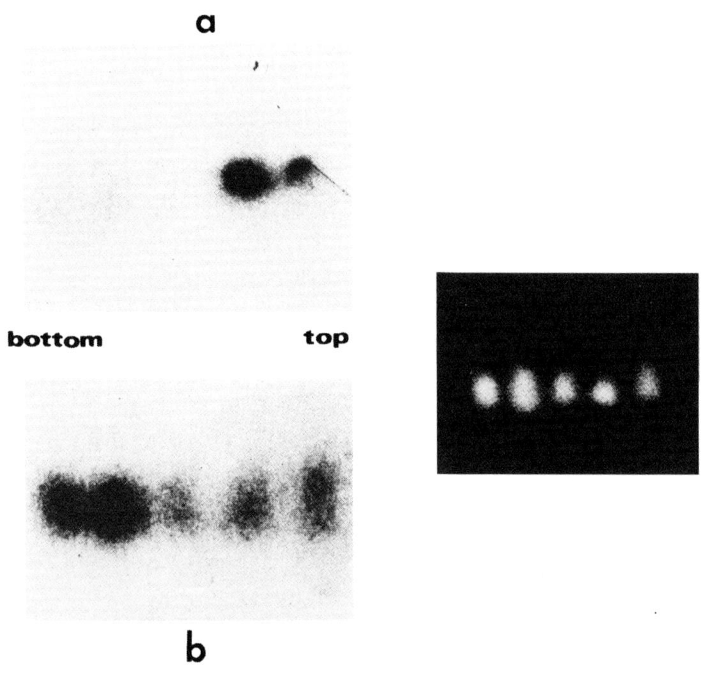

Fig. 9. Blot Hybridization of the DNA from the Different Monomer Fractions. Twenty micrograms from each of five monomer fractions was blotted onto nitrocellulose as described in Materials and Methods. The blots were hybridized to either (a) pHb 1001, a beta-globin cDNA, or (b) ovalbumin cDNA. Controls showed that under these conditions of hybridization, all of the DNA was successfully transferred to nitrocellulose and no more than 20% of the transferred DNA was lost during the hybridization and washing procedures; however, when detectable, this loss was the same in each fraction. Blot hybridization to RNA and small DNA fragments immobilized to nitrocellulose has been studied extensively by Dr. Patricia Thomas in this laboratory and the analysis will be presented elsewhere.

The availability of a reasonably pure population of precursor cells allows us to ask several questions about the switch from precursor cell to terminally differentiated cell. We have previously reported [11] that the globin genes are resistant to DNase I as assayed by solution hybridization in these precursor cells. More recently, we have been able to show that these genes are also resistant as assayed by blot hybridization. Most important, the blot hybridization data also showed that no discrete subfragments appeared after mild DNase I digestion, indicating the absence of hypersensitive regions before overt differentiation. These experiments suggest that the globin gene is in an inactive chromosomal configuration before the appearance of the first cytologically recognizable erythroid cells at 35 hours of development. This observation has been extended using the "run-off" transcription assay to measure the number of nascent RNA polymerases present on the globin genes in these cells. Finally, we have also investigated the degree of methylation of the globin genes in these precursor cells. There is a growing body of evidence suggesting that actively transcribed genes are undermethylated [12–16]. The degree of methylation of a gene is presently best assayed by digesting the DNA with methyl-sensitive restriction enzymes. MSP and HPA II both cut DNA at the same site (CCGG); however, HPA II is sensitive to a methyl group at CpG, the most common site of methylation, whereas MSP is not sensitive to methylation in this position. Figure 12 shows MSP and HPA II digests of erythroblast DNA and of erythroid precursor DNA hybridized to either an embryonic β-globin probe or a probe coresponding to the entire α domain (coding and noncoding region). As compared to the precursor DNA, erythroblast DNA is considerably undermethylated, both for the embryonic β-globin gene and for the coding regions of the α-genes. In nonerythroid differentiated cells (for example, brain and liver), these

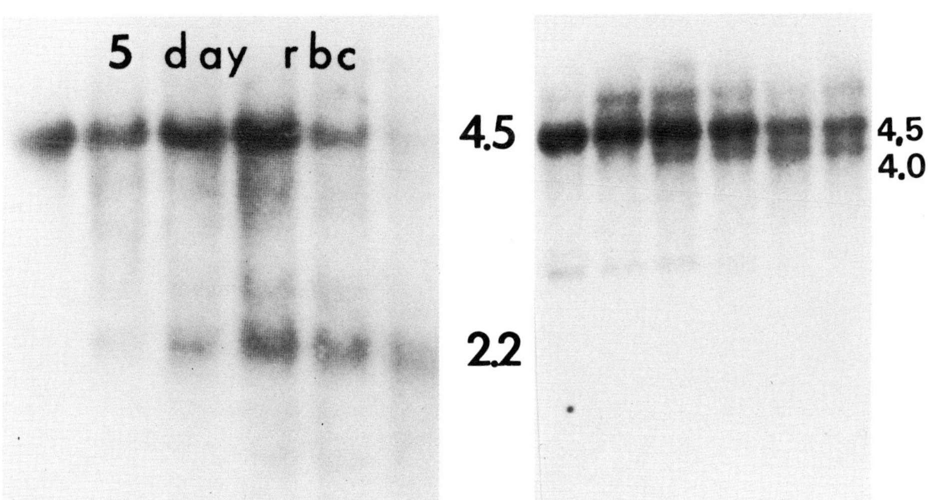

Fig. 10. Mapping the specific DNase cleavages in the embryonic and adult red cell β domain. Five-day (left) or adult (right) red cell nuclei were digested with DNase I (0; 0.1; 0.2; 0.4; 0.5; 0.8 μg/ml). All samples were restricted with Hind III and hybridized to pB2H2 (an embryonic β probe). Unlabeled bands probably represent hybridization to repetitive DNA sequences.

genes are highly methylated by this assay. Moreover, in embryonic red cells, the embryonic β-globin genes are unmethylated, but the hatching and adult genes are methylated; in definitive embryonic cells (synthesizing the adult and the hatching β globin chains), the embryonic β gene is now methylated, but the others are not; and finally, in adult erythrocytes, the adult β-globin gene is not methylated, but both the hatching and the embryonic genes are methylated [14]. Thus, in the chick β-globin domain, there seems to be a very tight correlation between the expressed gene, the gene that is transcribed, the gene that is undermethylated, and the gene that is sensitive to DNase I and capable of binding to an HMG column. In addition, where it has been analyzed most completely, the coding region of the α-globin domain possesses all these same properties; however, the adjacent, noncoding region on both the 5′ and 3′ sides show the opposite properties.

We next wished to understand what events were required for turning on the globin chromatin structure. It was previously proposed that the assembly of a new chromosome structure about the globin domain required the construction of a new chromosome at the time of DNA replication [17]. This type of proposal is clearly very compatible with the concept of a quantal mitosis previously put forward by Holtzer [18]. In order to test this notion, DNA synthesis was inhibited in 25-hour blastoderms by addition of FUdR to individual embryos. No hemoglobin appeared at 35 hours, and not surprisingly, the embryos died. The principal problem with this type of experiment is the difficulty in distinguishing an effect that is due to a coupling of differentiation to DNA replication from one that is secondary to the obvious cell death that is occurring in the presence of FUdR.

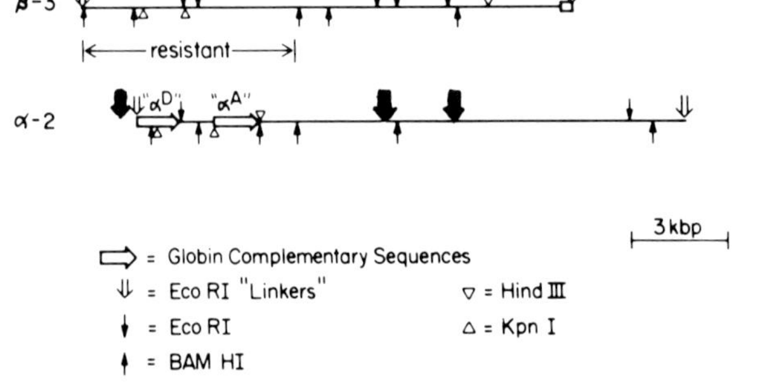

Fig. 11. Restriction maps for four λ Charon 4a recombinant chicken clones carrying globin genes. "a" and "e" denote adult and embryonic β-globin genes extending over the distance indicated by the thick arrows. The direction of transcription (5′ → 3′) was determined by Dodgson et al [1978] and M. Dolan, J.B. Dodgson, and J.D. Engel [unpublished results]. The thick vertical arrows point to sites of preferred cutting by DNase I as referred to the text.

By lowering the FUdR concentration to inhibit DNA replication by only 60%–70% and by reversing the FUdR (by adding thymidine to the embryo) after only five to eight hours of inhibition, we think we have been able to define conditions that cause no detectable signs of cell death, but significantly *delay* the appearance of Hb. Figure 13 shows the time course of accumulation of Hb per embryo for the first three days of incubation, as well as the time course for embryos treated with low doses of FUdR (60%–70% inhibition of H3-deoxyadenosine incorporation) for increasing periods of time before Hb synthesis begins. The data shows that a partial FUdR inhibition of replication, followed by reversal with thymidine, results in a lag in the accumulation of Hb. Additional experiments have demonstrated that the decreased Hb levels persist through the fifth day of incubation and

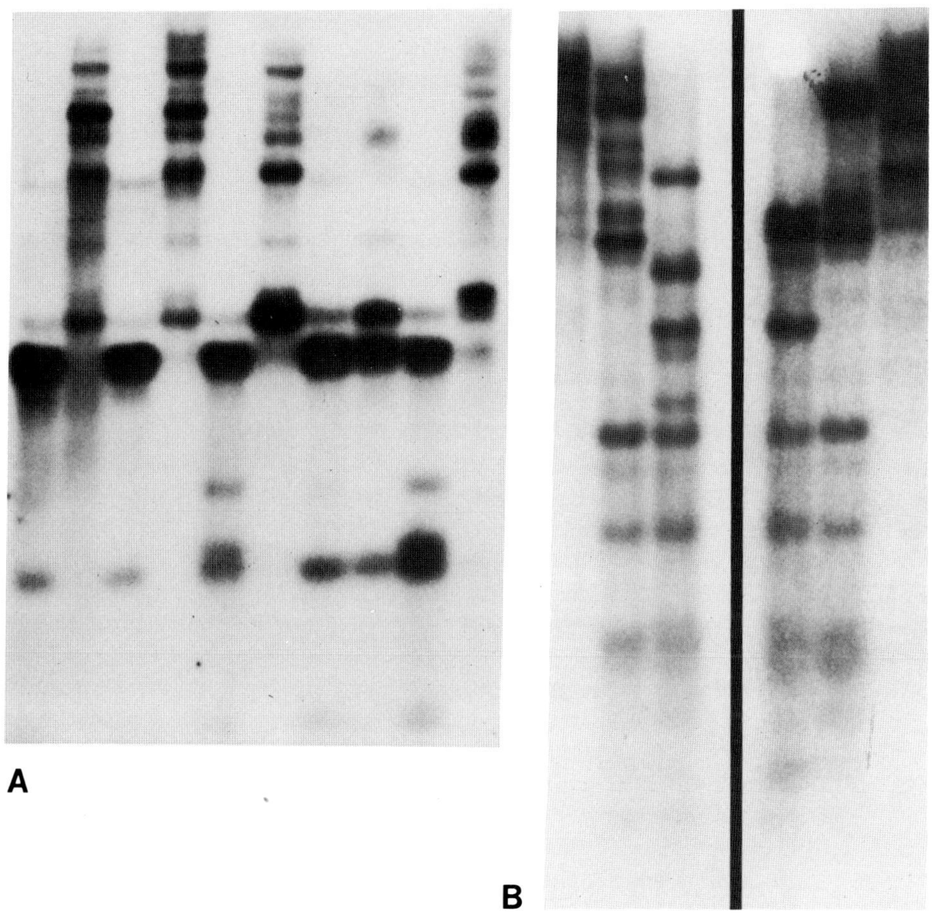

Fig. 12. Methylation of globin genes in red cell precursors. DNA from red cell precursors or red cells was digested with MSP or HPA II and hybridized to an embryonic β-globin probe (A) or to $\alpha 5$ and $\alpha 2$ (B), an entire α-domain probe (see Fig. 6). A. Left to right: MSP and HPA II (adult erythrocytes); MSP and HPA II (reticulocytes); MSP and HPA II (14-day definitive red cells); MSP and HPA II (5-day primitive embryonic red cells); MSP and HPA II (precursor red cells from 20-hour embryos). B. Left: Hybridization to 2. Right: Hybridization to 5. Left to right: precursor DNA, HPA II; red cell DNA, HPA II; red cell DNA, MSP; red cell, MSP; red cell DNA, HPA II; precursor DNA, HPA II.

that even lower concentrations of FUdR (20% inhibition of deoxyadenosine incorporation for five hours) produce a significant, though somewhat shorter, delay in the accumulation of Hb.

We do not think that these results can be explained in terms of cell death for the following reasons: 1) Over 80% of the embryos treated with 10^{-6} M FUdR or less for six hours or less will hatch some 20 days later. 2) During treatment with FUdR, there is no inhibition of H3-leucine incorporation. Autoradiography using labeled leucine or uridine showed all cells to be labeled with either isotope. If "thymidine-less" death had been occurring, the characteristic presence of unlabeled cells would have been expected. 3) FUdR treatment several generations earlier, between five and ten hours of incubation, results in no detectable delay in the appearance of Hb (data not shown).

Is the dependence on DNA synthesis a property of Hb synthesis per se or is it only a requirement for its initiation? To answer this question we monitored the effect of FUdR on the synthesis of Hb in cells which had already started to make Hb. Four-day erythroblasts, which make Hb and actively synthesize DNA, were incubated for ten hours in vitro and in vivo in the presence of doses of FUdR that block DNA synthesis by over 98%. Under these conditions, no inhibition of Hb synthesis was observed. Lower doses of FUdR gave the same results. Longer exposures to high doses of FUdR led to the accumulation of unlabeled cells and obvious signs of cell death. Consequently experiments were not continued beyond ten hours. In contrast, up to 30 hours of exposure to low doses of FUdR—doses

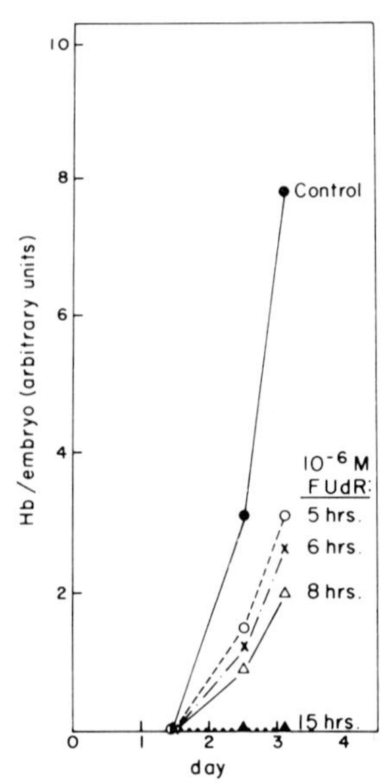

Fig. 13. Delayed appearance of Hb after transient exposure to submaximal doses of FUdR. Twenty-five-hour embryos were treated with 10^{-6} M FUdR (3 ml per embryo) for the stated periods of time. Thymidine (1.0 ml at 5 mg/ml) was then added to reverse the FUdR block to DNA synthesis. Hb was determined at various times thereafter. These assays were done on 25–30 embryos for each point according to methods previously described [10]. The 10^{-6} M FUdR inhibits H3-deoxyadenosine incorporation in OVO into DNA by about 60%–70%. Controls were embryos treated simultaneously with FUdR and TdR. Uridine does not reverse the effects of FUdR.

which cause a delay when given for only five hours at 25 hours of incubation—result in no signs of cell death and no decrease in Hb accumulation when administered at 4 days of incubation. These experiments show that a partial and transient inhibition of DNA replication inhibits only the initiation of Hb synthesis and not synthesis that has already been initiated. The results are compatible with, but clearly do not prove, the hypothesis that the replication-dependent assembly of a new daughter chromosome is required for the construction of an active chromatin structure and subsequent activation of the globin genes.

SUMMARY

We review our work dealing with the association of HMG 14 and 17 with actively transcribed genes. We next describe the existence of specific regions of the globin chromosomal domain that are cleaved by DNase I. These regions are located at specific DNA sequences; they are tissue-specific; in the β-globin domain, they are different in embryonic and in adult red cell nuclei; and while they tend to be located near the 5′ end of transcription units, they are undetected in some transcription units and they are also present in regions not known to be transcribed. Finally, we describe the preliminary biochemical characterization of a population of avian erythrocyte precursor cells. We show that the globin genes in these cells are not preferentially sensitive to digestion by DNase I, that they are not transcribed in an in vitro nuclear transcription system, that they are methylated, and that they lack the specific DNase I cleavages associated with actively transcribed globin genes.

ACKNOWLEDGMENTS

This work was supported by the National Science Foundation and the National Institutes of Health.

REFERENCES

1. Weisbrod S, Weintraub H: Isolation of a subclass of nuclear proteins responsible for conferring a DNase I sensitive structure on globin chromatin. Proc Nat Sci USA 76:630–634, 1979.
2. Weisbrod S, Groudine M, Weintraub, H: Interaction of HMG 14 and 17 with actively transcribed genes. Cell 19:289–301, 1980.
3. Gazit B, Panet A, Cedar H: Reconstitution of a DNase I sensitive structure on active genes. Proc Nat Acad Sci USA 77:1787–1790, 1980.
4. Goodwin GH, Johns EW: A new group of chromatin associated proteins with a high content of acidic and basic amino acids. Eur J Biochem 401:215–219, 1973.
5. Brown JL, Ingram U: Structural studies on chick embryonic Hb. J Biol Chem 249:3960–3972, 1974.
6. Stalder J, Groudine M, Dodgson J, Engel JD, Weintraub H: Hb switching in chickens. Cell 19: 973–980, 1980.
7. Levy WB, Dixon GH: Partial purification of transcriptionally active nucleosomes from trout testis cells. Nuceic Acids Res 5:4155–4163, 1978.
8. Albanese I, Weintraub H: Electrophoretic separation of a class of nucleosomes enriched in HMG 14 and 17 and actively transcribed globin genes. Nuceic Acids Res 8:2978–2805, 1980.
9. Stalder J, Larsen A, Engel JD, Dolan M, Groudine M, Weintraub H: Tissue-specific DNA cleavages in the globin chromatin domain introduced by DNase I. Cell 20:451–460, 1980.
10. Weintraub H, Cambell G, Holtzer H: Primitive erythropoiesis in early chick embryogenesis. J Cell Biol 50:652–668, 1971.
11. Weintraub H, Groudine M: Chromosomal subunits in active genes have an altered conformation. Science 93:848–858, 1976.
12. Bird AP, and Southern EM: Use of restriction enzymes to study eukaryotic DNA methylation. J Mol Biol 118:27, 1978.

13. Bird AP, Taggart MH, Smith BA: Methylated and unmethylated DNA compartments in sea urchin genome. Cell 17:889, 1979.
14. McGhee JD, Ginder GD: Specific DNA methylation sites in the vicinity of the chicken β-globin genes. Nature 280:419–420, 1979.
15. Mandel JL, Chanbar P: DNA Methylation: Organ specific variations within and around the ovalbumin genes. Nucleic Acids Res 7:2081, 1979.
16. Waalwijk C, Flavell RA: DNA methylation at a CCGG sequence in the large intron of the rabbit β-globin gene: Tissue-specific variations. Nucleic Acids Res 5:4631–4641, 1978.
17. Weintraub H, Flint SJ, Leffak M, Groudine M, Grainger R: The generation and propagation of variegated chromosome structures. Cold Spring Harbor Symp Quant Biol 42:401–407, 1978.
18. Holtzer H, Rubinstein N, Fellini S, Yeoh G, Chi J, Birnbaum J, Okayama M: Quantal and proliferative cell cycles. Q Rev Biophys 8:523–557, 1975.

Organization and Expression of Globin Genes, pages 191–201

DNA Methylation in the Chicken Adult β-Globin Gene: A Relationship With Gene Expression

Gordon D. Ginder and James D. McGhee

Department of Internal Medicine, Division of Hematology-Oncology, University of Iowa Hospitals and Clinics, Iowa City (G.D.G.) and Laboratory of Molecular Biology, National Institute of Arthritis, Metabolism and Digestive Diseases, National Institutes of Health, Bethesda, Maryland (J.D.M.)

DNA METHYLATION IN HIGHER EUKARYOTES

Methylation at the 5-position of cytosine is the only postsynthetic modification detected in higher eukaryotic DNA [2]. Such a modification provides a mechanism whereby daughter and parental DNA strands could be distinguished, and hence DNA methylation has been made the central feature in numerous models and hypotheses concerning the control of eukaryotic cell differentiation [3–5].

Initial studies of DNA methylation in higher eukaryotes were limited to measurement of the overall 5-methyl cytosine content of DNA from various tissues. Most of these studies have shown that 5-methyl cytosine comprises about 1%–3% of the total base composition in various animal cell DNAs [6, 7]. While some early reports suggested a difference in bulk 5-methyl cytosine content in different tissues of the same organism and moreover that the 5-methyl cytosine content changed during development [2, 7]; more precise measurements done with high-pressure liquid chromatography and mass spectroscopy have recently shown no detectable differences in the overall 5-methyl cytosine content between sea urchin embryo and germ cell DNA, between fish sperm and fish somatic (liver) DNA [8], nor any differences between various chicken tissue DNAs [9].

The availability of a large number of bacterial restriction endonucleases with defined cleavage recognition sequences [10] coupled with the fact that nearly all of 5-methyl cytosine occurs in the dinucleotide 5′ CpG 3′ [1, 8] has provided a new tool for studying DNA methylation at sites within or near specific genes. Bird and Southern [11] first showed that type II restriction endonucleases which contain the dinucleotide 5′ CpG 3′ within their recognition-cleavage sequence could be used to probe for 5-methyl cytosine by virtue of the fact that several of these endonucleases, including the enzyme HpaII, which cleaves at the sequence CCGG, are unable to cut DNA in which the cytosine in the CpG sequence is methylated [12]. This approach was made much more powerful when Waalwijk and Flavell [13] and

 Published 1981 by Alan R. Liss, Inc., 150 Fifth Avenue, New York, NY 10011

Singer et al [14] described an enzyme, MspI, which also cleaves at CCGG, but appears to cleave regardless of the methylation state of the internal cytosine. Cedar et al [15] and Bird et al [16] have provided confirmatory evidence that the difference in cleavage pattern between MspI and HpaII is in fact due to cytosine methylation.

We have used these two endonucleases, HpaII and MspI, to demonstrate a correlation between methylation at six specific CCGG sites in the vicinity of the chicken β-globin genes, and the expression of the chicken adult β-globin gene.

THE CHICKEN GLOBIN SYSTEM

The chicken globin system provides a valuable model of globin gene expression during differentiation in that there is a developmental switch between embryonic β-globin expression and adult β-globin expression [17]. (For further discussion see paper by Weintraub in this volume.) One technical advantage of the chicken globin system for examining in vivo DNA methylation patterns lies in the fact that the circulating nucleated erythrocytes and reticulocytes provide a population of pure erythroid cells for analysis. Of particular relevance to the current studies, it appears that in 5-day embryos there is little or no detectable synthesis of adult β-globin protein or mRNA [18]. This has allowed us to examine the methylation pattern of DNA isolated from embryonic erythroid cells in which the adult β-globin gene is not active and compare the result with DNA isolated from adult erythroid tissues, in which the gene is active or has been active.

METHYLATION AT CCGG SITES WITHIN THE CHICKEN ADULT β-GLOBIN GENE: CORRELATION WITH GENE ACTIVITY

The experimental details and controls used in these studies are described elsewhere [1; and submitted manuscript]. Briefly, DNA was extracted from various chicken tissues, digested with either HpaII or MspI, electrophoresed on a 1% agarose gel, and transferred to nitrocellulose filters by the method of Southern [19]. The nitrocellulose filters were then hybridized to ^{32}P-labeled probes made by nick translation either of the cDNA plasmid pHb1001 [18] or the chicken adult β-globin clone CAβG-1 [20]. The filters were autoradiographed and the fragment band patterns were analyzed. Those CCGG sites which are cleaved by MspI and/or are present in the unmethylated clone CAβG-1, but which are not cleaved by HpaII, are interpreted as being methylated at the internal cytosine. Sites which are cleaved by both MspI and HpaII are interpreted as being unmethylated.

Figure 1 depicts the overall restriction map and detailed CCGG site map of the chicken adult β-globin gene determined by analysis of the clone CAβG-1, as described elsewhere [20]. It should be noted that in Figures 2–4, in which the cDNA clone (pHb 1001) was used as a hybridization probe, detailed mapping and various controls [data not shown] have demonstrated that only the bands of length 0.19 kb (not usually detected for technical reasons) and 1.4 kb, as well as the sum of these two bands at 1.6 kb, arise from sequences within the adult β-globin gene. Other bands are due to cross-reaction with embryonic β-like gene(s) [1, 21].

The digestions shown in Figure 2 were hybridized against probes made from the cDNA β-globin plasmid pHb1001 (a gift from Winston Salser). As shown in Figure 2, when both oviduct DNA (Lane D) and adult erythrocyte DNA (Lane B) are digested with MspI, a 1.4 kb fragment containing mRNA coding sequences is generated; this same band appears in the HpaII digest of adult erythrocyte DNA (Lane A) and, as seen from the clone map (Fig. 1b) is the band which contains the majority of the adult β-globin gene sequence. Figure 2, Lane A, shows that the same pattern is observed on HpaII digestion of adult reticulocyte DNA (obtained from phenylhydrazine-induced anemic adults).

On the other hand, when DNA from adult oviduct is digested with HpaII (Fig. 2, Lane C), a coding fragment of 1.6 kb is generated in addition to the 1.4 kb fragment.

Figure 3, Lane B, shows the results using the same probe against 5-day embryo erythrocyte DNA digested with HpaII. The 1.4 and 1.6 bands are both present as in the case of the oviduct DNA. A similar pattern containing these two bands has been observed in all of the nonerythroid adult somatic chicken tissues thus far examined. These results immediately suggest that the CCGG sites surrounding the 1.4 kb adult β-gene fragment are completely unmethylated in adult erythroid cells, while at least one of these sites (site 5, Fig. 1b) is at least partially methylated in other somatic tissues and 5-day embryo erythroid cells. Sperm DNA appears to be highly methylated, at least at CCGG sites, and neither 1.4 nor 1.6 kb bands are detected after

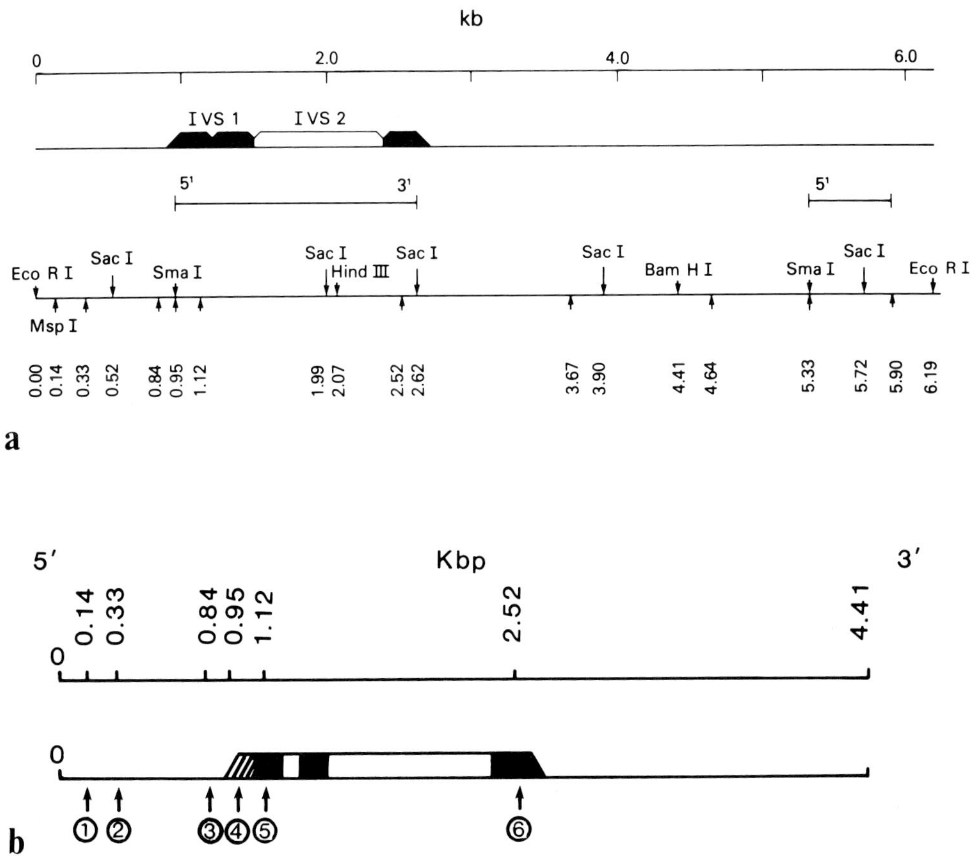

Fig. 1. The general restriction map of the clone CAβG-1 (a) with detailed map of CCGG sites (b) is depicted. The map was derived previously [20].

HpaII digestion [data not shown]. The higher molecular weight bands in the HpaII digests of Figures 2–4 indicate changes occuring in the methylation patterns of the embryonic β-globin genes as well.

The digestions in Figure 4 are essentially the same as in Figure 3 except that a probe highly enriched for the 5′ end of the β-globin cDNA sequence was made by isolating a 100 base pair 5′ end fragment of globin cDNA from pHb1001. In both, the MspI digest (Lane A) and HpaII digest of adult reticulocyte DNA (Lane B), there are bands at both 1.4 kb and 0.19 kb while in the 5-day embryo erythrocyte DNA HpaII digest (Lane C) there are bands at 1.6 kb and 1.4 kb but no band at 0.19 kb. These digests again localize the position of the variable methylation pattern to the CCGG site number 5 at map position 1.12 (in Fig. 1) while sites 4 and 6 appear to be unmethylated in all tissues examined.

An additional interesting observation was made in this experiment; namely that the 0.55 kb band which is bounded by map positions 5.33 and 5.90 on CAβG-1 (see

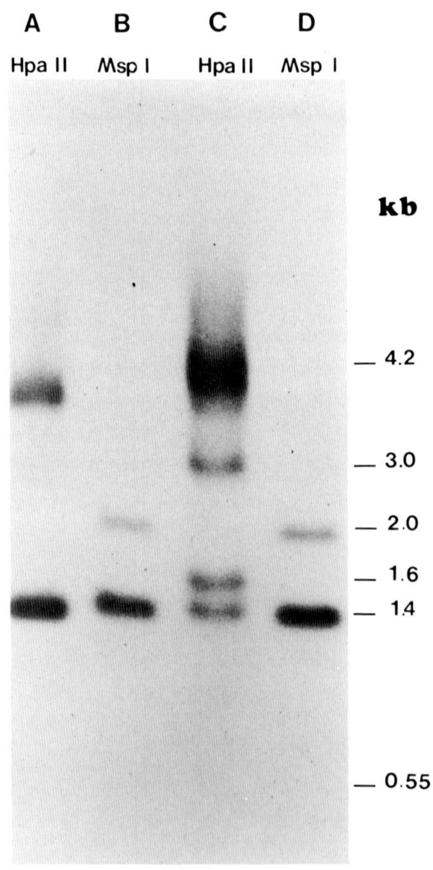

Fig. 2. DNA from adult erythrocyte (Lanes A and B) adult oviduct (Lanes C and D) were digested with either HpaII (Lanes A and C) or MspI (Lanes B and D) and transferred to nitrocellulose filters. Resultant filters were hybridized against the α^{32P} cDNA probe pHb 1001.

Fig. 1a) and corresponds to an expressed embryonic β-globin gene [21] is seen in the 5-day embryo erythrocyte DNA (Lane C) but not in the adult reticulocyte DNA (Lane B). Thus these two CCGG sites in the embryonic β-globin gene appear to be methylated in nonexpressing adult red cells, but unmethylated in embryonic erythrocytes which presumably express the gene.

The probe derived from the cDNA clone only allows one to analyze methylation sites within or close to the gene coding region. To examine the multiple CCGG sites which lie on the 5′ side of the gene, as well as to eliminate cross-reacting fragments derived from embryonic genes, the following method was employed. DNA from a given tissue was extracted, digested with both EcoRI and HindIII, and electrophoresed on a preparative gel. The 1.8–2.2 kilobase pair size class DNA was isolated from the gel by the method of Vogelstein [22]. This DNA, which should

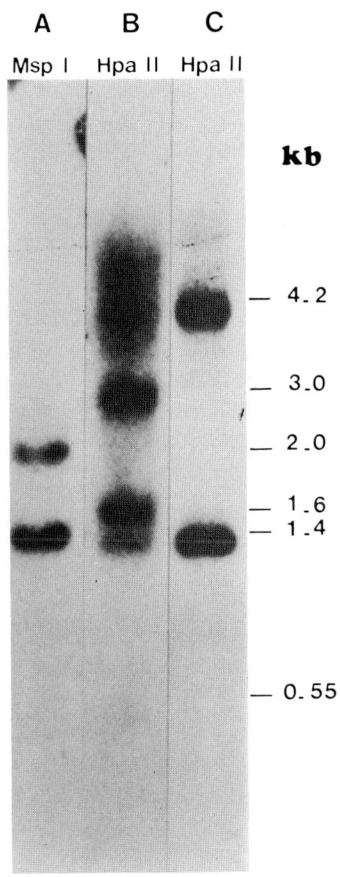

Fig. 3. DNA from 5-day-old embryonic erythrocytes (Lanes A and B) or adult reticulocytes were digested with either MspI (Lane A) or HpaII (Lanes B and C). Filters were hybridized with pHb 1001 as in Figure 1.

contain the 5′ end of the adult β-globin gene (see Fig. 1a), was then further digested with either MspI or HpaII. As a hybridization probe, we used the analogous 2.1 kb Eco RI/HindIII fragment isolated from the adult β-globin clone, CAβG-1.

The results of such an experiment are shown in Figure 5. The MspI digest of 5-day embryonic erythrocyte DNA (Lane B) generates all the expected fragment sizes from the map of CAβG-1, if one assumes that the 0.1 kb fragments are below the limit of detection. The diffuse hybridization pattern at approximately 2 kb is apparently due to a moderately repeated sequence contained in our probe solutions [W.I. Wood, personal communication]. Since site 4 at 0.95 map units is always unmethylated (see above), no adult β-globin gene band can be greater than 1.2 kb and hence this 2 kb hybridization does not concern the present analysis.

The HpaII digest of adult reticulocyte DNA (Lane A) has an additional band at 0.7 kb which, judging from the lengths of the possible fragments, indicates partial methylation at site 2, map position 0.33; otherwise no sites are methylated. In contrast, the HpaII digest of 5-day embryo erythrocyte DNA (Lane C), as well as

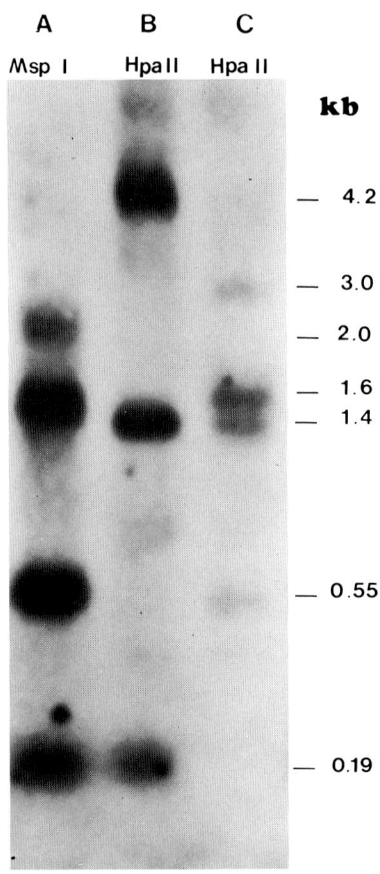

Fig. 4. DNA from adult erythrocytes (Lane A) and embryonic erythrocytes (Lane B) were digested with HpaII, electrophoresed, and transferred to nitrocellulose filter and hybridized to a 5′ end fragment of pHb 1001 made by cleaving the plasmid with MspI and isolating the 5′ end globin coding fragment.

the digests of DNA from adult oviduct and brain (not shown), reveal no bands at 0.5 kb or 0.19 kb. Instead, there are prominent new bands at 0.85 kb and 1.2 kb. Consideration of the possible fragment sizes (and recalling that site 4 is completely unmethylated) suggests that there is a high degree of methylation (greater than or equal to 90%) at CCGG sites 1 and 2. There is at least partial methylation at site 5 (which we knew from the previous hybridizations to pHb1001) as well as site 3.

Figure 6 shows the composite map of the methylation patterns of CCGG sites in and near the adult β-globin gene in the different tissues examined. It appears that in adult erythroid tissues, all but one of the CCGG sites in the coding region and in the 5′ flanking region of the adult β-globin gene are unmethylated. In contrast, four of these sites are highly methylated in adult nonerythroid tissues as well as in 5-day-old embryo erythrocytes, which are not expressing this gene. It may be of further interest that in all somatic tissues examined, two CCGG sites seem to be maintained

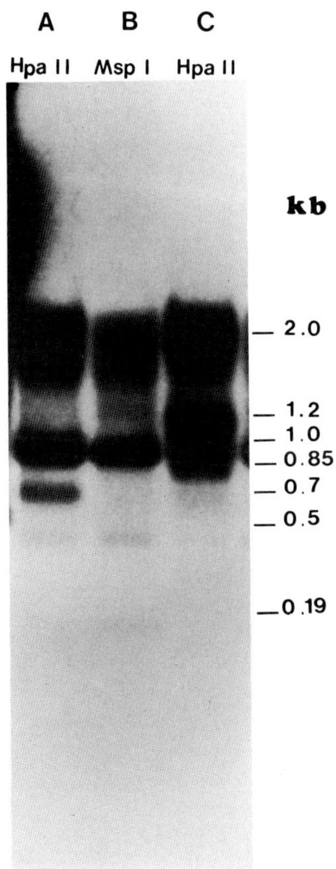

Fig. 5. DNA from adult reticulocytes (Lane A) and embryonic RBCs (Lanes B and C) were digested with both EcoRI and Hind III. The 1.8 to 2.2 kb size class (which includes the 5′ end of the adult β-globin gene) was isolated from an agarose gel and digested with either HpaII (Lanes A and C) or MspI (Lane B) and transferred to a filter. The hybridization probe in this experiment was an α^{32}P-labelled 2.1 kb Hind III/EcoRI fragment of the chicken adult β-globin gene clone CAβG-1 (see map Fig. 1a).

in the unmethylated state; in sperm DNA, however, these same sites appear to be methylated. We have also found the same relationship between DNA methylation and gene expression at CpGs within other restriction endonuclease sites in the adult β-globin gene [Ginder and McGhee, in preparation]. In addition, preliminary evidence, as discussed above and as shown in Figure 4, suggests that at least two CCGG sites in an embryonic β-globin gene are unmethylated in embryonic red cells, but are at least partially methylated in adult reticulocytes. Thus the same relationship between gene expression and DNA methylation appears to hold true in the expected reciprocal fashion for at least one embryonic β-globin gene as well as the adult β-globin gene.

THE CORRELATION OF DNA METHYLATION WITH GENE EXPRESSION AND POSSIBLE ROLE IN DIFFERENTIAL GLOBIN GENE SWITCHING

Our results demonstrating a correlation between the absence of DNA methylation at certain specific sites in the chicken adult β-globin gene and expression of the gene [1] have been accompanied by similar evidence in a variety of eukaryotic and viral gene systems. Mandel and Chambon [23] have reported the same correlation in the chicken ovalbumin, ovomucoid, and conalbumin genes as well as confirmatory data on the chicken adult β-globin gene. Van der Ploeg and Flavell [24] have looked at a large number of CpG sites in the human non-α-globin gene cluster and have reported a similar correlation between lack of methylation of the sites in or near each specific globin gene and the expression of that gene. Most importantly the methylation pattern in the intergenic flanking regions does not vary with gene expression. Other authors have found that in integrated viruses such as Herpes virus saimiri [25] and Adenovirus 12 [26] specific regions of these genomes which

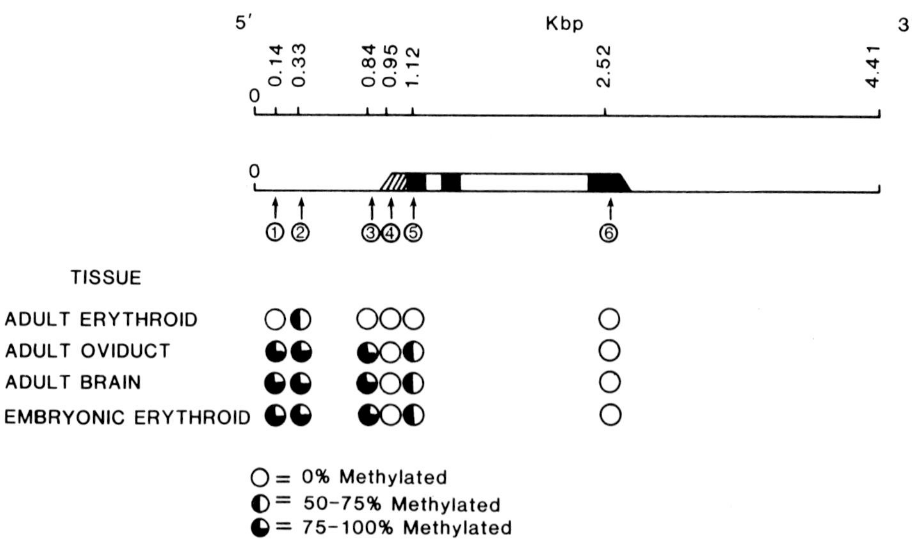

Fig. 6. Map showing the methylation pattern of internal cytosines in six CCGG sites within and in the 5′ upstream region of the chicken adult β-globin gene in various tissues examined.

are being expressed in a given cell or cell line are either undermethylated or unmethylated, while the same regions are highly methylated in cell lines in which the viral genes are inactive.

Despite all the above correlations, the role of DNA methylation in gene expression remains speculative. As suggested by van der Ploeg and Flavell [24], and as is certainly consistent with the "active" methylation pattern for which we find the chicken adult β-globin gene, absence of methylation may be a necessary but not sufficient condition for gene transcription. If so, the methyl group could presumably exert a negative effect upon transcription although the possible mechanism is not at all clear. A direct steric blockage of the polymerase by a methyl group seems unlikely since at least bacterial RNA polymerase can transcribe methylated DNA [12]. A second mechanism might be to alter the local chromatin structure; however, at least gross effects on chromatin structure seem unlikely since perfectly normal nucleosome core particles can be formed on DNA which is completely substituted with glycosylated hydroxymethylcytosine residues [McGhee and Felsenfeld, in preparation]. On the other hand, the finding of a correlation between low levels of DNA methylation and increased DNaseI sensitivity [23] is at least suggestive that chromatin structure and DNA methylation may be related in some way.

An alternative mechanism by which a methyl group could exert negative control is by acting as a binding site for an inhibitor protein. Although the testing of such a model must obviously await more direct studies of transcription in highly reconstituted or native cell systems, two observations must be borne in mind. Specific in vitro initiation of transcription has been reported using DNA which has been cloned in bacteria [27, 28, and see papers in this volume by Luse, Proudfoot, and Flavell] and which hence is abnormally methylated. However, as yet these in vitro transcription systems have not demonstrated tissue specificity. Secondly, certain organisms such as insects [29] can apparently complete their normal course of development without the benefits of differential DNA methylation.

Nonetheless, the accumulating evidence of strong correlations between specific DNA methylation and gene expression in globin systems and other genes argues for a potential role for DNA methylation in differentiation and raises caution in the use of unmethylated transcriptional templates cloned in bacterial cells. Further studies in which cloned eukaryotic genes are specifically methylated in vitro and studies in which DNA methylation is manipulated in vivo such as the recent study of Jones and Taylor [30] should help to elucidate the potential role of DNA methylation in gene switching.

SUMMARY

We have previously reported a correlation between site-specific DNA methylation and expression of the chicken β-globin genes [1]. We have now mapped six CCGG sites within and around the chicken adult β-globin gene and determined the methylation pattern of the internal cytosines (CC*GG) of these sites using the bacterial restriction endonucleases HpaII, which cleaves at CCGG but not C5meCGG, and MspI, which cleaves at both CCGG and C5meCGG. We find that three of these sites are unmethylated only in the DNA from adult erythroid cells which are expressing or have expressed the adult gene, but are methylated in DNA from adult nonerythroid cells as well as 5-day embryonic erythrocytes. Of the remaining three sites, one appears to be at least partially methylated and two appear to be unmethylated in all of the somatic tissues examined. A preliminary description of the methylation patterns of two CCGG sites in an embryonic β-globin gene is also described. From these data and similar correlations reported in other gene systems, we suggest a possible role for DNA methylation in differential globin gene expression.

ACKNOWLEDGMENTS

This work was supported in part by a professional enrichment grant to G.D.G. from the American Cancer Society, Iowa Division.

REFERENCES

1. McGhee JD, Ginder GD: Specific DNA methylation sites in the vicinity of the chicken β-globin genes. Nature 280:418–420, 1979.
2. Grippo P, Iaccarino M, Parisi E, Scarano E: Methylation of DNA in developing sea urchin embryos. J Mol Biol 36:195–208, 1968.
3. Riggs A: X inactivation, differentiation, and DNA methylation. Cytogenet Cell Genet 14:9–25, 1975.
4. Holliday R, Pugh JE: DNA modification mechanisms and gene activity during development. Science 187:226–232, 1975.
5. Taylor JH: Enzymatic methylation of DNA: Patterns and possible regulatory roles. In "Molecular Genetics," part III, New York: Academic Press, 1979, pp 89–115.
6. Sneider TW, Potter VR: Methylation of mammalian DNA: Studies on Novikoff hepatoma cells in culture. J Mol Biol 42:271–284, 1969.
7. Vanyushin BE, Mazin AL, Vasilyev VK, Belozersky AN: The content of 5-methyl cytosine in animal DNA: The species and tissue specificity. Biochim Biophys Acta 299:397–403, 1973.
8. Pollack JM, Swihart M, Taylor JH: Methylation of DNA in early development: 5-methyl cytosine content of DNA in sea urchin sperm and embryos. Nucleic Acids Res 5:4855–4863, 1973.
9. Razin A, Cedar H: Distribution of 5-methyl cytosine in chromatin. Proc Natl Acad Sci USA 74: 2725–2728, 1977.
10. Roberts RJ: Restriction and modification enzymes and their recognition sequences. Nucleic Acids Res 8:r63–r80, 1980.
11. Bird AP, Southern EM: Use of restriction enzymes to study eukaryotic DNA methylation. J Mol Biol 118:27–47, 1978.
12. Mann MB, Smith HO: Specificity of HpaII and HaeIII DNA methylases. Nucleic Acids Res 4: 4211–4221, 1977.
13. Waalwijk C, Flavell RA: MspI, an isoschizomer of HpaII which cleaves both methylated and unmethylated HpaII sites. Nucleic Acids Res 5:3231–3236, 1978.
14. Singer J, Roberts-Ems J, Riggs AD: Methylation of mouse liver DNA studied by means of the restriction enzymes MspI and HpaII. Science 203:1019–1020, 1979.
15. Cedar H, Solage A, Glaser G, Razin A: Direct detection of methylated cytosine in DNA by use of the restriction enzyme MspI. Nucleic Acids Res 6:2125–2132, 1979.
16. Bird AP, Taggart MH, Smith BA: Methylated and unmethylated DNA compartments in the sea urchin genome. Cell 17:889–901, 1979.
17. Brown JL, Ingram WM: Structural studies on chick embryonic hemoglobins. J Biol Chem 249: 3960–3972, 1974.
18. Salser WA, Cummings I, Liu A, Strommer J, Padayatty J, Clark P: Analysis of chicken globin cDNA clones: Discovery of a novel chicken alpha globin gene induced by stress. In "Cellular and Molecular Regulation of Hemoglobin Switching," Stamatoyannopoulos G, Nienhuis A (eds): New York: Grune and Stratton, 1979, pp 621–647.
19. Southern EM: Detection of specific sequences among DNA fragments separated by gel electrophoresis. J Mol Biol 98:503–517, 1975.
20. Ginder GD, Wood WI, Felsenfeld G: Isolation and characterization of recombinant clones containing the chicken adult β-globin gene. J Biol Chem 254:8099–8102, 1979.

21. Dodgson JB, Strommer J, Engel JD: Isolation of the chicken β-globin gene and a linked embryonic β-like globin gene from a chicken DNA recombinant library. Cell 17:879–887, 1979.
22. Vogelstein B, Gillespie D: Preparative and analytical purification of DNA from agarose. Proc Natl Acad Sci USA 76:615–619, 1978.
23. Mandel JL, Chambon P: DNA methylation: Organ specific variations in the methylation pattern within and around ovalbumin and other chicken genes. Nucleic Acids Res 7:2081–2103, 1979.
24. van der Ploeg LHT, Flavell RA: DNA methylation in the human $\gamma\delta\beta$-globin locus in erythroid and non-erythroid tissues. Cell 19:947–958, 1980.
25. Desrosiers RC, Mulder C, Fleckenstein B: Methylation of Herpes virus Saimiri DNA in lymphoid tumor cell lines. Proc Natl Acad Sci USA 76:3839–3843, 1979.
26. Sutter D, Doerfler W: Methylation of integrated adenovirus type 12 DNA sequences in transformed cells is inversely correlated with viral gene expression. Proc Natl Acad Sci USA 77: 253–256.
27. Weil PA, Luse DS, Segall J, Roeder RG: Selective and accurate initiation of transcription at the Ad2 major late promotor in a soluble system dependent on purified RNA polymerase II and DNA. Cell 18:469–484, 1979.
28. Wasylyk B, Kedinger C, Corden J, Brison O, Chambon P: Specific in vitro initiation of transcription of conalbumin and ovalbumin genes and comparison with adenovirus-2 early and late genes. Nature 285:367–372, 1980.
29. Bird AP: DNA methylation and the frequency of CpG in animal DNA. Nucleic Acids Res 8: 1499–1504, 1980.
30. Jones PA, Taylor SM: Cellular differentiation, cytidine analogs and DNA methylation. Cell 20: 85–93, 1980.

mRNA Metabolism

Organization and Expression of Globin Genes, pages 205–217

Processing of the β-Globin mRNA Precursor in Mouse and Human Cells

Jeffrey Ross, Lynne E. Maquat, Alan J. Kinniburgh, Larry R. Beach, George R. Honig, Jack Lazerson, and William B. Ershler

The McArdle Laboratory for Cancer Research, University of Wisconsin, Madison (J.R., L.E.M., A.J.K., L.R.B.), The Children's Hospital, Chicago (G.R.H.), The Milwaukee Children's Hospital, Milwaukee (J.L.), and the Department of Medicine, Center for the Health Sciences, Madison (W.B.E.)

INTRODUCTION

The transfer of information in eukaryotes from genes to functional proteins takes place via complex enzymatic and transport steps, most of which are not well understood. We know still less at this point about how these steps are controlled. It seems likely that some genes are regulated primarily at the transcriptional level [1, 2], and the expression of other genes can be modulated at the translational level [3, 4]. Since many eukaryotic genes are probably transcribed to large precursor RNA molecules, RNA processing is an essential step in the information transfer process, but the extent, if any, to which genes are regulated by processing events is unknown. The expression of some viral genes might be controlled at the RNA processing level [5–7]. Other data indicate that RNA transcripts from some nuclear [8–11] and mitochondrial [12, 13] structural genes are not processed to mRNA under certain conditions.

We and others have studied the synthesis and processing of the β-globin mRNA precursor in mouse and human erythroid cells [14–20]. The emphasis of our experiments has been to describe the steps by which the precursor is transformed into mature mRNA, with the ultimate goal of understanding the enzymology of RNA processing and its potential role in gene regulation. During the course of these studies, we obtained some evidence for a hypothesis discussed for several years by hematologists and biochemists investigating the etiology of β^+ thalassemia: The phenotype of a cell can be changed by mutations that affect RNA processing. These experiments are discussed below. Therefore, although there is no proof that RNA processing events can modulate the expression of nonribosomal nuclear genes, the β^+ thalassemia data do demonstrate that the rate of information transfer can be retarded if processing of mRNA precursor molecules occurs inefficiently.

INTERMEDIATE-SIZED RNAs DERIVED FROM THE MOUSE β-GLOBIN mRNA PRECURSOR

The mouse β-globin genes are transcribed into an mRNA precursor containing approximately 1,600 to 1,800 nucleotides [14–16, 19, 20]. The precursor contains one copy of the β-globin mRNA (700–800 nucleotides), but the mRNA sequences are separated into three noncontiguous segments by two intervening RNA sequences (IVS) [14, 21, 22]. In addition to the transcribed nucleotides, the precursor contains 150 nucleotides of poly A which are added posttranscriptionally to the 3′ terminus [15, 20]. The precursor is processed to mRNA by removal of the IVS and ligation of the globin-specific segments. RNA splicing was first described for adenovirus RNA processing [23–25].

Initial experiments were performed to investigate the reaction sequence involved in processing of the mouse β-globin mRNA precursor. Specifically, we wished to determine if the larger IVS was removed and the RNA spliced in a single step, or if two or more reactions were required. If processing occurred in a stepwise fashion, then we would expect to observe intermediate-sized RNAs, that is, RNAs that were smaller than the precursor but larger than mature mRNA. Fifteen-day-old mouse fetal liver erythroid cells were cultured for several hours and were then pulse-labeled with ^3H nucleosides for eight to ten minutes [20, 26]. (The liver of the embryonic mouse is the major erythropoietic organ from approximately 12 to 18 days of gestation; these cells can be incubated in primary stationary cultures, and they synthesize large quantities of adult globin mRNA and globin protein [27].) Total cell RNA was then electrophoresed in a denaturing polyacrylamide gel [20], and the RNA eluted from each fraction was assayed for β-globin mRNA nucleotide sequences by molecular hybridization to cloned, filter-bound, β-globin cDNA [17]. After an eight-minute labeling period, four size classes of β-specific RNA were observed: the precursor (1,600–1,800 nucleotides), mature mRNA (700–800 nucleotides), and two intermediate-sized RNAs containing approximately 1,000 and 900 nucleotides (Fig. 1A). Even after labeling times of only 5 minutes, the majority of the pulse-labeled mRNA sequences was detected in precursor or mature mRNA molecules. This result indicates that the intermediate-sized RNAs turn over quite rapidly. Pulse-chase kinetic analyses were not feasible, because of the rapid turnover time. However, there is no evidence that any globin mRNA sequences are wasted within the nucleus [26]. Therefore, it seems likely that the intermediate-sized RNAs shown in Figure 1 are obligate intermediates involved in the processing of this precursor.

THE GENERATION OF MOUSE β-GLOBIN mRNA BY PARTIAL RNA SPLICING REACTIONS

To determine how the intermediates are generated, each of the intermediate-sized RNA fractions was purified by eluting appropriate molecular weight RNAs from gel slices and by annealing the RNA with filter-bound β-globin DNA. The hybridized RNA was eluted from the filters and analyzed by RNAase protection experiments. This technique has been described [22]. Briefly, radioactive RNA (precursor, intermediates, or mRNA) is incubated with full-length globin cDNA, which is prepared from the mature mRNA. The DNA-RNA hybrids are then treated with RNAase A, under conditions in which the enzyme hydrolyzes only single-

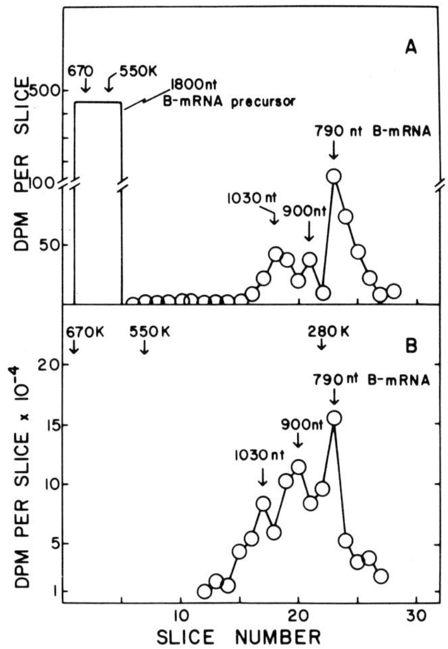

Fig. 1. Analysis of pulse-labeled β-globin mRNA sequences from mouse fetal liver cells. A. Radioactive RNA was prepared from mouse fetal liver cells that had been incubated for eight minutes with ^{3}H uridine [26]. The RNA was electrophoresed in a denaturing polyacrylamide gel (98% formamide) [20]. The gel was then sliced, RNA was eluted from each fraction, and the RNA was hybridized to cloned β-globin cDNA immobilized on nitrocellulose filters [17]. The filters were then treated with RNAase A, to hydrolyze nonannealed RNA, and the annealed RNA was eluted and counted. The first five gel fractions correspond to unprocessed precursor molecules that were pooled and hybridized to filter-bound DNA. Arrows indicate the location of molecular weight markers 18S rRNA, MW 670,000, 670K; 16S rRNA, 550,000, 550K. B. Radioactive RNA was prepared and the RNA was electrophoresed and hybridized as described in A, but the RNAase step was omitted. The 280K marker is BMV-4 RNA (MW 280,000). (© Cell, The MIT Press.)

stranded regions of the hybrid. A perfectly matched DNA-RNA hybrid (cDNA-mRNA) will be unaffected by the RNAase. When a hybrid between cDNA and an IVS-containing RNA is treated with RNAase, the annealed mRNA-cDNA segments will be protected from digestion, but the IVS will be hydrolyzed, because they do not hybridize to the cDNA. The RNAase-resistant RNA is then separated from the cDNA and electrophoresed in a gel. This RNA should resolve into two or more subfractions, depending on the number of IVS in the intermediate. If the RNA contained a single IVS, the RNAase protection experiment will yield two RNA pieces. If the RNA contained two IVS, three RNA pieces will be generated, and so forth.

When this experiment was performed with the purified 1,000-nucleotide intermediate, three RNA fragments were generated, designated I–III (Fig. 2A). They contained approximately 455, 225, and 130 nucleotides, respectively, and were in equimolar amounts. They were approximately the same size as the fragments generated in identical experiments with the β-globin mRNA precursor [22]. Furthermore, the total size of the RNAase-protected fragments was 810 nucleotides, which is similar to that of newly synthesized β-globin mRNA (700–800

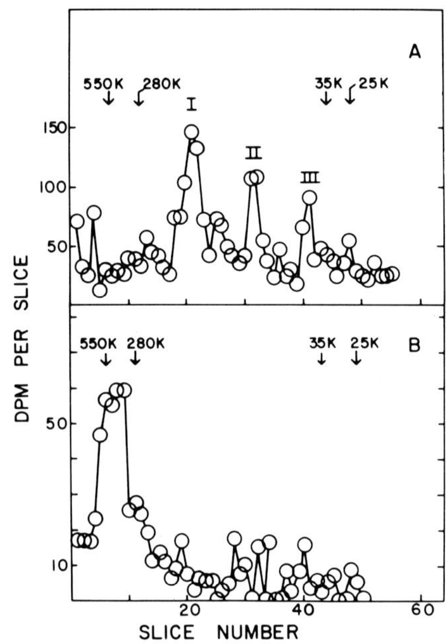

Fig. 2. Analysis of intervening sequences in the 1,000-nucleotide β-globin mRNA processing intermediate. The 1,000-nucleotide intermediate was prepared from the gel of Figure 1B. The RNA was hybridized in solution to unlabeled, full-length globin cDNA [22]. A portion of this DNA-RNA hybrid was treated with RNAase A; the remainder was untreated. Digested and undigested samples were extracted with phenol, denatured, and electrophoresed in denaturing gels. The gels were sliced, and the radioactivity in each slice was measured. Additional molecular weight markers include 5S rRNA, MW 35,000, 35K and 4S tRNA, MW 25,000, 25K. A. The 1,000-nucleotide intermediate annealed to cDNA and treated with RNAase. B. The 1,000 nucleotide intermediate annealed to cDNA and not treated with RNAase. (© Cell, The MIT Press.)

nucleotides). Controls indicated that the RNA itself was full-length (Fig. 2B). Therefore, the molecules generated by treating the hybrids with RNAase are mRNA-specific fragments and are not produced by random degradation.

This result indicates that the 1,000-nucleotide intermediate-sized RNA has the same mRNA fragment organization as the precursor—that is, it contains two intervening sequences but is 600–700 nucleotides smaller than the precursor. To account for both the smaller size and the two IVS in the intermediate, we propose that only a portion of the larger IVS is removed from the precursor molecule during the initial processing reaction(s). The remaining precursor fragments are joined together, creating a processing intermediate with two IVS. Additional partial splicing reactions are required to excise the IVS nucleotides that remain in the intermediate.

The organization of the 900-nucleotide RNA was analyzed in a similar manner. Five RNA fragments were protected by RNAase (Fig. 3A). RNAs I–III corresponded to the mRNA-specific segments observed with the precursor or 1,000-nucleotide intermediate RNAs. The 800 nucleotide molecule (V) corresponded to mature β-globin mRNA that contaminated the 900 nucleotide RNA preparation. Peak IV contained approximately 370 nucleotides. There were approximately two moles of peak I per mole of peaks II, III, and IV (Table I). These data indicate that

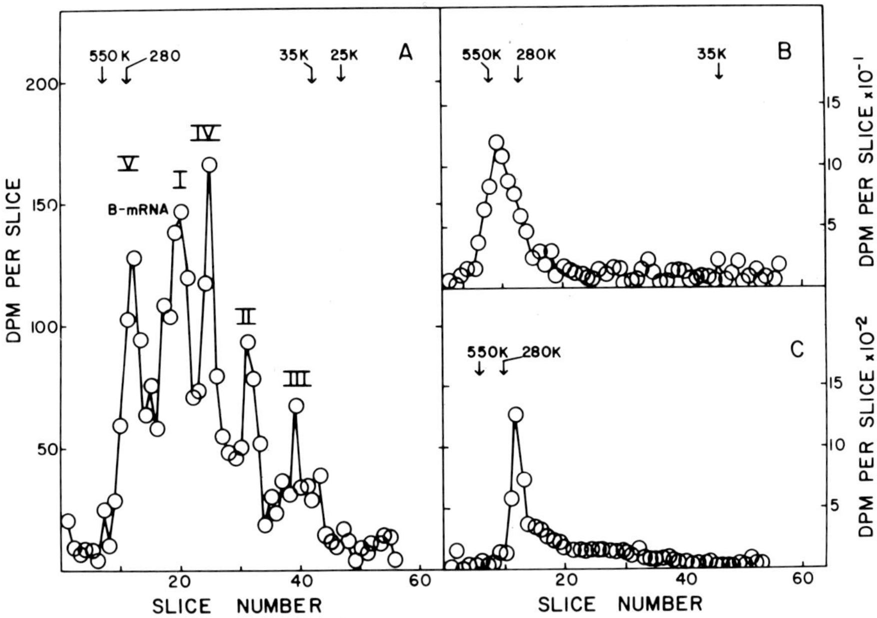

Fig. 3. Analysis of intervening sequences in the 900-nucleotide β-globin mRNA intermediate. Purified intermediate was prepared as described in Figure 1B, hybridized, treated with RNAase A, and electrophoresed as described in Figure 2. A. RNA hybridized to cDNA and treated with RNAase. B. RNA hybridized to cDNA but not treated with RNAase. C. ^3H β-globin mRNA hybridized to cDNA and treated with RNAase. (© Cell, The MIT Press.)

TABLE I. Quantitation of RNA Fragments Derived After Cleavage of Globin cDNA-900 Nucleotide RNA Hybrids With RNAase A[1]

Globin RNA fragment	Length (nucleotides)	Radioactivity per fragment (dpm)	Moles of fragment	900 Nucleotide molecules (%)[2]	
				I-II-III	I-IV
I	410	744	1.00	48	52
II	222	274	0.68		
III	144	132	0.47		
IV(II-III fusion)	366	437	0.64		

[1]The data are from the experiment shown in Figure 3.
[2]The percentage of 900 nucleotide intermediate RNA molecules with the II-III fusion, to form fragment IV (see Fig. 4), is derived as follows:

$$\frac{437}{274 + 132 + 437.}$$

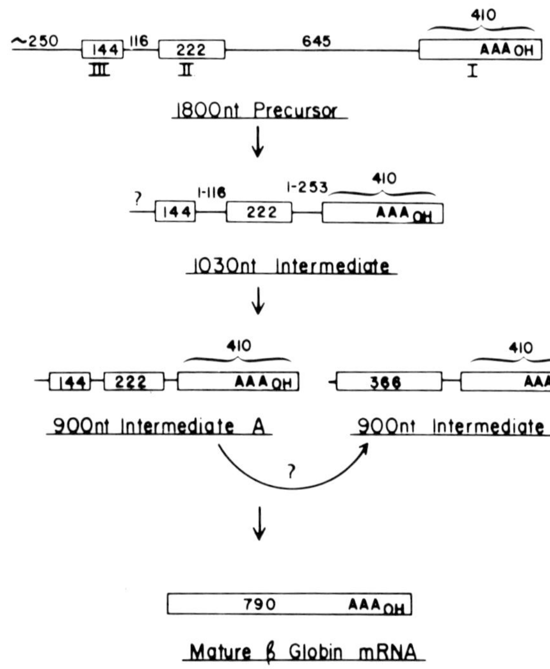

Fig. 4. Model for the processing of the mouse β-globin mRNA precursor. (© Cell, The MIT Press.)

the 900-nucleotide intermediate actually consists of two distinct RNAs, one with two IVS and three mRNA segments, the other with one IVS, one mRNA segment (I), and a fusion product of two mRNA segments (II + III = IV). Therefore, approximately half of the 900-nucleotide fraction contains the II–III fusion.

A theoretical model that illustrates how the precursor might be processed is shown in Figure 4. The first detectable step is a partial splicing reaction, in which a portion, but not all, of the larger IVS is excised, and the remainder of the molecule is ligated to generate the 1,000-nucleotide intermediate. A portion of the smaller IVS may also be removed at this step. Our data do not allow us to determine precisely how many nucleotides are removed from each IVS. One of the two 900-nucleotide intermediates (designated A in Fig. 4) contains two IVS and probably arises by additional partial splicing reactions involving the IVS of the 1,000-nucleotide intermediate. The other 900-nucleotide RNA (designated B in Fig. 4) contains a fused 5′ terminal-middle RNA segment, approximately 100 nucleotides of the larger IVS, and the 3′ terminal containing mRNA fragment. It is not clear if intermediate B is derived from intermediate A or if each is generated independently from the 1,000-nucleotide intermediate.

PROCESSING OF THE HUMAN β-GLOBIN mRNA PRECURSOR

These results on the mouse precursor indicated that nucleotides within IVS can provide signals for RNA processing. We reasoned that mutations involving these signal nucleotides would either block or retard the rate of processing of a precursor. (Experiments with SV40 or SV40-globin hybrids have indicated that intact mRNA-

IVS junctions are necessary for proper splicing of mRNA precursors [28, 29].) Furthermore, it seemed reasonable to propose that the etiology of β^+ thalassemia involves mutations in IVS that specifically affect the rate of processing of the β-globin mRNA precursor. Initial experiments with nucleated erythroid cells were performed to determine if human α and β-globin mRNA precursors could be identified and to measure the efficiency with which these RNAs are processed to mRNAs. Bone marrow aspirate cells from nonthalassemic individuals were pulse-labeled for 12 minutes with ^3H nucleosides. At this time actinomycin D was added, a portion of the cells (the pulse fraction) was harvested immediately, and the remainder (the chase fraction) was incubated for 30 minutes in the actinomycin D. RNA was prepared from the pulsed and chased cells and was analyzed by gel electrophoresis and hybridization as described for the mouse experiments except that human α and β-globin cDNA clones were used in the hybridization assays.

Discrete peaks of β-globin RNA corresponding to approximately 1,900, 1,600, 1,400, 1,200, 1,000, and 900 nucleotides, as well as mRNA-sized molecules (700 to 800 nucleotides) were observed in RNA from nonthalassemic individuals (Fig. 5C and data from two other nonthalassemic individuals not shown). In some of these gels, one or more of the intermediate-sized RNAs were not resolved as discrete peaks but migrated as shoulders of adjacent RNA peaks. Since β-globin RNAs larger than 1,900 nucleotides were not detected, the 1,900-nucleotide RNA is designated as the human β-globin mRNA precursor. After a 30-minute incubation with actinomycin D, 85% of the β-globin mRNA-specific radioactivity was in mature mRNA molecules (Fig. 5D and Table II). Few, if any, precursor or intermediate-sized molecules remained after the chase. Therefore, the pulse-labeled precursors and intermediates (Fig. 5C) are substrates in a series of efficient processing reactions which lead to the production of β-globin mRNA. Appropriate hybridization controls have been performed to ensure that the molecules we have designated as β are neither γ nor δ-globin RNAs (L.E.M. and J.R., unpublished observations). Peaks of pulse-labeled α-globin RNA were also observed (Fig. 5A). A presumed precursor (approximately 1,100 nucleotides) and two classes of intermediate-sized molecules (approximately 900–1,000 nucleotides) were processed efficiently to α-globin mRNA during the 30-minute chase period (Fig. 5B and Table II). We conclude that the α and β-globin mRNA precursors in nonthalassemic human erythroid cells are processed efficiently to mature mRNAs.

INEFFICIENT PROCESSING OF THE β-GLOBIN mRNA PRECURSOR IN CELLS OF PATIENTS WITH β^+ THALASSEMIA

In order to test the hypothesis that the rate of processing of β-globin mRNA is retarded in β^+ thalassemia, similar experiments were performed with bone marrow cells from several patients with homozygous disease. One experiment is shown in Figure 6. The sizes of the β-globin mRNA precursor and intermediate-sized molecules from pulse-labeled cells corresponded to those observed in the nonthalassemic individuals (compare Figs. 5C and 6D). The 1,900-nucleotide precursor was not detected after a 10-minute chase, indicating that it had turned over (Fig. 6E). However, 55% of the total β-globin mRNA-specific radioactivity was accounted for by intermediate-sized RNAs that had not been completely processed (Fig. 2E and Table II). Little, if any, of these intermediate-sized RNAs was processed to mRNA during the subsequent 20 minutes of the total 30-minute chase

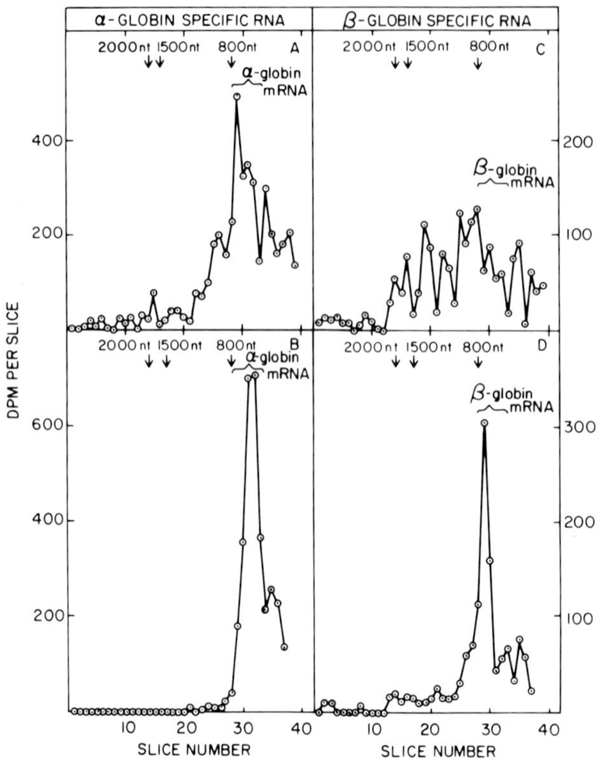

Fig. 5. Kinetic analysis of processing of α and β-globin RNAs from bone marrow cells of the nonthalassemic individual. Nucleated bone marrow cells from a patient with hemolytic anemia were cultured for one hour, at which time ^{3}H nucleosides were added [18]. After 12 minutes actinomycin D (10 μg/ml) was added, a portion of the cells was immediately harvested, and the remainder was incubated for 30 minutes with the actinomycin D. RNA was harvested and electrophoresed as described in Figure 1, and the eluted RNA was hybridized to cloned human α or β-globin cDNA immobilized on nitrocellulose filters. As a control, each hybridization reaction contained a PMB9 filter which provided a background (0–20 dpm) for nonspecific binding of RNA to the filters. A. Hybridization of α-globin cDNA to RNA from pulse-labeled cells. B. Hybridization of α-globin cDNA to RNA from cells pulse-labeled for 12 minutes and then chased for 30 minutes with actinomycin D. C. Hybridization of β-globin cDNA to RNA from cells pulse-labeled for 12 minutes. D. Hybridization of β-globin cDNA to RNA from cells pulse-labeled for 12 minutes and then chased for 30 minutes with actinomycin D.

(Fig. 6F). These results indicate that the β-globin mRNA processing intermediates in this patient are not converted efficiently to mRNA, in contrast to the situation with nonthalassemic individuals. Similar data have been obtained with four other patients with homozygous β^+ thalassemia (data not shown). It should be noted, however, that the nature of the processing defect seems to vary from patient to patient. That is, in some patients, such as the one shown above, the precursor and some of the larger intermediate-sized RNAs turn over during the 30-minute chase. In other patients a portion of the precursor and the larger intermediates do not turn over during the chase. In each case deficient β-globin mRNA production is the result of a defect in the RNA processing mechanism, but the defect is not the same in each patient.

TABLE II. Quantitation of Pulse-Chase Experiments With Nucleated Bone Marrow Cells From One Patient With Hemolytic Anemia and One Patient With β^+ Thalassemia[1]

	Hemolytic anemia		β^+ thalassemia	
	Pulse	Chase	Pulse	Chase
Radioactivity in α-globin mRNA sequences:				
Distribution (%)				
Precursor	5	0	20	0
Intermediates	15	0	25	5
mRNA-size	80	100	55	95
Total (dpm)	3,190	2,450	2,140	2,880
Radioactivity in β-globin mRNA sequences:				
Distribution (%)				
Precursor	5	0	5	0
Intermediate	55	15	45	55
mRNA-size	40	85	50	45
Total (dpm)	1,210	910	690	1,260
Total radioactivity ratio: α/β	2.6	2.7	3.1	2.3

[1]These data are derived from the experiments shown in Figures 5 and 6.

Since the lesion in this disease affects the quantity of β but not α-globin protein, it was important to determine if processing of the α-globin precursor and intermediates was also defective. Four α-globin RNAs were detected in pulse-labeled cells of the thalassemic patient and were similar in size to those in normal individuals (compare Figs. 5A and 6A). Approximately 95% of the α-globin mRNA specific radioactivity was processed to mRNA by the end of the 30-minute chase (Fig. 6C and Table II). These results indicate that the α-globin mRNA precursor and intermediates are processed to mRNA efficiently and without significant wastage in cells from patients with β^+ thalassemia.

We propose that the mutation(s) in at least some of the β^+ thalassemias occur in one or both IVS of the β-globin gene. Based on experiments with the mouse precursor, these mutations could occur within the IVS and/or at the mRNA-IVS junctions. Our data also account for the clinical and biochemical heterogeneity of the β^+ thalassemias [30], since we have observed qualitatively different processing defects in different patients. This result implies that the β^+ thalassemia phenotype can result from mutations at more than one site in the IVS and that mutations at different RNA splicing signal sites can affect processing in different ways.

RNA PROCESSING, INTERVENING SEQUENCES, AND GENE EXPRESSION

These and other experiments in a variety of cell systems demonstrate that RNA processing is a complex series of enzymatic and transport steps. The thalassemia data indicate that abnormalities in this pathway can dramatically alter the cell phenotype. Therefore, processing of RNA plays an essential role in gene expression. The critical unanswered question concerns whether any genes are regulated at

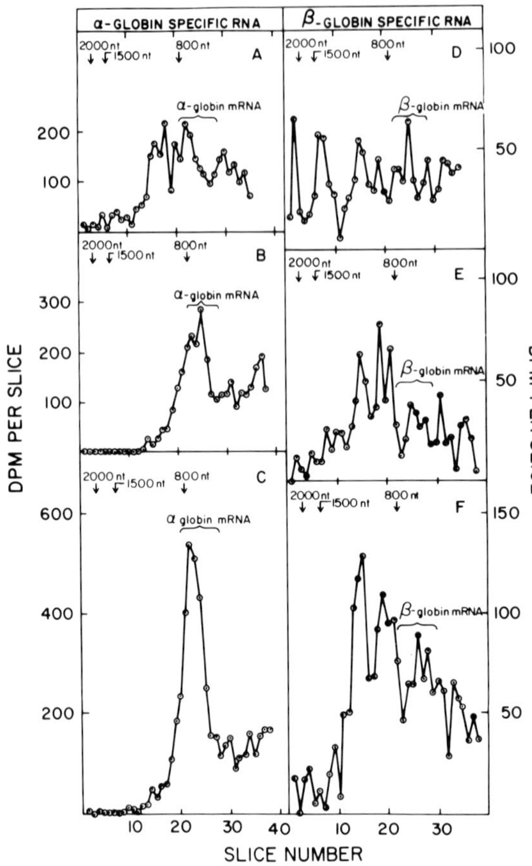

Fig. 6. Kinetic analysis of processing of α and β-globin RNAs from cells of a patient with homozygous β^+ thalassemia. Nucleated cells were analyzed as described in the legend to Figure 1, using hybridization probes noted in Figure 5. Hybridization of cloned α-globin cDNA filters to RNA from (A) cells pulse labeled for 12 minutes, (B) cells incubated for 10 minutes with actinomycin D, and (C) cells incubated for 30 minutes with actinomycin D. Hybridization of cloned β-globin cDNA filters to RNA from (D) cells pulse-labeled for 12 minutes, (E) cells incubated for 10 minutes with actinomycin D, (F) cells incubated for 30 minutes with actinomycin D.

the RNA processing level. As noted above, most of the evidence pertaining to this question has come from viral systems or from yeast mitochondria. It is not clear at this point if nuclear genes that specify proteins can be regulated in a similar fashion. It will be interesting to determine, for example, if hemoglobin switching is regulated in some way by RNA processing events.

It is also not clear why the IVS are excised in stepwise fashion (partial splicing) rather than in a single step. Perhaps partial splicing is the result of the evolution of IVS. If the size of the IVS has expanded during evolution, as a result of the sequential addition of blocks of nucleotides, then the enzyme system responsible for IVS excision might have evolved as well. Thus, a simple IVS in an evolutionarily primitive β-globin gene might have been excised in one step. When the IVS enlarged, this enzyme no longer recognized the whole IVS but was able to excise portions of it in multiple steps. Alternatively, a second splicing enzyme might have evolved to excise the newly added block of nucleotides.

While most theories concerning the function of IVS in DNA and RNA concern recombination, evolution, or splicing phenomena [31], we suggest that the IVS of at least some genes might function as regulatory signals. For example, the inappropriate transcription of certain genes, such as those coding for hormones, could be lethal to the organism. Therefore, it is critical that the transcription of these genes be repressed in most cells. One mechanism for insuring that genes are fully repressed is to have a backup shutoff system, analogous in some ways to attenuators in the amino acid biosynthetic systems of some prokaryotes [32]. The IVS could provide such a backup system, if they contained transcriptional terminators or signals for the destruction of the RNA precursors. We have looked for and not found significant levels of globin mRNA in lymphocytes (J.R., unpublished data), although other laboratories have reported small quantities of globin mRNA in nonerythroid cells [33].

The thalassemia experiments indicate, but do not prove, that thalassemia, like sickle cell disease, results from mutations in the structural globin gene. In the hemoglobinopathies the mutations occur in the globin-coding or mRNA region of the gene; in thalassemia we hypothesize that the mutations occur in IVS. These results have encouraged us to study the mechanism of RNA processing with in vitro systems, in order to characterize the enzymatic machinery and the processing signal sites in the RNA precursor. One goal of these experiments might be to design therapeutic agents that would affect the rate of processing of an abnormal precursor RNA.

SUMMARY

The processing of β-globin mRNA precursor molecules in nucleated mouse and human erythroid cells was investigated. To determine the mechanism by which the two intervening sequences in the mouse precursor RNA are excised and the globin mRNA segments are joined, processing intermediates approximately 600–700 nucleotides smaller than the precursor were purified and analyzed by RNAase protection experiments. Two of these intermediates each contain two small intervening sequences. This result indicates that at least one of the intervening sequences is excised from the precursor in two or more steps (partial splicing). At each step only a portion of the intervening sequence is excised and discarded, and the remaining segments are then ligated, creating a processing intermediate with a partially spliced intervening sequence. Pulse-chase kinetic analyses of human bone marrow cells revealed that the human α and β-globin genes are transcribed into precursor RNAs that probably undergo complex processing reactions, similar to those in the mouse. These reactions occur efficiently in cells of nonthalassemic individuals. In cells from patients with β^+ thalassemia, 85%–95% of the α-globin precursor RNA is processed to α-globin mRNA; in contrast, the β-globin precursor is processed inefficiently. Fifty percent or less of the thalassemic β-globin mRNA precursor is processed to mRNA during a 30-minute chase with actinomycin D. These data indicate that the phenotype of β^+ thalassemia results from inefficient processing of the β-globin mRNA precursor to mRNA. β^+ thalassemia is thus the first documented example of an RNA processing disease.

ACKNOWLEDGMENTS

We thank Lorri Steffen and Deborah Hass for expert technical assistance. We are grateful to the following individuals for their support. Drs. B. Forget and S. Weissman for human globin cDNA clones, Drs. H. Faber, O. Smithies, and F. Blattner for mouse globin cDNA clones, Dr. T. Maniatis for human globin genomic clones, Drs. E. Prendergast and M. Borzy for performing some of the bone marrow aspirations, and Dr. J. Pellett for providing surgical bone chip specimens.

This work was supported by grants CA-07175 and CA-23076 from the N.I.H. L.E.M., A.J.K., L.R.B., and G.R.H. were supported by grants CA-09230, CA-23076, CA-91935, and AM-19046, respectively, from the N.I.H.

REFERENCES

1. Rhoads RE, McKnight GS, Schimke RT: Quantitative measurement of ovalbumin messenger ribonucleic acid activity. J Biol Chem 248:2031–2039, 1973.
2. Ross J, Ikawa Y, Leder P: Globin messenger RNA induction during erythroid differentiation of cultured luekemia cells. Proc Natl Acad Sci USA 69:3620–3623, 1972.
3. Bruns GP, London IM: The effect of hemin on the synthesis of globin. Biochem Biophys Res Commun 18:236–242, 1965.
4. Gross KW, Jacobs-Lorena M, Baglioni C, Gross PR: Cell-free translation of maternal messenger RNA from sea urchin eggs. Proc Natl Acad Sci USA 70:2614–2618, 1973.
5. Nevins JR, Darnell JE: Steps in the processing of Ad2 mRNA:poly(A)$^+$ nuclear sequences are conserved and poly(A) addition precedes splicing. Cell 15:1477–1493, 1978.
6. Aloni Y: Extensive symmetrical transcription of simian virus 40 DNA in virus-yielding cells. Proc Natl Acad Sci USA 69:2404–2409, 1972.
7. Chiu NH, Radonovich MF, Thoren MM, Salzman NP: Selective degradation of newly synthesized nonmessenger simian virus 40 transcripts. J Virol 28:590–599, 1978.
8. Denis H, Wegnez M: Biochemical research on oogenesis:oocytes of Xenopus laevis synthesize but do not accumulate 5S RNA of somatic type. Dev Biol 58:212–217, 1977.
9. Jacq B, Jourdan R, Jordan BR: Structure and processing of precursor 5S RNA in Drosophila melanogaster. J Mol Biol 117:785–795, 1977.
10. Johnson LF, Levis R, Abelson HT, Green H, Penman S: Changes of RNA in relation to growth of the fibroblast. J Cell Biol 71:933–938, 1976.
11. Beltz GA, Flint SJ: Inhibition of HeLa cell protein synthesis during adenovirus infection: Restriction of cellular messenger RNA sequences to the nucleus. J Mol Biol 131:353–373, 1979.
12. Church GM, Slonimski PP, Gilbert W: Pleiotropic mutations within two yeast mitochondrial cytochrome genes block mRNA processing. Cell 18:1209–1215, 1979.
13. Halbreich A, Pajot P, Foucher M, Grandchamp C, Slonimski P: A pathway of cytochrome b mRNA processing in yeast mitochondria: Specific splicing steps and an intron-derived circular RNA. Cell 19:321–329.
14. Smith K, Lingrel JB: Sequence organization of the β-globin mRNA precursor. Nucleic Acids Res 5:3295–3301, 1978.
15. Bastos RN, Aviv H: Globin RNA precursor molecules: Biosynthesis and processing in erythroid cells. Cell 11:641–650.
16. Curtis PJ, Weissman C: Purification of globin messenger RNA from dimethyl-sulfoxide-induced Friend cells and detection of a putative globin messenger RNA precursor. J Mol Biol 105: 1061–1075, 1976.
17. Kinniburgh AJ, Ross J: Processing of the mouse β-globin mRNA precursor: At least two cleavage-ligation reactions are necessary to excise the larger intervening sequence. Cell 17:915–921, 1979.
18. Maquat LE, Kinniburgh AJ, Beach LR, Honig GR, Lazerson J, Ershler WB, Ross J: Processing of human β-globin mRNA precursor to mRNA is defective in three patients with β^+-thalassemia. Proc Natl Acad Sci USA 77:4287–4291, 1980.
19. Kwan S-P, Wood TG, Lingrel JB: Purification of a putative precursor of globin messenger RNA from mouse nucleated erythroid cells. Proc Natl Acad Sci USA 74:178–182, 1977.

20. Ross J, Knecht DA: Precursors of alpha and beta globin messenger RNAs. J Mol Biol 119:1–20, 1978.
21. Tilghman SM, Curtis PJ, Tiemeier DC, Leder P, Weissman C: The intervening sequence of a mouse β-globin gene is transcribed within the 15S β-globin mRNA precursor. Proc Natl Acad Sci USA 75:1309–1313, 1978.
22. Kinniburgh AJ, Mertz JE, Ross J: The precursor of mouse β-globin messenger RNA contains two intervening RNA sequences. Cell 14:681–693, 1978.
23. Klessig DF: Two adenovirus mRNAs have a common 5′ terminal leader sequence encoded at least 10kb upstream from their main coding regions. Cell 12:9–21, 1977.
24. Chow LT, Gelinas RE, Broker TR, Roberts RJ: An amazing sequence arrangement at the 5′ ends of adenovirus 2 messenger RNA. Cell 12:1–8, 1977.
25. Berget SM, Moore C, Sharp PA: Spliced segments at the 5′ terminus of adenovirus 2 mRNA. Proc Natl Acad Sci USA 74:3171–3175, 1977.
26. Ross J: A precursor of globin messenger RNA. J Mol Biol 106:403–420, 1976.
27. Cole RJ, Paul J: The effects of erythropoietin on haem synthesis in mouse yolk sac and cultured foetal liver cells. J Embryol Exp Morphol 15:245–260, 1966.
28. Lai C-J, Khoury G: Deletion mutants of simian virus 40 defective in biosynthesis of late viral mRNA. Proc Natl Acad Sci USA 76:71–75, 1979.
29. Hamer DH, Leder P: SV40 recombinants carrying a functional RNA splice junction and polyadenylation site from the chromosomal mouse β^{maj} globin gene. Cell 17:737–747, 1979.
30. Bank A, Ramirez F: The molecular biology of the thalassemia syndromes. CRC Crit Rev Biochem 5:343–367, 1978.
31. Crick F: Split genes and RNA splicing. Science 204:264–271, 1979.
32. Zurawski G, Elseviers D, Stauffer GV, Yanofsky C: Translational control of transcription termination at the attenuator of the Escherichia coli tryptophan operon. Proc Natl Acad Sci USA 75: 5988–5992, 1978.
33. Humphries S, Windass J, Williamson R: Mouse globin gene expression in erythroid and non-erythroid tissues. Cell 7:267–277, 1976.

Organization and Expression of Globin Genes, pages 219–230

Processing of γ, δ, and β-Globin mRNA Precursors in Human Fetal, Normal Adult, and β^+ Thalassemia Erythroid Cells

Judith A. Kantor, Russel E. Kaufman, Patricia H. Turner, Margaret Farquhar, George Stamatoyannopoulos, and Arthur W. Nienhuis

Clinical Hematology Branch, National Heart, Lung, and Blood Institute, Bethesda, Maryland (J.A.K., P.H.T., A.W.N., R.E.K.) and University of Washington, Seattle (M.F., G.S.)

INTRODUCTION

The switch in hemoglobin phenotype from Hb F ($\alpha_2\gamma_2$) to Hb A ($\alpha_2\beta_2$) during the perinatal period reflects a change in the predominant non-α globin mRNA species that accumulates in erythroblasts [1]. Therefore the selective expression of the two fetal globin genes followed by the adult δ and β-globin genes occurs because of regulation exerted at the level of transcription or RNA processing rather than at the level of mRNA translation.

The basic structure of the globin genes consists of three exons, with two introns at positions corresponding to amino acids 29–31 and 104–105 [2–5]. In the case of the mouse β-globin gene the initial product of transcription appears to be a high molecular weight RNA that represents a transcript of the three exons and the two introns [6–8]. Subsequent processing of the RNA includes the stepwise removal of the intervening sequences by a mechanism which includes nucleolytic cleavage of the RNA molecule followed by ligation of the exons to form the mature RNA. The demonstration that intervening sequence RNA is removed piecemeal by a series of excision and religation steps was first reported by Ross and co-workers for the mouse β-globin mRNA precursor [9, also see paper in this volume]. It is known that the primary transcript of the rabbit β-globin is also processed in a stepwise fashion [see Flavell, this volume].

The complex biochemical mechanisms related to the processing of globin mRNA precursor to mature mRNA provides a potential level for the control and regulation of these genes. To explore this hypothesis, we have compared the concentration of the precursors to γ and β mRNA to those of the mature mRNA species in human fetal and adult erythroid cells and have also measured the concentration of δ-globin mRNA precursor in adult cells. In an effort to define the molecular lesion in β^+ thalassemia, the β-globin mRNA precursor concentration was also measured in bone marrow cells from patients with β^+ thalassemia.

Russel E. Kaufman's present address is Duke University, Durham, North Carolina.

EXPERIMENTAL PROCEDURES

The human β-like globin gene cluster includes the embryonic (ϵ), fetal ($^{G}\gamma$,$^{A}\gamma$) and adult (δ,β) genes. These have been cloned and characterized [5]. Each gene contains a 850–900 base pair (bp) fragment defined by the restriction endonucleases Bam HI and Eco RI. This fragment contains the entire large intron of the gene as well as 12 bp of coding sequence at the 5′ end and 50 bp of coding sequence at the 3′ end (Fig. 1). Prior studies have shown that the sequences of the large introns of these genes have diverged considerably and therefore do not cross react with one another [2, 12]. Therefore, the intron from each gene should anneal selectively to the precursor mRNA transcribed from that particular gene.

To measure the concentration of the various globin mRNA precursors, we have devised a hybridization assay which allows the use of a double-stranded probe (Fig. 1). The large intron of the several globin genes have been subcloned [13] using the plasmid vector, pBR322 [14]. These recombinant plasmids provide a source of pure intron fragments which are made radioactive by a standard nick-translation reaction [15]. Each probe was incubated in the presence of total cellular RNA for various periods of time and the rate of duplex formation was quantitated by use of the single-strand specific nuclease S_1. In these hybridization reactions, two types of duplexes may be formed. The first are RNA:DNA duplexes involving one strand of the probe and the complementary RNA sequences in the globin mRNA precursor. The second are DNA:DNA duplexes formed by the reannealing of the complementary sequences of the double-stranded probe. Two approaches were used to

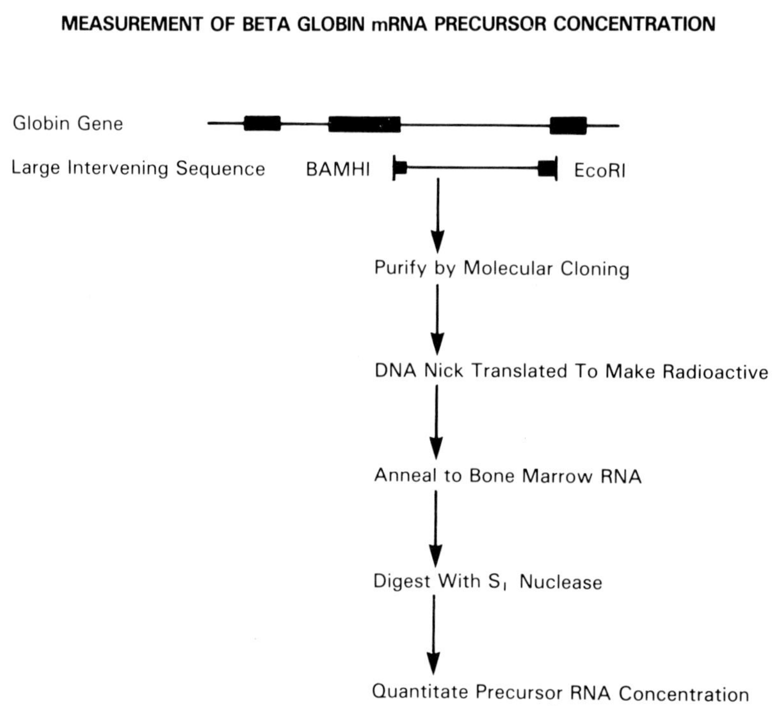

Fig. 1. Schematic representation of the hybridization assay for the measurement of β, δ, and γ-globin mRNA precursor concentrations.

estimate the relative amounts of each type of duplex formed in the hybridization reaction. An equivalent amount of hybridization probe was incubated in the presence of E coli RNA and the rate of duplex formation was compared to that in the presence of bone marrow mRNA. The second assay involved the use of RNase A in low salt to destroy RNA including that present as RNA:DNA duplex molecules. Two identical series of reactions were incubated for variable periods of time; one was treated with RNase and the fraction of single-stranded or double-stranded probe (DNA:DNA duplex) was determined by batch chromatography on hydroxylapatite. The second series of reactions was analyzed on hydroxylapatite without RNase A treatment so that both RNA:DNA and DNA:DNA duplexes were scored, giving a measure of total duplex formation. The difference between total and DNA:DNA duplex at each time point gave a measure of RNA:DNA duplex formation.

A quantitative estimate of the amount of reactive RNA species was obtained by comparing the rate of DNA:DNA duplex formation in the control reaction to that of RNA:DNA duplex formation in the reaction with bone marrow cell RNA. Various mathematical equations were used to make duplex formation linear with respect to time, thereby facilitating the comparison of these rates (see figures). One assumption implicit in these comparisons is that under identical reaction conditions, the rate constants for DNA:DNA and RNA:DNA duplex formation are the same. This may not be strictly true. However the data derived by this method provided an estimate of the relative concentrations of specific RNA molecules, even though absolute quantitation may not be possible.

RESULTS

Nature of Globin mRNA Precursors in Human Erythroid Cells

Electrophoresis of nuclear or polyadenylated RNA in agarose gels under denaturing conditions may be used to define the molecular size of precursor RNA molecules. Following resolution of the RNA in the gel, it is transferred to diazotized aminobenzyloxymethyl (ABM) cellulose paper by the Northern blotting technique and annealed to a radioactive probe chosen to detect particular RNA molecules [16]. For example, use of the large intron from a globin gene as a probe displays precursor RNA molecules containing complementary sequences whereas use of a probe specific for the coding sequences for a globin will demonstrate mature mRNA and all precursor molecules present in sufficient concentration to allow detection.

One such experiment is shown in Figure 2. Nuclear RNA from bone marrow cells of an individual with sickle cell anemia was resolved electrophoretically and hybridized to either the large intron (Lane B) or coding sequence probe (Lanes D and E). The intron probe annealed to an RNA species of approximately 1,800 nucleotides (17S) and also to an RNA species which migrated at the position of mature mRNA. We believe that this latter signal is derived from the annealing of the intron-associated coding sequences (Fig. 1) to the considerable amount of mature mRNA in the nuclear sample since a signal of similar intensity was observed with an equivalent amount of purified reticulocyte globin mRNA (Lane A). Shown in Lane D are the RNA species which anneal to the globin coding sequence probe; only the 17S precursor and 10S mature mRNA are visualized, although the intensity of the

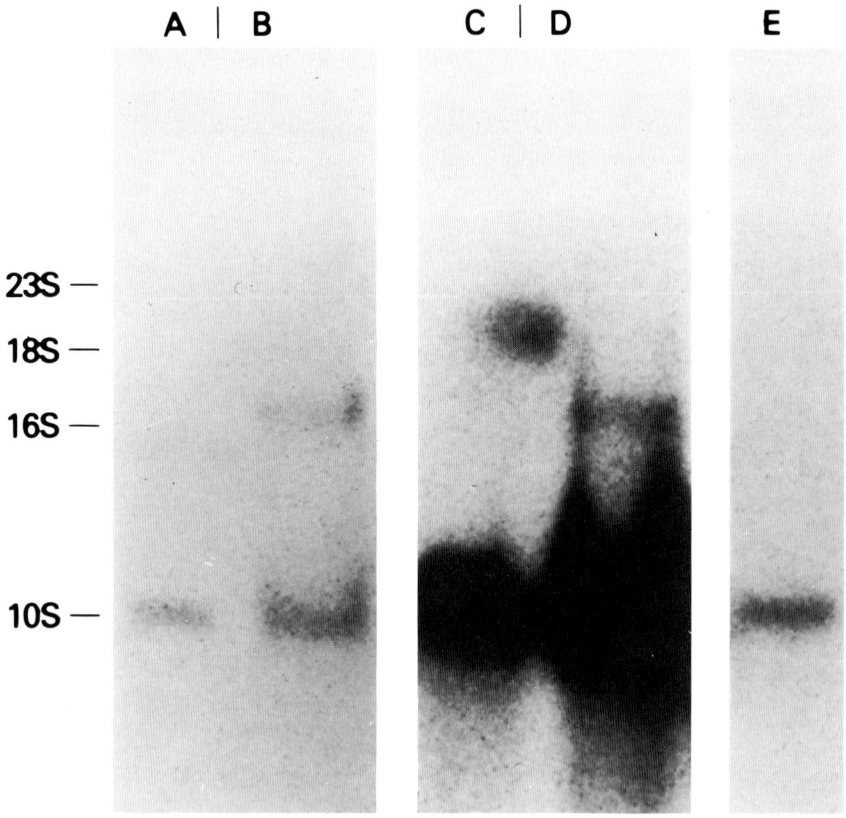

Fig. 2. Demonstration of a 17S β-globin mRNA precursor in human bone marrow cell nuclei. RNA was fractionated by methylmercury hydroxide gel electrophoresis and transferred to ABM-cellulose paper by the method of Alwine et al [16]. A. Ten nanograms of globin mRNA annealed to the β intron probe. B. Three micrograms of nuclear RNA annealed to the β intron probe. C. Ten nanograms of globin mRNA annealed to the β-globin coding sequence probe (Hha I fragment of JW102 [19]). D, E. Three micrograms of nuclear RNA annealed to the β-globin coding sequence probe. Radioautographic exposure for lanes A-D was 24 hours, while lane E was exposed for 6 hours. RNA markers were run simultaneously and their positions are indicated in the margins on the left.

signal caused by the relatively high concentration of mature mRNA would obscure any RNA species which migrated close to 10S but which was present in relatively low concentration. The shorter exposure of this same track is shown in Lane E in this figure.

Thus human bone marrow cells contain an 1,800-nucleotide precursor which anneals to both a β-globin coding sequence probe and the large intron probe [17]. The weak radioautographic signal compared to that of mature RNA suggests that it is present in a low concentration. This 1,800-nucleotide precursor is of sufficient length to contain an entire transcript of the gene beginning at the CAP site which corresponds to the 5′ end of the mRNA and including a transcript of the two introns and also 150–200 nucleotides of poly A at the 3′ end.

Smaller RNA species intermediate in size between this precursor and mature mRNA were not demonstrated although such molecules might be present in a concentration so low that they would not be visualized in these blotting experiments, eg, less than ten to 20 molecules cell. By using 30 μg of nuclear RNA rather than 3 μg used in each lane of the experiments shown in Figure 2, we have been able to demonstrate a 1,700–1,800 nucleotide RNA species which hybridized to the large intron probe derived from the δ-globin gene. Others have performed analogous experiments to define a comparably sized γ mRNA precursor molecule in human fetal erythroid cells [12].

In summary, current data suggests that the primary transcription products of the γ, δ, and β-globin genes are 1,700–1,800-nucleotide precursor molecules. These are processed to mature γ, δ, or β-globin mRNA by the removal of intervening sequence RNA without accumulation of a significant concentration of intermediate-sized RNA species.

The Concentrations of γ and β-Globin mRNA Precursor in Fetal and Adult Erythroid Cells

Total cellular RNA was recovered from hepatic erythroblasts of three fetuses of variable gestational age. Aliquots of these RNA samples were annealed to the γ and β intron probes and the rate of duplex formation was determined by analysis of individual aliquots of the reaction mixture which had been incubated for variable periods of time, using single-strand-specific nuclease. These results were compared to data obtained using RNA extracted from adult bone marrow cells of patients with sickle cell anemia. Figure 3 shows the results obtained in a direct comparison of RNA from fetal and adult erythroblasts while Table I contains all the data obtained in these series of analysis. In general, the γ mRNA precursor is present in a considerably lower concentration in adult erythroid cells than it is in fetal erythroblasts while the β-globin mRNA precursor is present in a very low concentration in fetal erythroblasts compared to its concentration in adult bone marrow cells.

By using single-stranded cDNAs complementary to the individual globin mRNAs, prepared synthetically from globin mRNA [1] or from recombinant plasmids which contain cloned cDNA sequences [19], the concentration of mature γ and β mRNA have been determined. These results are also given in Table I. The concentrations of precursor and mature mRNA species are best compared as the ratio of the number of molecules of each per cell. The ratio for γ mRNA sequences is approximately 1:100 in fetal erythroid cells and 1:15 in adult erythroid cells. Similarly, the ratio of β-globin mRNA precursor to mature mRNA is 1:500 in adult bone marrow cells and approximately 1:19–1:70 in fetal erythroblasts. These results might be interpreted to suggest that there is a change in the efficiency of processing of the β and γ-globin mRNA precursors during development and that this change contributes to the switch from predominantly γ to predominantly β-globin mRNA accumulation in fetal and adult erythroblasts, respectively. However, this interpretation must be regarded as highly tentative given our reservations, outlined above, about the precise quantitation of precursor, particularly when it is present in low concentration.

β-Globin mRNA Precursor Concentration in Erythroblasts From Patients With β^+ Thalassemia

In β^+ thalassemia there is a quantitative decrease in the amount of β-globin mRNA causing a deficiency in β-globin production [20, 21]. We have attempted to define the mechanism by which mutations affect β-globin gene expression leading

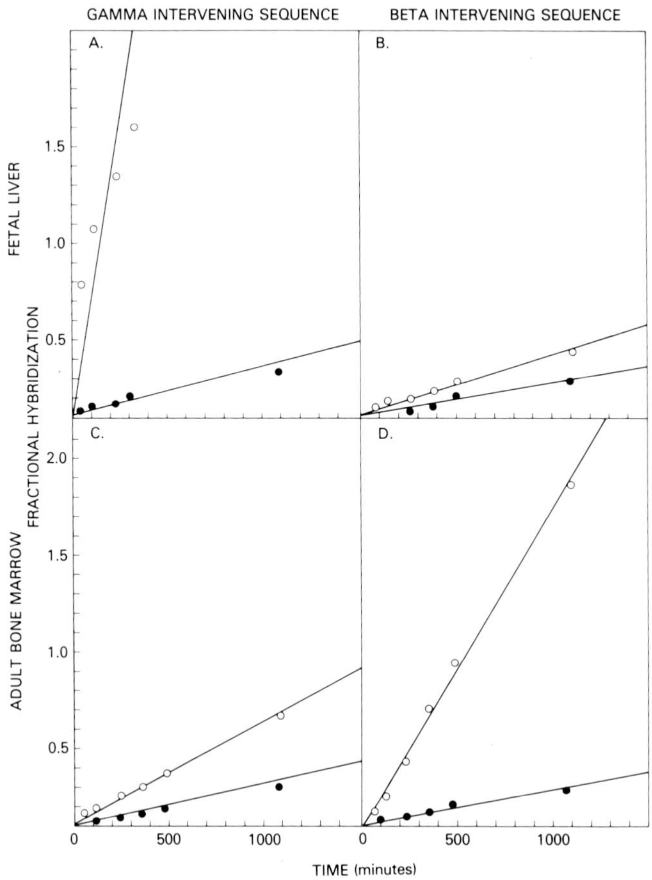

Fig. 3. Quantitation of γ and β-globin mRNA precursor concentration in human fetal liver and adult bone marrow cells. Each experimental reaction (○) contained 180 μg of total cellular RNA and 240 pg of [^{32}P] intron probe (specific activity = 6×10^7 cpm/μg) in a total volume of 100 μl divided among ten capillaries. The control reaction (●) contained 180 μg of Escherichia coli RNA. Background S_1-resistant probe present at zero time represented 5%–10% of the total and was subtracted from each data point. The control value at each time point was subtracted from the experimental value; the difference was doubled and then added to the control value to compensate for the fact that only one strand in RNA:DNA duplexes is labeled while in the control reaction both strands in DNA duplexes are labeled. The experimental data is plotted as ln 1/1-H while the control data is plotted as H/1-H where H is the fractional hybridization. Precursor concentration is estimated from the slope of the control versus experimental data.

to this disease state. Total cellular RNA from the bone marrow of four patients, all of which are homozygous for β^+ thalassemia, was hybridized to the β-globin gene intron probe. The concentration of β-globin mRNA precursor was compared to that found in control cells of patients with sickle cell anemia. Shown in Figure 4 is the analysis of RNA from two such patients compared to a control. In general, we found that the β-globin mRNA precursor concentration was two to threefold higher in the cells of patients with β^+ thalassemia compared to the controls. The average for the four patients studied was 22.5 ± 4.0 pg/μg and for the four controls it was 10.3 ± 1.0 pg/μg, corresponding to approximately 220 and 100 precursor molecules per cell,

TABLE I. Summary of γ- and β-Globin Precursor and mRNA Concentrations in Erythroid Cells

	DNA Probe				Ratio	DNA Probe				Ratio
	Gamma intron		Gamma coding			β intron		β coding		
Fetal liver	pg/μg of RNA	Molecules cell	pg/μg of RNA	Molecules cell	Coding precursor	pg/μg of RNA	Molecules cell	pg/μg of RNA	Molecules cell	Coding precursor
112 Day	13.0	130	854	12,727	97.9	2.4	24	54	801	33.4
128 Day	11.4	114	1474	21,859	191.7	2.8	28	131.5	1950	69.6
168 Day	16.9	169	258	3,826	22.6	2.5	25	31.6	469	18.8
Adult bone marrow										
Control (SCA) 10% Hb F	3.0	30	30.8	457	15.2	9.9	99	3400	50422	509.0
β^+ thalassemia	2.4	24	29.1	432	18.0	28.9	289	549	8142	28.2
β^+ thalassemia	1.6	16	22.3	331	20.7	18.9	189	472	6999	37.0

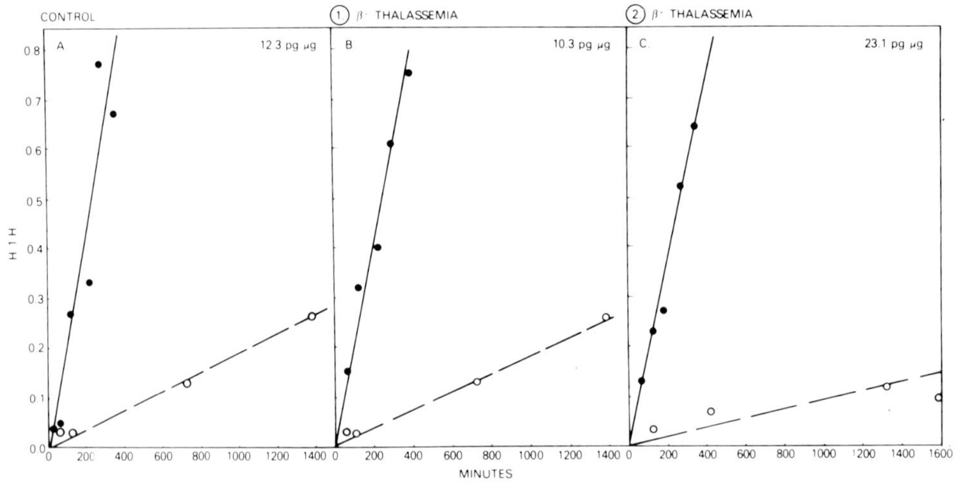

Fig. 4. Quantitation of the β-globin mRNA precursor concentration in total cellular RNA from bone marrow cells of a control patient and two patients with β^+ thalassemia. The rate of annealing of the double-stranded intron probe in a control reaction containing 100 μg of E coli RNA (○) in the presence of 80 μg of RNA from a patient with sickle cell anemia (● Panel A) or in the presence of 90 μg of RNA from two patients with β^+ thalassemia (● panels B and C). The data was analyzed as outlined in the legend to Figure 3.

respectively [17]. These data suggested that transcription of the β-globin gene was normal in erythroblasts of these patients with β^+ thalassemia but that either inefficient processing of the β-globin mRNA precursor or instability of mature β-globin mRNA accounted for the deficiency in β-globin mRNA accumulation.

Direct measurement of RNA synthesis and stability of two patients revealed that they had mutations that affected the processing of their β-globin mRNA [17]. Bone marrow cells were pulse-labeled in [^{3}H]nucleosides for two hours and then analyzed immediately or chased in cold nucleosides for an additional 20 hours before RNA extraction. Newly synthesized α or β-globin mRNA was quantitated by a series of hybridization reactions [17] which utilize DNA fragments isolated from recombinant plasmids which contain either the α or β-globin mRNA coding sequences [22]. This pulse-chase data showed that the ratio of newly synthesized β to α mRNA sequences was higher in nuclear RNA from β^+ thalassemic cells at two hours than in nuclear RNA from control cells but that the β to α ratio in newly synthesized sequences which reached the cytoplasm was considerably lower than in control cells. The β-globin mRNA sequences which reached the cytoplasm of the β thalassemic cells appeared to be stable during the 20-hour chase.

Northern blot analysis of these patients' RNA in analogous experiments to those shown in Figure 2 allowed us to demonstrate abnormal RNA species in the cells of these two patients with β^+ thalassemia. The bone marrow cells of a Greek patient contained a 1,300-nucleotide RNA species at a concentration of approximately 100 molecules/cell. This RNA annealed to both the β-globin gene intron and coding sequence probes. We hypothesized that it represents an RNA species from which part but not all of the RNA transcribed from the large intron has been removed. Presumably a mutation has affected the efficiency which a particular excision splicing site can be used. In contrast, the cells of an American black patient contained an RNA species complementary to some portion of both the large intron and coding sequence probes; it comigrated with mature mRNA and thus was approximately 600 nucleotides long. We hypothesize that this RNA species is derived from a processing step in which the splice is not made efficiently.

The Relative Concentration of δ and β-Globin mRNA Precursors in Adult Erythroid Cells

Hb A_2 ($\alpha_2\delta_2$) is a minor hemoglobin accounting for approximately 3% of the total found in adult red cells. Indirect evidence suggests that the concentration of globin mRNA is very low in human erythroblasts [11]. Even though δ-globin is made in bone marrow cells, no δ-globin synthesis occurs in reticulocytes, suggesting that δ-globin mRNA may be relatively unstable [11]. To obtain further insight into the basis for the large discrepancy in the amounts of δ and β-globin mRNA, we measured the concentration of δ and β-globin mRNA precursors in human bone marrow erythroblasts (Fig. 5). These hybridization reactions were analyzed as outlined above to individually quantitate the total (RNA:DNA and DNA:DNA) and DNA:DNA duplex formation.

The bone marrow cells of a patient with sickle cell anemia contained considerably higher concentrations of β than of δ-globin mRNA precursor. An estimate from the hybridization data suggests that there are approximately 100 β and only eight to 15 δ-globin mRNA precursor molecules/cell. Thus the ratio of δ to β-globin mRNA precursor is roughly 1:10 while the concentration of δ-globin to β-globin in red cells is 1:40. Again some reservation must be applied to the precise quantitative interpretation of the precursor concentration measurements. The concentration of δ-globin mRNA precursor was markedly lower than that of the β-globin mRNA precursor in RNA from cells of a patient homozygous for β^+ thalassemia but slightly higher than that found in normal bone marrow cells, in keeping with modest elevation of Hb A_2 in this patient's red cells.

DISCUSSION

We have found that the concentration of γ, δ, and β-globin mRNA precursors in fetal and adult erythroblasts is roughly proportional to the concentration of the mature mRNA species derived from these precursors and therefore these concentrations reflect the pattern of globin synthesis and accumulation within human red cells. Only in the cells of patients with β^+ thalassemia was there a disproportionally high concentration of precursor compared to the mature mRNA, but these patients are known to have mutations which affect mRNA production [17, 18]. In normal cells, then, the steady-state concentration of globin mRNA precursors seems predictive of the predominant globin mRNA species. The simplest explanation for these findings is that primary regulation of expression of the individual globin genes is based on modulation of the rate of their transcription. Thus one might suppose that in fetal erythroid cells the γ gene is more frequently (and/or rapidly) transcribed than in adult erythroid cells, leading to marked accumulation of γ mRNA in fetal cells and neglible amounts in adult cells. Conversely, the β-globin gene may be more frequently (and/or rapidly) transcribed in adult compared to fetal erythroblasts.

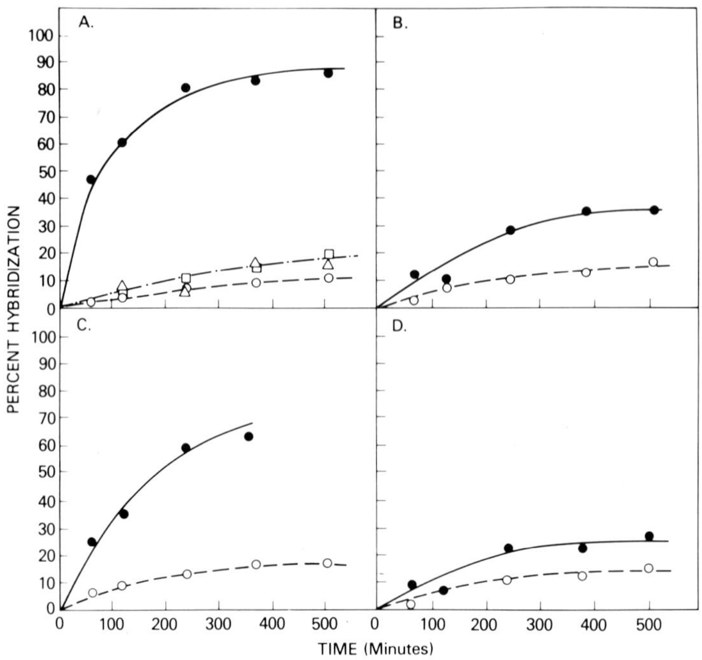

Fig. 5. Annealing of bone marrow total cellular RNA from a patient with homozygous β^+ thalassemia and from a patient with sickle cell anemia to the β or δ intron probes. One hundred twenty-five μg of total cellular RNA from a β^+ thalassemia patient (A and B) or from a control patient with sickle cell anemia (C and D) was annealed to 300 ng of [^{32}P] nick-translated β intron probe (panel A or C respectively) or δ intron probe (panel B or D respectively) in a total reaction volume of 160 μl. Individual reactions containing 10 μl were incubated for the times specified. Duplicates were removed and the total duplex (●) (RNA [^{32}P] DNA-[^{32}P]-DNA) and [^{32}P]DNA-[^{32}P]DNA duplex (○) were scored as described by Kantor et al [17]. A control experiment containing 125 μg of E coli RNA instead of human bone marrow was analyzed in an identical fashion; total duplex (△) or DNA-DNA duplex (□).

An alternative explanation is that the rate and efficiency of processing of specific globin mRNA precursors may vary in erythroid cells during different developmental periods. For example, the γ mRNA precursor might be efficiently and rapidly processed in fetal erythroid cells, leading to accumulation of large amounts of γ-globin mRNA, whereas in adult bone marrow cells processing might be inefficient, leading to a low concentration of mature γ mRNA. To suggest that the rate of transcription of the γ-globin genes is identical in both fetal and adult erythroblasts and yet to account for the relatively low steady-state concentration of γ mRNA precursor in adult cells, one would have to suppose that the half-life of the precursor is considerably shorter in adult compared to fetal erythroblasts. Thus processing of the γ-globin mRNA precursor would of necessity be both more rapid but highly inefficient in adult erythroblasts compared to fetal cells to account for all the data. Such a change in the kinetics of processing of a particular RNA species might well be accompanied by a change in the processing pathway. Modifications of the Burke and Sharp techniques for detecting and quantitating precursor species [23] might, therefore, be profitably applied to test this hypothesis.

Some data may already be brought to bear to argue that processing mechanisms are fairly general. In monkey kidney cells infected with a hybrid SV40 virus containing the mouse β-globin gene, the large β-globin gene intron is efficiently removed [24, 25]. Similarly, transformation of mouse fibroblasts with rabbit globin genes leads to transformants which contain mRNA species from which the intervening sequence RNA has been correctly excised [26]. However, these experiments provide little insight into the kinetics of processing and, therefore, do not bear directly on the question as to whether alterations in the processing of a particular globin mRNA precursor during development might account for selective accumulation of specific globin mRNAs in human erythroid cells.

SUMMARY

The concentration of globin mRNA precursors has been measured in fetal, normal adult, and β thalassemic erythroblasts. In normal cells, the concentration of γ, δ, and β-globin mRNA precursors is roughly proportional to the amounts of these mature mRNAs. These data seem most compatible with regulation at the level of gene transcription as being the most important modulation in determining the pattern of accumulation of globin mRNA species in cells from different developmental periods. In contrast, in bone marrow cells from patients with homozygous β^+ thalassemia, the β-globin mRNA precursor is present at a higher than normal concentration despite the deficiency in β-globin mRNA. These results may be interpreted to indicate that mutations affecting mRNA processing may cause β^+ thalassemia.

REFERENCES

1. Farquhar MN, Papayannopoulou T, Brice M, Kan YW, Stamatoyannopoulos G: Cellular regulation of fetal hemoglobin synthesis in man: Investigation of γ and β mRNA accumulation in clonal erythroid cultures initiated from erythroid progenitors derived from fetuses, neonates and adult individuals. Dev Biol 80:64–78, 1980.
2. Lawn RM, Fritsch EF, Parker RC, Blake G, Maniatis T: The isolation and characterization of linked δ and β-globin genes from a cloned library of human DNA. Cell 15:1157–1174, 1978.
3. Jeffreys AJ, Flavell RA: The rabbit β-globin gene contains a large insert in the coding sequence. Cell 12:1097–1108, 1977.
4. Tilghman SM, Tiemeier DC, Seidman JG, Peterlin BM, Sullivan M, Maizel JV, Leder P: Intervening sequence of DNA identified in the structural portion of a mouse β-globin gene. Proc Natl Acad Sci USA, 75:725–729, 1978.
5. Fritsch EF, Lawn RM, Maniatis T: Molecular cloning and characterization of the human β-like globin gene cluster. Cell 19:959–972, 1980.
6. Tilghman SM, Curtis PJ, Tiemeier DC, Leder P, Weissman C: The intervening sequence of a mouse β-globin gene is transcribed within the 15S β-globin mRNA precursor. Proc Natl Acad Sci USA 75:1309–1313, 1978.
7. Kinniburgh AJ, Mertz JE, Ross J: The precursor of mouse β-globin mRNA contains two intervening RNA sequences. Cell 14:681–693, 1978.
8. Ross J, Knecht DA: Precursors of α and β-globin mRNAs. J Mol Biol 119:1–20, 1978.
9. Kinniburgh AJ, Ross J: Processing of the mouse β-globin in RNA precursor: At least two cleavage-ligation reactions are necessary to excise the large intervening sequence. Cell 17:915–921, 1979.
10. Hardison RC, Butler JT, Lacy C, Maniatis T, Rosenthal N, Efstatiadis A: The structure and transcription of four linked rabbit β-like globin genes. Cell 18:1285–1297, 1979.
11. Wood WG, Old JM, Roberts AVS, Clegg JB, Weatherall DJ, Quattrin N: Human globin gene expression: Control of β, δ, and $\delta\beta$ chain production. Cell 15:437–446, 1978.
12. Courtney M, Williamson R: A nuclear precursor to human γ-globin mRNA. Nucleic Acids Res 7:1121–1129, 1979.

13. Kretschmer PJ, Kaufman RE, Coon HC, Chen MJ, Geist CE, Nienhuis AW: Hemoglobin switching in sheep: Cloning and characterization of the β^A and β-like embryonic globin genes from genomic DNA. J Biol Chem 255:3204–3211, 1980.
14. Bolivar F, Rodriguez RL, Greene PJ, Betlach MC, Heyneker HL, Boyer HL: Construction and characterization of new cloning vehicles. New multipurpose cloning system. Gene 2:95–113, 1977.
15. Rigby PWJ, Dieckman M, Rhodes C, Berg P: Labeling deoxyribonucleic acid to high specific activity in vitro by nick-translation with DNA polymerase I. J Mol Biol 113:237–251, 1977.
16. Alwine JC, Kemp JD, Stark GR: Method for detecting specific RNAs in agarose gels by transfer to diazobenzyloxymethyl paper and hybridization with DNA probes. Proc Natl Acad Sci USA 74: 5350–5354, 1977.
17. Kantor JA, Turner PH, Nienhuis AW: β thalassemia: Mutations which affect processing of the β-globin mRNA precursor. Cell 8:148–157, 1980.
18. Maquat LE, Kinniburgh AJ, Beach LR, Honig GR, La Zerson J, Ershler WB, Ross J: Processing of the human β-globin mRNA precursor to mRNA is defective in three patients with β^+ thalassemia. Proc Natl Acad Sci USA 77:4287–4291, 1980.
19. Willing MC, Nienhuis AW, Anderson WF: Selective activation of human β- but not γ-globin genes in human fibroblast $\times$ mouse erythroleukemia cell hybrids. Nature 277:534–538, 1979.
20. Kacian DL, Gambino R, Dow LW, Grossbard E, Natta C, Ramirez F, Speigelman S, Marks PA, Bank A: Decreased globin mRNA in thalassemia detected by molecular hybridization. Proc Natl Acad Sci USA 70:1886–1890, 1973.
21. Housman D, Forget BG, Skoutchi A, Benz EJ Jr: Quantitative deficiency of β chain specific globin mRNA in the thalassemia syndromes. Proc Natl Acad Sci USA 70:1809–1813, 1973.
22. Wilson JT, Wilson LB, deRiel JK, Villa-Komaroff L, Efstratiadis A, Forget BG, Weissman SM: Insertion of synthetic copies of human genes into bacterial plasmids. Nucleic Acids Res 5:563–581, 1978.
23. Berk AJ, Sharp PA: Spliced early mRNAs of Simian virus 40. Proc Natl Acad Sci USA 75: 1274–1278, 1978.
24. Hamer D, Leder P: SV40 recombinants carrying a functional RNA splice junction and polyadenylation site from the chromosomal mouse βmaj globin gene. Cell 17:737–747, 1979.
25. Hamer D, Leder P: Expression of the chromosomal mouse βmaj globin gene cloned in SV40. Nature 281:35–40, 1979.
26. Mantei N, Bell W, Weissman C: Rabbit β-globin mRNA production in mouse L cells transformed with cloned rabbit β-globin chromosomal DNA. Nature 281:40–46, 1979.

Organization and Expression of Globin Genes, pages 231–239

Transcription of the β Major and β Minor Genes in Mouse Erythroleukemia Cells

Peter Curtis, A.C. Finnigan, and G. Rovera

The Wistar Institute, Philadelphia

INTRODUCTION

The chromosomal genes for mouse α, β major, and β minor globin have been cloned and extensively characterized [1–4]. The β-globin genes, like most eukaryotic genes characterized so far, contain intervening sequences (IVS) [5] which are not found in the mature mRNA. One IVS is located at the codon for amino acids (30–31) consisting of 116 nucleotides, the other is located at the codon for amino acids (104–105) consisting of 650 base pairs (bp). The large IVS of the β major globin gene has been shown to be unique to β major globin DNA, ie, there is no other homologous sequence in total mouse DNA [6]. When the large IVS of the β major and β minor globin genes are compared, only a small sequence at the 5′ end is homologous, the rest being nonhomologous [7, 8].

DNA and histones are complexed in chromatin to form nucleosomes which are randomly distributed with respect to the DNA sequences. There would appear to be a change in chromatin structure of the globin genes found in cells where the genes are expressed. Globin DNA in chromatin from erythroid cells (ie, actively expressed) was more sensitive to DNase I digestion than globin DNA in chromatin from nonerythroid cells [9]. However, on the basis of this criterion, the globin genes of uninduced as well as induced mouse erythroleukemia (MEL) cells are in an active state but are not expressed so that the active state is not the only requirement for transcription [10–13].

Pulse-label experiments and labeling of RNA by isolated nuclei have shown that induction of MEL cells results in activation of transcription of the globin genes, though the molecular mechanism of this activation is not known [14–16].

However, transcription would not appear to be the only level at which control of globin gene expression occurs. Several studies imply that post-transcriptional processing plays an important role. Globin RNA has been detected, localized to the nucleus, in several nonerythroid cells [12, 17, 18] and ovalbumin mRNA sequences are represented in liver and spleen nuclear RNA [19, 20], although contradictory results have appeared [21]. In induced MEL cells, globin RNA is selectively processed and transported to the cytoplasm. The percentage of newly transcribed RNA that is globin-specific is ~ 0.005%, while its steady level in the cell is 0.2% of total cell RNA [22, 23].

From studies, mainly using the inducible MEL cells, the earliest RNA transcript from the β-globin gene is a 15S molecule ~ 1,600 nucleotides long containing both the intervening sequences but no other additional sequences [22–28]. It possesses both the 5′ and 3′ termini of mature mRNA, ie, the cap structure $N^7mGppN^6mA_mpCp$ at the 5′ end and poly A at the 3′ end. Kinetic studies showed that this precursor has a half-life of 2 minutes, so that capping and polyadenylation must be very rapid events. The 15S precursor is synthesized in induced MEL cells at a rate of six to ten molecules/minute/cell, which is the same as the rate of accumulation of mature 10S globin mRNA in the cell, implying that processing is an efficient step [23]. A number of globin transcripts of intermediate sizes have been detected [29], including a β-globin RNA which still possesses two intervening sequences and yet is greatly reduced in size compared with the full-length transcript [30].

DIFFERENTIAL EXPRESSION OF THE ADULT β-GLOBIN GENES IN FRIEND CELLS

MEL cells can be induced by a variety of compounds to synthesize hemoglobin. The mechanism(s) of induction are not well understood. A variety of data suggest that butyric acid, hemin, and DMSO very likely act by three separate mechanisms, while it is not clear whether other inducers are like DMSO or not [for discussion see 31]. Induction of MEL cells with DMSO results in the synthesis of α-globin and predominantly β major globin; the ratio of β major:β minor globin is approximately 4:1 in normal mouse DBA/2. It has been reported to vary from 2:1 to 1.4:1 in DMSO-treated clone 745 cells [32–36], while hemin-treated MEL cells contain essentially β minor globin [34–37]. In vitro translation of poly A RNA from such cells showed only the presence of β minor globin, indicating that it was not due to translational inhibition of β major mRNA. Two possible mechanisms could explain the differential expression of the two β-globin genes. The first possibility is that in every cell the two genes are coexpressed to different extents. The second possibility is that subsets of cells could express at any given time only one of the two genes: The different ratio of β major and β minor synthesizing cells would generate the final ratio of the two hemoglobins present in the total population. In favor of this last hypothesis Alter and Goff [42] have reported the possibility of isolating clones of clone 745 cells responding to DMSO treatment with the synthesis of only β minor globin chains but not β major. Similar types of clones were also isolated by Rovera and Surrey [31]. So far clones of 745 cells responsive to DMSO with the synthesis of only β major globin chain, however, have not been reported, even if theoretically they should be more easily isolated. A more direct analysis of the types of hemoglobin present in single cells would give a more direct answer to this problem. The differential expression of the β-globin genes must occur either at the level of transcription, nuclear processing, or mRNA stability, whether or not a cellular mechanism exists.

ANALYSIS OF THE β-GLOBIN mRNAs

The β major and β minor globin differ by only six to nine amino acids [38], and this homology makes it difficult to study the kinetic parameters of the mRNAs since they cannot be distinguished by hybridization using globin cDNA probes. However, as has been noted earlier [7, 8], the large IVS of β major and β minor genes have been shown to be nonhomologous for a large part of their sequence, so probes prepared from this region of the cloned chromosomal gene will be specific for either β major or β minor. Therefore, it is feasible to use the IVS of the globin genes as probes to determine if the transcripts of β major gene are detectable in hemin-treated cells and whether there is differential expression of the two β genes in the DMSO-treated cells.

In order to detect the β major and minor globin transcripts specifically we have used the approach of Weaver and Weissmann's [39] modification of the Berk and Sharp method [40]. Within the chromosomal β major gene there is a restriction site Pst 1, which is located in the nonhomologous region of the large intervening sequence. Therefore a probe that is specific for β major globin nuclear transcripts was constructed as outlined in Figure 1. The chromosomal β major globin DNA was digested with the restriction enzyme Pst 1, giving three fragments 5.5 kilobases (kb), 5.0 kb, and 2.5 kb. To improve the efficiency of labeling the 5′ termini, the

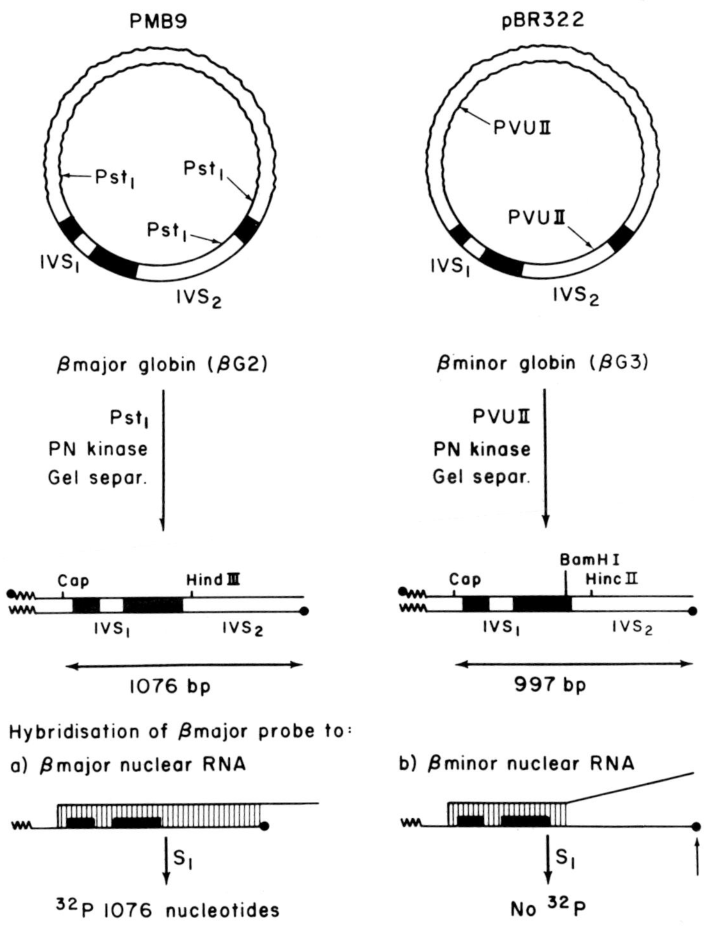

Fig. 1. Specific probes for the β major and β minor globin nuclear precursors were prepared by digestion of the plasmids pMB9-βG2 and pBR322-βG3 with Pst 1 or PVU II and then 5′ ends labeled after treatment with exonuclease III (to expose 5′ ends) and bacterial alkaline phosphatase. [^{32}P]-2.5 kb fragment from βG2 and [^{32}P]-7.1 kb fragment from βG3 were isolated by agarose gel electrophoresis.

fragments were treated briefly with Exonuclease III followed by bacterial alkaline phosphatase, after which they were labeled with [γ-^{32}P]ATP by polynucleotide kinase. The labeled fragments were separated by gel electrophoresis on 1% agarose from which the 2.5 kb fragment was eluted by electrodialysis. This fragment contained one labeled end located close to the 5′ end of the cloned chromosomal β major globin DNA. The other labeled end is located in the nonhomologous region of the large intervening sequence and can be protected by hybridization with full-length RNA transcripts of the β major globin gene or transcripts that have retained this part of the large intervening sequence. Mature mRNA and β minor transcripts would not be expected to protect the labeled end. A probe with similar specificity was constructed from the chromosomal β minor globin gene using the restriction enzyme PVUII whose site is also located in the large IVS within a sequence of 24 nucleotides which are totally absent from the β major gene.

To confirm the specificity of the probes, they were hybridized back to the chromosomal β major and β minor globin DNA by Southern blot analysis. However, the probes, prepared as described above, contained coding sequences which

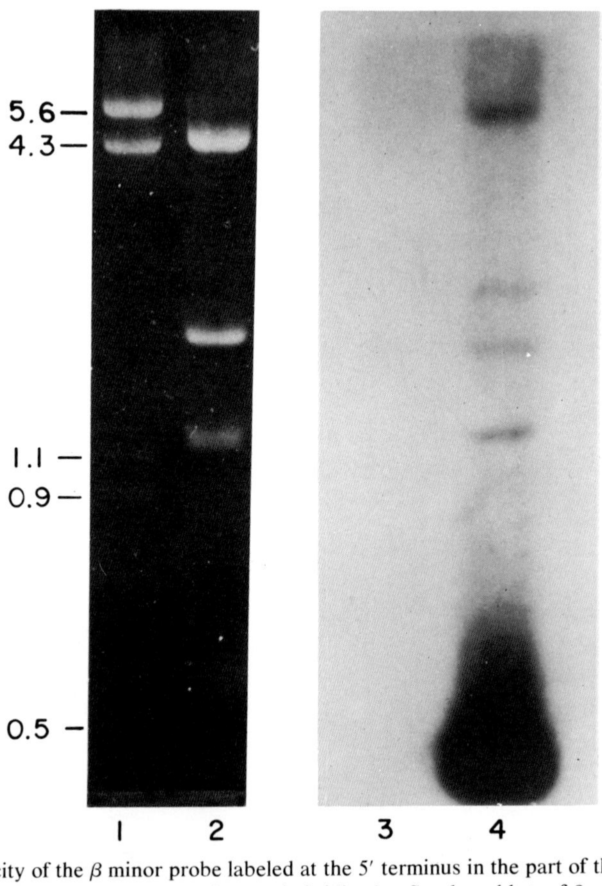

Fig. 2. Specificity of the β minor probe labeled at the 5′ terminus in the part of the large intervening sequences distal to the cap site. The probe was hybridized to Southern blots of β major gene restricted with Pst 1 and Hind III and β minor genes restricted with PVU II and Bam HI. Lane 1: Ethidium bromide staining of restricted β major gene; Lane 2: ethidium bromide staining of β minor gene; Lane 3: hybridization of β major gene to β minor probe; Lane 4: hybridization of β minor gene to β minor probe.

for this test were removed by digestion of the β major and β minor probes with restriction enzymes Hind III and Hinc II, respectively. Hybridization of the shortened probes to nitrocellulose filters containing β major and β minor globin DNA restriction fragments transferred from an agarose gel demonstrated a complete specificity for homologous DNA under conditions 50% formamide, 0.75 M NaCl (Fig. 2), which are less stringent than those used for the RNA-DNA hybridizations 80% formamide, 0.4 M NaCl.

The aim of the subsequent experiments was to determine the effect of hemin treatment on the transcription of the β major globin gene. RNA was prepared from isolated nuclei of untreated, DMSO- and hemin-treated MEL cells. After heat denaturation, the β major globin probe, the labeled 2.5 kb Pst 1 fragment, was hybridized with nuclear RNA in 80% formamide, 0.4 M NaCl, 20 mM Hepes buffer, pH 6.8, at 57°C for 16 hours, which are conditions that favor RNA-DNA hybrid formation [41]. The annealed nucleic acids were digested subsequently with S1 nuclease to remove single-stranded ends, followed by electrophoresis in an urea-polyacrylamide gel electrophoresis. Protected fragments were visualized by autoradiography.

A prominent labeled band of approximately 1,070 nucleotides resulted from hybridization of nuclear RNA from DMSO-treated cells with the β major globin probe, with minor amounts of bands corresponding to 800 and 300 nucleotides (Fig. 3). In contrast, no bands were observed using nuclear RNA from untreated or

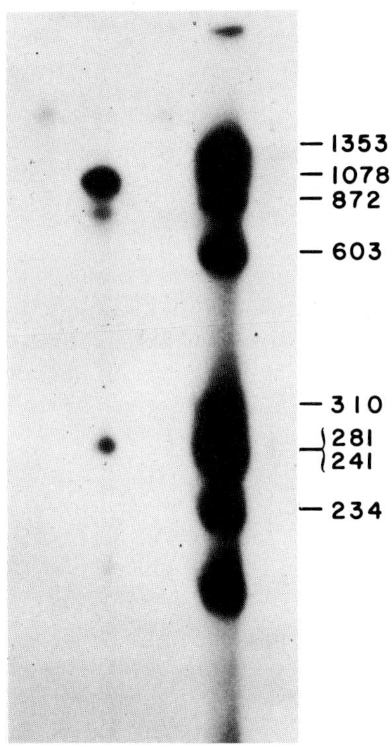

Fig. 3. Urea-polyacrylamide gel electrophoresis (PAGE) of S1 nuclease-resistant β major-specific probe after hybridization with nuclear RNA. A. untreated MEL; B. DMSO-induced MEL cells; C. hemin-induced MEL cells; D. marker of ϕX 174 DNA cut with Hae III.

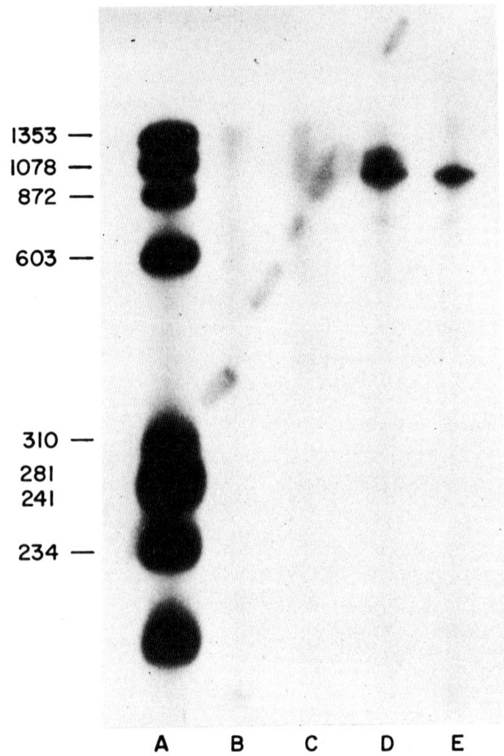

Fig. 4. Urea-polyacrylamide gel electrophoresis (PAGE) of S1 nuclease-resistant β minor specific probe after hybridization with nuclear RNA. A. marker of ϕX 174 DNA cut with Hae III; B. yeast RNA; C. untreated MEL; D. DMSO-induced MEL cells; E. hemin-induced MEL cells.

hemin-treated cells, and the same negative result was obtained with cytoplasmic RNA from untreated and DMSO- and hemin-treated cells and yeast RNA.

To determine that hemin treatment of MEL cells had affected the β minor globin gene, the β minor probe was hybridized with nuclear RNA from untreated and DMSO- and hemin-treated cells. A protected fragment of approximately 1,000 nucleotides was observed in untreated, DMSO-, and hemin-treated cells (Fig. 4). Using the same amount of RNA in each hybridization, more radioactivity was seen when the RNA was obtained from DMSO-treated cells, somewhat less from hemin-treated cells, and least, though still detectable, from untreated cells. No band was observed by hybridizing the probe with MEL cytoplasmic RNA or yeast RNA.

DISCUSSION

Kinetic studies have established that β-globin mRNA is derived from a 15S precursor, which possesses a 5′ cap structure identical to that of mature mRNA, a poly A tail, and both intervening sequences [22–28]. The 5′ terminus of the 15S precursor coincides with that of mature mRNA [39], and therefore in the case of the β major mRNA precursor is located 1,076 nucleotides from the Pst 1 site which is in the large intervening sequence.

Hybridization of the labeled β major probe with nuclear RNA of DMSO-treated cells resulted in the protected fragment approximately 1,070 nucleotides long. The assay therefore has detected the 15S precursor as the predominant β major nuclear transcript in agreement with the earlier results [see 23, 24]. By the same assay no 15S precursor was detected in hemin-treated cells, implying that the lack of β major globin expression is the result of no transcription, unless the β major transcripts are selectively degraded so rapidly that its steady-state level cannot be detected. The absence of a 15S β major precursor in untreated cells agrees with the results of Rovera and Surrey [31], who detected by immunoprecipitation only β minor but not β major globin polypeptides in untreated cells.

The amounts of the 15S β minor precursors in untreated, DMSO-, and hemin-treated cells correlates approximately with the levels of β minor globin polypeptide detected in these cells [37].

In our study we observed from the hybridization of the β major globin probe with nuclear RNA from DMSO-treated cells minor bands of 800 and 300 nucleotides. Protection of the labeled 800-nucleotide DNA strand implies excision of part or all of the small intervening sequence from the β major nuclear transcript, since the 3′ end of the small intervening sequence is located 815 nucleotides from the labeled Pst 1 site. The labeled 300-nucleotide DNA could result from a β major nuclear transcript where sequences located 300 nucleotides from the Pst 1 site were excised; presumably the excised sequences extended to the 5′ end of the large intervening sequence. Alternatively the 300-nucleotide DNA could arise by hybridization of the probe with RNA excised from the large intervening sequence of the β major nuclear transcript.

A number of globin RNA molecules of intermediate size between the 15S species and the mature mRNA have been detected [29]; in particular, Kinniburgh and Ross [30] identified an intermediate, 1,030 nucleotides long, which possessed two intervening sequences. Since in our assay splicing of sequences from the large intervening sequence would result in a small labeled DNA or none at all (except for removal of the small region between the Pst 1 site and the 3′ end of the large intervening sequence), the splicing of a part of the small intervening sequence must be the first step in the processing of the 15S precursor followed by removal of part of the large intervening sequence.

SUMMARY

Induction of Friend erythroleukemic cells clone 745 by DMSO results in appearance of β major:β minor globin polypeptides in a ratio of 2:1, while induction by hemin produces only β minor globin together with a α-globin. To determine the level at which control is exerted the presence of β major and β minor globin nuclear RNA transcripts has been investigated. The β major and β minor genome DNA sequences were selectively labeled in the nonhomologous region of the large intervening sequence and hybridized to nuclear RNA of untreated and treated Friend cells.

The nuclear RNA of DMSO-treated cells contained both β major and β minor globin nuclear RNA transcripts whereas hemin-treated and untreated cells contained only β minor globin transcripts. Nuclear transcripts with the size of partially processed β major globin RNA were detected also in DMSO-treated cells.

ACKNOWLEDGMENTS

This work was supported from grants CA 10815 and CA 24273 from the National Cancer Institute.

REFERENCES

1. Tilghman SM, Tiemeier DC, Polsky F, Edgell MH, Seidman JG, Leder A, Enquist LW, Norman B, Leder P: Cloning specific segments of the mammalian genome: Bacteriophage λ containing mouse globin and surrounding gene sequences. Proc Natl Acad Sci USA 74:4406–4410, 1977.
2. Tilghman SM, Tiemeier DC, Seidman JG, Peterlin BM, Sullivan M, Maizel JV, Leder P: Intervening sequence of DNA identified in the structural portion of a mouse β-globin gene. Proc Natl Acad Sci USA 75:725–729, 1978.
3. Leder A, Miller HI, Hamler DH, Seidman JG, Norman B, Sullivan NM, Leder P: Comparison of cloned mouse α- and β-globin genes: Conservation of intervening sequence locations and extragenic homology. Proc Natl Acad Sci USA 75:6187–6191, 1978.
4. Konkel DA, Tilghman, SM, Leder P: The sequence of the chromosomal mouse β-globin major gene: Homologies in capping, splicing and poly (A) sites. Cell 15:1125–1132, 1978.
5. Dawid, IB, Wahli W: Application of recombinant DNA technology to questions of developmental biology: A review. Dev Biol 69:305–328, 1979.
6. Miller, HI, Konkel DA, Leder P: An intervening sequence of the mouse, β-globin major gene shares extensive homology only with β-globin genes. Nature 275:772–774, 1978.
7. Tiemeier DC, Tilghman, SM, Polsky FI, Seidman JG, Leder A, Edgell MH, Leder P: A comparison of two cloned mouse β-globin genes and their surrounding and intervening sequences. Cell 14:237–245, 1978.
8. Konkel DA, Maizel JV Jr, Leder P: The evolution and sequence comparison of two recently diverged mouse chromosomal β-globin genes. Cell 18:865–873, 1979.
9. Weintraub H, Groudine M: Chromosomal subunits in active genes have an altered conformation. Science 193:848–856, 1976.
10. Wallace RB, Dube SK, Bonner J: Localization of the globin gene in the template active fraction of chromatin of Friend leukemia cells. Science 198:1166–1168, 1977.
11. Miller DM, Turner P, Nienhuis AW, Axelrod DE, Gopalakrishnan TV: Active conformation of the globin genes in uninduced and induced mouse erythroleukemia cells. Cell 14:511–521, 1978.
12. Gottesfeld JM, Partington GA: Distribution of messenger RNA-coding sequences in fractionated chromatin. Cell 12:953–962, 1977.
13. Lau AF, Ruddon RW, Collett MS, Faras AJ: Distribution of the globin gene in active and inactive chromatin fractions from Friend erythroleukemia cells. Exp Cell Res 111:269–276, 1978.
14. Aviv H, Veloch Z, Bastos R, Levy S: Biosynthesis and stability of globin mRNA in cultured erythroleukemic Friend cells. Cell 8:495–503, 1976.
15. Orkin SH, Swerdlow PS: Globin RNA synthesis in vitro by isolated erythroleukemic cell nuclei: Direct evidence for increased transcription during erythroid differentiation. Proc Natl Acad Sci USA 74:2475–2479, 1977.
16. Lowenhaupt, K, Trent C, Lingrel JB: Mechanisms for accumulation of globin mRNA during dimethyl sulfoxide induction of murine erythroleukemia cells: Synthesis of precursors and mature mRNA. Dev Biol 63:441–454, 1978.
17. Humphries S, Windass J, Williamson R: Mouse globin gene expression in erythroid and non-erythroid tissues. Cell 7:267–277, 1976.
18. Perlman SM, Ford PJ, Rosbash MM: Presence of tadpole and adult globin RNA sequences in oocytes of xenopus laevis. Proc Natl Acad Sci USA 74:3835–3839, 1977.
19. Axel R, Feigelson P, Schutz G: Analysis of the complexity and diversity of mRNA from chicken liver and oviduct. Cell 7:247–254, 1976.
20. Roop DR, Nordstrom JL, Tsai SY, Tsai MJ, O'Malley B: Transcription of structural and intervening sequences in the ovalbumin gene and identification of potential ovalbumin mRNA precursors. Cell 15:671–685, 1978.

21. Groudine M, Weintraub H: Rous sarcoma virus activates embryonic globin genes in chicken fibroblasts. Proc Natl Acad Sci USA 72:4464–4468, 1975.
22. Curtis PJ, Weissmann C: Purification of globin messenger RNA from dimethylsulfoxide-induced Friend cells and detection of a putative globin messenger RNA precursor. J Mol Biol 106:1061–1075, 1976.
23. Curtis PJ, Mantei N, Weissmann C: Characterization and kinetics of synthesis of 15S β-globin RNA, a putative precursor of β-globin mRNA. Cold Spring Harbor Symp Quant Biol 42:971–984, 1977.
24. Curtis PJ, Mantei N, Van Den Berg J, Weissmann C: Presence of a putative 15S precursor to β-globin mRNA but not to α-globin mRNA in Friend cells. Proc Natl Acad Sci USA 74:3184–3188, 1977.
25. Tilghman SM, Curtis PJ, Tiemeier DC, Leder P, Weissmann C: The intervening sequence of a mouse β-globin gene is transcribed within the 15S β-globin mRNA precursor. Proc Natl Acad Sci USA 75:1309–1313, 1978.
26. Ross, J: A precursor of globin messenger RNA. J Mol Biol 106:403–420, 1976.
27. Kwan, SP, Wood TG, Lingrel JB: Purification of a putative precursor of globin messenger RNA from mouse nucleated erythroid cells. Proc Natl Acad Sci USA 74:178–182, 1977.
28. Bastos RN, Aviv H: Globin RNA precursor molecules: Biosynthesis and processing in erythroid cells. Cell 11:641–650, 1977.
29. Farace MG, Ullu, E, Fantoni A, Rossi GB, Cioe L, Dolei A: Nuclear RNA sequences coding for α and β globins in erythroid cells: Evidence for multiple intermediate molecules. Blood 53:134–141, 1979.
30. Kinniburgh AJ, Ross J: Processing of the mouse β-globin mRNA precursor: At least two cleavage-ligation reactions are necessary to excise the larger intervening sequence. Cell 17:915–922, 1979.
31. Rovera, G, Surrey, S: Use of resistant or hypersensitive variant clones of Friend cells in analysis of mode of action of inducers. Cancer Res 38:3737–3744, 1978.
32. Kabat, D, Sherton CC, Evans LH, Bigley R, Koler RD: Synthesis of erythrocyte-specific proteins in cultured Friend leukemia cells. Cell 5:331–338, 1975.
33. Nudel U, Salmon JE, Terada M, Bank A, Rifkind RA, Marks PA: Differential effects of chemical inducers on expression of β globin genes in murine erythroleukemia cells. Proc Natl Acad Sci USA 74:1100–1104, 1977.
34. Alter BP, Goff SC: Globin synthesis in mouse erythroleukemia cells in vitro: A switch in β chains due to inducing agent. Blood 50:867–876, 1977.
35. Alter, BP, Goff SC: Variable globin chain synthesis in mouse erythroleukemia cells. Blood 52: 1047–1057, 1978.
36. Rovera G, Abramczuk J, Surrey S: The effect of hemin on the expression of β globin genes in Friend cells. FEBS Lett 81:366–370, 1977.
37. Rovera G, Vartikar J, Connolly GR, Magarian C, Dolby TW: Hemin controls the expression of the β minor globin gene in Friend erythroleukemic cells at the pretranslational level. J Biol Chem 253:7588–7590, 1978.
38. Gilman JG: Mouse haemoglobin beta chains. Sequence data on embryonic chain and genetic linkage of the γ-chain locus to the adult γ-chain locus Hbb. Biochem J 155:231–241, 1976.
39. Weaver RF, Weissmann C: Mapping of RNA by a modification of the Berk-Sharp procedure: the 5′ termini of 15S β-globin mRNA precursor and mature 10S β-globin mRNA have identical map coordinates. Nucleic Acids Res 7:1175–1193, 1979.
40. Berk AJ, Sharp PA: Sizing and mapping of early adenovirus mRNAs by gel electrophoresis of S1 endonuclease-digested hybrids. Cell 12:721–732, 1977.
41. Casey J, Davidson N: Rates of formation and thermal stabilities of RNA:DNA and DNA:DNA duplexes at high concentrations of formamide. Nucleic Acids Res 4:1539–1552, 1977.
42. Alter BP, Goff SC: Hemoglobin switching. A cellular model. Science 207:647–649, 1980.

Globin Messenger RNA Metabolism in Normal and Thalassemic Erythroblasts

Edward J. Benz, Jr., Alphonse L. Scarpa, Jon K. deRiel, Barry L. Tonkonow, A. Kim Ritchey, and Bernard G. Forget

The Hematology Section, Department of Internal Medicine, Yale University School of Medicine, New Haven, Connecticut

INTRODUCTION

The β thalassemia syndromes are characterized by defective synthesis of structurally normal β-globin chains [1–4]. At the biochemical level, two major classes of β thalassemia can be distinguished: (1) in β^0 thalassemia, β-globin synthesis is entirely absent; (2) in β^+ thalassemia, β-globin synthesis is variably reduced, but readily detectable. The β-globin chains which are produced appear to be structurally normal. The molecular basis of these biosynthetic defects has been partially defined in several types of β^0 thalassemia. In some cases, absent β-globin synthesis can be attributed to deletion or rearrangement of the β-globin gene, or to point mutations within the coding regions of β-globin mRNA which render the mRNA nontranslatable [1–5]. However, in the majority of cases of β^0 thalassemia, and in all cases of β^+ thalassemia studied, the β-globin gene is present and intact. In these cases of β^0 thalassemia, β mRNA may be absent or present in reduced amounts, but nonfunctioning. In all cases of β^+ thalassemia studied thus far, β-globin mRNA appears to be reduced to a degree correlating well with the protein biosynthetic defect; moreover, β mRNA from these patients functions normally in mRNA-dependent cell-free protein synthesizing systems [cf 1–4]. Indeed, β^+ thalassemia (and β^0 thalassemias not attributable to gene deletions or rearrangements) accounts for the vast majority of patients with clinically disabling thalassemia syndromes [2–4]. Thus, further understanding of the molecular pathology of the common forms of β thalassemia will require direct examination of the transcription, processing, nuclear-cytoplasmic transport, and cytoplasmic stability of β-globin mRNA in maturing erythroblasts from individual patients.

β thalassemia results in decreased production of the normal major adult hemoglobin, Hb A ($\alpha_2\beta_2$). However, both thalassemic and nonthalassemic erythroblasts produce other hemoglobins in small quantities: the minor adult hemoglobin, Hb A_2 ($\alpha_2\delta_2$), accounting for 1.5–5% of the total intracellular hemoglobin, and fetal hemoglobin, Hb F ($\alpha_2\gamma_2$), which rarely comprises more than 1% of the total hemoglobin in normal individuals, but is increased in β thalassemia. Hb F production during postnatal life is restricted to a small subpopulation of cells called "F" cells [cf 1–4].

In general, the amounts of these minor hemoglobins produced in individual subjects seem to correlate well with the amounts of mRNA coding for δ and γ globin chains [2–5]. Enhanced synthesis of γ or δ-globin chains would be desirable in order to provide alternative non-α-globin chains as a source for hemoglobin tetramer formation in β thalassemic individuals. Thus, further understanding of the mechanisms which determine the levels of γ and δ-globin mRNA in human erythroblasts would clearly be of interest. The differential regulation of the non-α globin genes remains poorly understood, at least in part, because the dynamic determinants of globin mRNA accumulation are not yet elucidated. Analysis of δ and γ mRNA transcription, processing, transport, and stability will also be required if therapeutic manipulation of the γ and δ-globin genes is to be attempted in an effort to develop more rational therapies for β thalassemia.

Until recently, studies of the individual globin mRNAs in erythroid cells have been limited to measurements of so-called "steady-state" mRNA levels. The amount of each globin mRNA already accumulated in the erythroid cells being studied have been quantitated in various clinical conditions, but no reliable techniques for measuring the kinetics of mRNA production and degradation have been available. Previous methods have relied on the use of highly radioactive copy DNA (cDNA) preparations which can be partially purified so that the predominant component of the cDNA sample is complementary only to α, β, or γ mRNA [cf 2]. These cDNAs can be utilized as molecular hybridization probes for quantitative analysis of the relative amounts of each globin mRNA present in a specimen of RNA isolated from erythroid cells or subcellular compartments. Unfortunately, these measurements reflect only the static levels of each globin mRNA present, and inferences concerning the dynamics resulting in these steady-state levels are at best indirect.

Recent studies of the steady-state levels of mRNA in nuclei, erythroblast cytoplasm, and reticulocytes of β thalassemic patients suggest strongly that, in many cases, defective accumulation of β-globin mRNA cannot be attributed simply to a primary defect in the synthesis (transcription) of mRNA precursor from the mutant β-globin gene. In each of three patients studied by Nienhuis and co-workers [6], nuclear RNA from early erythroblasts contained considerably higher levels of β globin mRNA sequences than cytoplasmic mRNA, which, in turn, contained higher levels than reticulocytes. Similarly, in a case of β^0 thalassemia studied by Ottolenghi and co-workers [7], nuclear β-globin mRNA sequences were detectable, but no β mRNA was present in cytoplasm. These results provide strong indirect evidence for posttranscriptional defects as the major cause of defective β-globin mRNA accumulation. In contrast, a second patient with β^0 thalassemia studied by Ottolenghi and co-workers [7], and a patient studied by this laboratory [8], demonstrated severe and fixed defects in β-globin mRNA in all cell fractions at all stages of erythropoiesis. Thus, a transcriptional defect remains a possibility in these individuals.

In this communication, we describe our initial attempts to assess the dynamics of mRNA accumulation in thalassemic and nonthalassemic erythroblasts. As described in the Methods section, our approach has been to generate radioactive newly synthesized globin mRNA by incubation of erythroblasts in vitro with tritiated uridine, extraction of the RNA, and subsequent hybridization to pure nonradioactive DNA hybridization probes. Our results suggest that the reduced accumulation of δ-globin chains during erythropoiesis compared to β-globin chains is due primarily to differences in transcription and/or early intranuclear processing of the β and δ mRNA precursors. In contrast, β-globin mRNA accumulation appears to be normal during initial pulse incubations in β^+ thalassemia; a subsequent posttranscriptional decline, due to abnormal processing, transport, and/or stability, results in the net deficit of β mRNA sequences in the steady state.

METHODS

The detailed methodology employed for these studies is described in a separate communication [9]. Briefly, bone marrow erythroblasts were obtained at the time of diagnostic bone marrow aspiration from five nonthalassemic patients (chosen because of the presence of erythroid hyperplasia) and five patients with β^+ thalassemia. Cells were suspended at approximately $1\text{–}3 \times 10^7$ cells/ml, and incubated for 20 minutes with 0.5 millicuries/ml of ^{3}H-uridine (pulse incubation). In some cases, an aliquot of the cells was subsequently incubated for 18–24 hours, after exhaustive washing to remove ^{3}H-uridine and suspension of the cells in medium supplemented with 10 millimolar nonradioactive uridine (chase incubation). Total cellular RNA was extracted by the method of Lowenhaupt and Lingrel [10]. Approximately $4 \times 10^4\text{–}2 \times 10^5$ cpm of radioactive RNA were then incubated with a "stack" of nitrocellulose filters, to each of which had been bound a unique cloned recombinant DNA fragment containing only sequences complementary to one of the individual globin mRNA or mRNA precursor sequences. Care was taken to insure that sufficient DNA was bound to each filter to establish a four to sixfold molar excess of DNA sequences over total complementary globin mRNA or mRNA precursor sequences added. After incubation for 48 hours at 37–42°C in 50% formamide buffer [cf 9], the filters were exhaustively washed by a method essentially identical to that of Ross and co-workers [11]. The filters were then dried and counted. The amount of ^{3}H-radioactivity remaining bound to each filter after these procedures was taken as a measure of the amount of newly synthesized globin mRNA accumulated during the pulse incubation, or the amount of RNA synthesized in vitro and remaining intact after the chase incubation. Since absolute rates of globin mRNA synthesis are exceedingly difficult to assess in these systems, we have expressed our data in the form of relative amounts of radioactive mRNAs accumulating; that is, the data are expressed as ratios of one mRNA produced relative to others obtained from the same RNA preparation.

This experimental strategy utilizes the absolute purity of cloned DNA segments to permit detection of small amounts of radioactive mRNA under conditions of vast DNA excess for molecular hybridization. In this manner, no artefacts due to differences in the nonradioactive pools of preexisting mRNA are likely to distort the data. Moreover, the rate of the hybridization reaction is determined by the DNA probe, even though the reaction rate is slowed somewhat by immobilization on filters. Finally, the approach permits use of cloned DNA fragments which exist in vivo in double-stranded DNA form; by initially fixing the DNA to the filters in the denatured state, the DNA fragments are immobilized in single-stranded form and thus remain available for hybridization.

Data obtained from pulse-chase labeling experiments was compared to steady-state levels of mRNA measured by standard molecular hybridization techniques mentioned in the introduction and described in detail elsewhere [cf 12].

RESULTS AND DISCUSSION

Measurement of Pulse-Chase-Labeled Radioactive Globin mRNA Sequences

In order to detect newly synthesized globin mRNA sequences, it was necessary to detect the relatively low levels of radioactivity incorporated into newly synthesized globin mRNAs during a short pulse incubation, under conditions which would not be affected by the much larger pools of nonradioactive globin mRNA already present in the cell at the time of the incubation with radioactive RNA precursors. In order to accomplish this aim, we employed excess amounts of absolutely pure cloned globin gene-specific DNA sequences attached to nitrocellulose filters in denatured form. The excess amounts of DNA included on the filters allowed complete hybridization of all RNA sequences, both radioactive and nonradioactive, without artefacts due to binding of other radioactive RNA sequences complementary to nonglobin DNA contaminants present in previously available molecular hybridization probes. Representative data are summarized in Table I and discussed in the next paragraph.

Typically, we obtained approximately 5×10^5–5×10^6 cpm of radioactive RNA from 10^7–10^8 nucleated erythroblasts after a 20-minute pulse-labeling experiment. Specific activities ranged from 8×10^3–2×10^5 cpm/mcg total cellular RNA; variability in the specific activity was due in part, but not entirely, to variable contamination of the RNA preparation with cellular DNA, despite extensive efforts to remove DNA from these preparations. Fortunately, in preliminary experiments [data not shown], we were able to demonstrate that the ratios of individual radioactive globin mRNAs measured were not affected by DNA contamination. Table I

TABLE I. Hybridization of ^3H Bone Marrow Erythroblast RNA to Filter-Bound Globin DNA Probes[1]

	Type of plasmid DNA bound to filter				Ratio	
Source of RNA	α	β	γ	pMB9	β/α	γ/α
β^+ thal #1	0.11%[2]	0.14%	0.013%	0.005%	1.33	0.11
Nonthalassemic	0.087%	0.083%	0.006%	0.006%	0.95	0

[1]Cells were incubated with ^3H-uridine, and RNA extracted as described in the text. Each ^3H RNA was hybridized to nitrocellulose filters, each containing one of the following DNA preparations: a) "α" — 25 micrograms of bacterial plasmid "JW101," containing an α-globin cDNA insert; b) "β" — 25 micrograms "JW102," containing a β-globin cDNA insert; c) "γ"—25 micrograms "JW151," containing a γ-globin cDNA insert; d) "pMB9"—bacterial plasmid pMB9, 25 micrograms, containing no eukaryotic DNA inserts [cf 14]. Essentially identical results were obtained by using smaller amounts of plasmid fragments purified to enrich each DNA preparation for the desired globin cDNA insert. RNA inputs were 200,000 cpm. Thus, 0.1% binding represents 200 cpm remaining attached to the filters after posthybridization washing.

[2]$\% = \dfrac{\text{cpm } ^3\text{H RNA bound to filter}}{\text{cpm } ^3\text{H RNA added to filter}} \times 100.$

describes typical results obtained from a nonthalassemic patient and a β^+ thalassemic patient. In these, and other, experiments, we determined the amount of total globin mRNA by preliminary molecular hybridization experiments using highly specific globin cDNA probes. These measurements enabled us to determine the input of total globin mRNA sequences which would result by addition of a given number of cpm of ^{3}H-RNA to the filters. In this manner, we could ensure that sufficient amounts of the complementary cloned globin DNA sequence were added to the nitrocellulose filters to ensure DNA excess. Invariably, 1–2 mcg of DNA sequence complementary to the mRNA sequences were sufficient to ensure a four to fivefold excess of globin DNA sequences in each experiment. In the experiment outlined in Table I, 200,000 cpm of each mRNA was added to the filters. The number of cpm remaining bound to the filters after the washing procedures described in Methods are expressed as percentages of the input radioactivity. As indicated in Table I, filters containing only bacterial plasmid DNA invariably bound only 10–25 cpm. Filters containing globin gene DNA sequences, as shown in Table I, typically bound 0.05%–0.2% of the input radioactivity. As indicated in the Table, the washing procedures are sufficiently stringent to permit discrimination of closely related globin sequences, such as the β and γ mRNAs. Although considerable variability was encountered from patient to patient, results of repeated measurements with the same patient were reasonably reproducible. Thus, as indicated by Table I, this approach permits measurement of radioactive mRNA sequences synthesized in vitro in intact cells during a 20-minute pulse incubation.

Accumulation of Newly Synthesized Globin mRNA in Normal and β^+ Thalassemic Erythroblasts

Tables II and III indicated results obtained from five patients with nonthalassemic disorders causing erythroid hyperplasia (hereditary spherocytosis (two patients), polycythemia vera (two patients), and sickle cell anemia (one patient)), and four patients "homozygous" for β^+ thalassemia. The thalassemic group included patients of Italian and Greek, but not black, origin, and individuals with transfusion-dependent thalassemia major as well as thalassemia intermedia.

TABLE II. ^{3}H Globin mRNA Content of Pulse-Chase-Labeled mRNA in Normal Erythroblasts[1]

	^{3}H β/α mRNA Ratio	
Patient	Pulse	Chase
#1	1.24	—
#2	0.93	—
#3	0.86	0.78
#4	0.83	0.81
#5	1.00	1.25

[1]Cells were "pulse"-labeled for 20 minutes with ^{3}H-uridine, and then "chased" for 24 hours in the absence of ^{3}H-uridine. The β/α ratios were determined as discussed in Table I and the text. Steady-state β/α mRNA ratios ranged from 0.78 to 1.05. RNA inputs ranged from 8×10^4 to 2×10^5 cpm, and the percent ^{3}H cpm bound ranged from 0.03 to 0.12%.

TABLE III. Steady-State and ^{3}H Globin mRNA in Pulse-Chase-Labeled β^+ Thalassemia Erythroblasts[1]

	β/α Ratio			
Patient	Globin synthesis	Steady-state mRNA	^{3}H mRNA (pulse)	^{3}H mRNA (chase)
#1	0.16	0.16	1.3-1.4	—
#2	0.09	0.11	0.80-1.1	—
#3	0.14	0.41	0.84-0.91	0.44
#4	0.06	0.12	1.57	—

[1]Pulse-chase incubation conditions, filter hybridization analysis, and hybridization analysis of steady-state mRNA were as described in the text. Globin synthesis was performed by incubating erythroblasts with ^{3}H-leucine and analyzing radioactive globins by carboxymethyl cellulose column chromatography as described elsewhere [15]. RNA inputs ranged from 8×10^4 to 4×10^5 cpm and only results showing binding to filters at least twice background are shown. Ranges of values shown for patients #1–#3 represent ranges of three to five measurements of each RNA sample. Ratios were determined as in Table I.

TABLE IV. γ and δ ^{3}H mRNA Accumulation in β^+ Thalassemia Erythroblasts[1]

	Globin synthesis		^{3}H Pulse mRNA		^{3}H Chase mRNA	
Patient	γ/α	δ/α	γ/α	δ/α	γ/α	δ/α
#1	0.139	<0.1[2]	0.12	—	—	—
#2	0.067	<0.1	0.20	—	—	—
#3	0.211	<0.1	0.22	0.08	0.11	0.04
#4	0.07	<0.1	0.08	—	—	—
Nonthalassemic	0	<0.1	0	0.11	—	0.036

[1]Pulse-chase incubations and measurements of globin synthesis are described in the text and Tables I–III. The δ-globin DNA probe consisted of physically purified Eco RI fragment "F" from λ phage recombinant HβGl [cf 12] which contains only the more divergent (compared to β) 3′-mRNA coding sequence and large intervening sequence of the δ-globin gene. Eco RI fragment F was first subcloned in plasmid pBR322, and the insert subsequently purified by alkaline sucrose density gradient centrifugation. Ten micrograms of insert was bound to each filter. Patient numbering is the same as in Table III.

[2]<0.1, δ-globin peak too small to be quantitated. RNA inputs were as in Tables II and III, except that 1.2×10^6 cpm were analyzed to determine ^{3}H δ mRNA. Ratios were determined as in Table I.

As shown in Table II, nonthalassemic erythroblasts synthesized roughly equal amounts of β and α mRNA during a 20-minute pulse incubation. The β/α ^{3}H-mRNA ratio remained stable during a 24-hour chase incubation in the absence of ^{3}H-uridine. Both levels of radioactive mRNAs correlated well with steady-state mRNA levels determined by standard liquid hybridization techniques. To our surprise, as shown in Table III, β^+ thalassemic erythroblasts synthesized approximately equal amounts of β and α-globin mRNA during the pulse incubation. These results indicate clearly that a primary transcriptional defect is unlikely to account for the defective β mRNA accumulation in the steady state in these disorders. In each of the four cases, the newly synthesized radioactive β/α mRNA ratio was considerably higher than the steady-state β/α mRNA ratio determined with the same RNA samples using liquid hybridization techniques. In the one patient (patient # 3) in whom sufficient cells were available for a "chase" incubation, the β/α mRNA ratio declined to 0.44, a level correlating well with steady-state β/α mRNA ratios in these erythroblasts. These results indicate a clearcut posttranscriptional decline in β mRNA accumulation, compatible with a defect in either mRNA processing or long-term cytoplasmic stability. Further studies of the posttranscriptional defect in this patient are described in a subsequent section.

The unexpectedly abundant amounts of newly synthesized β mRNA detected in the above experiments cannot be due to artefacts resulting from γ-δ-β cross hybridization, as indicated in Table IV. These results indicate that γ/α, and, where measured, δ/α ratios of newly synthesized RNA, were clearly distinguishable by the use of the appropriate cloned DNA sequence probes from β/α mRNA ratios, and were, in any event, too low to contribute significantly to the measured β/α mRNA ratios.

Production of δ mRNA in Normal and β Thalassemic Erythroblasts

Table IV indicates that in both normal and β thalassemic erythroblasts, the initial accumulation of δ mRNA sequences is considerably lower than the accumulation of α or β mRNAs. These results indicate that the primary cause of differential δ and β-globin gene expression encountered in adult erythroblasts is mediated either at the transcriptional level, or during early intranuclear processing events which occur too quickly to be distinguished in the 20-minute pulse incubation employed. Our results are not compatible with previous hypotheses [5] which suggest the long-term cytoplasmic stability of δ mRNA may be the primary factor contributing to the low rates of δ-globin gene expression. However, as indicated in Table IV, a further decline in the δ/α mRNA ratio was noted during prolonged chase incubations, suggesting that further relative declines in δ mRNA accumulation may increase the discrepancy between δ and β-globin gene expression. Unfortunately, the number of cpm binding to δ DNA filters in these "chase" experiments was exceedingly low; we are reluctant to conclude that the further declines noted during chase incubations are statistically significant.

Apparent Persistence of Intervening Sequence Transcripts in a Patient With Homozygous β^+ Thalassemia

As noted earlier, only one patient with β thalassemia, patient # 3, yielded sufficient bone marrow erythroblasts for studies with both pulse and chase-type incubations. As indicated in Table III, this patient produced normal levels of β and α mRNA during the pulse incubation; however, the β/α mRNA ratio declined during a 24-hour chase incubation. An interesting finding in this patient, noted in Table III, is the relatively high level of β and α mRNA sequences detected in steady-state RNA from bone marrow erythroblasts. These findings are similar to those reported by Nienhuis, Turner, and Benz [6] in three other patients with β^+ thalassemia. As indicated in Table V, peripheral blood mRNA and β/α globin synthesis by peripheral blood reticulocytes yielded somewhat lower β/α ratios than expected on the basis of results obtained during the chase incubation or by analysis

TABLE V. Persistence of ^3H β Intron RNA in Pulse-Chase-Labeled β^+ Thalassemic Erythroblasts[1]

RNA sample	β/α	β Intron/β
β^+ Thalassemia, Pt. #3		
Pulse	0.91	1.08
Chase	0.44	0.26
Peripheral blood	0.27	—
Nonthalassemic, Pt. #5		
Pulse	1.00	0.71
Chase	1.25	<0.01

[1]The probe for the large β gene intervening sequence intron was obtained by isolating the 1,000 bp Eco RI-Bam HI double digestion fragment from a Pst I subclone of HβGI [13]; the Pst I clone contained only the β-globin gene. The Eco RI-Bam HI fragment contained the entire large intervening and approximately 60 bp of mRNA coding sequences flanking the ends of the intron. The "intron" fragment was cloned in pBR322 and used as probe (10 μg/filter). β intron/β = ^3H cpm bound to β intron probe/^3H cpm bound to JW102 β cDNA probe, after subtraction of background cpm as described in Table I. The β/α ratios are derived from Tables II and III. The β intron/β ratios correspond to: 51.9% (β^+ thal pulse), 20.4% (β^+ thal chase), 41.5% (nonthal pulse), and 0% (nonthal chase), of total ^3H cpm in β gene transcripts present as intervening sequence RNA.

of steady-state RNA from erythroblasts. Thus, we analyzed pulse- and chase-labeled radioactive RNA from this patient for the presence of persistent RNA sequences complementary to the intervening sequences of the gene. As discussed by other reviewers in these proceedings, the intervening sequences of globin genes are apparently transcribed into the initial mRNA precursors, but subsequently excised, and presumably degraded. In order to measure newly synthesized RNA containing the intervening sequence, or intron, we purified the large intervening sequence of the β-globin gene from the λ phage clone HβGI, which contains the entire human δ and β-globin genes [13]. The purified intervening sequence fragments were then "subcloned" in plasmid pBR322, and utilized as nonradioactive DNA probes by the nitrocellulose filter technique described earlier.

As shown in Table V, normal erythroblasts contained approximately 40% of total β globin gene transcripts in the form of intervening sequences. A similar level of β intron radioactivity was observed with the thalassemic erythroblasts. During prolonged 24-hour chase incubations, intervening sequence radioactivity essentially disappeared from nonthalassemic erythroblasts. In contrast, erythroblasts from patient # 3 retained small but readily detectable levels of β intron radioactivity. As shown in Table V, β introns still comprised approximately 20% of the reduced level of total β mRNA sequences remaining after the chase incubation.

The above results constitute preliminary evidence for persistence of transcripts of the intervening sequence in this patient with β thalassemia, and might account, at least partially, for the higher level of β mRNA detected in total cellular RNA from nucleated erythroblasts as compared to reticulocytes. It is tempting to speculate that this individual may fail to process the initial mRNA precursor properly, with the result that precursor "hangs up" in the nucleus in high molecular weight form. We are currently studying RNA from this patient with RNA gel blotting methods in order to assess these possibilities.

SUMMARY AND CONCLUSIONS

We have studied globin mRNA metabolism in nonthalassemic and β^+ thalassemic erythroblasts. Our findings suggest that in both normal and β thalassemic erythroblasts, initial levels of β and α mRNA production are equal. These data rule out a primary transcriptional defect in the most common forms of β thalassemia. Newly synthesized β mRNA appears to remain relatively stable in normal patients during prolonged chase incubations, whereas, in at least one patient with β^+ thalassemia, a posttranscriptional decline of β mRNA to thalassemic levels occurs. In this individual, persistence of β intervening sequence transcripts is also apparent during the chase incubation, suggesting that faulty β mRNA processing may be the primary molecular defect. In contrast, normal δ-globin gene expression appears to be reduced at the level of transcription or early intranuclear processing.

REFERENCES

1. Weatherall DJ, Clegg JB: "The Thalassemia Syndromes," 2nd ed. Oxford: Blackwell Scientific Publications, 1972.
2. Benz EJ, Forget BG: Pathogenesis of the thalassemia syndromes. Pathobiology Annual, 10:1–33, 1980.
3. Nienhuis AW, Benz EJ: Regulation of hemoglobin synthesis during development of the red cell. N Eng J Med 297:1316, 1977.
4. Benz EJ, Forget BG: Molecular genetics of the thalassemia syndromes. Progr Hemat 9:107, 1975.
5. Wood WG, Old JM, Roberts AVS, Clegg JB, Weatherall DJ, Quattuin N: Human globin gene expression: Control of β, δ, and $\delta\beta$ chain production. Cell 15:437, 1978.
6. Nienhuis AW, Turner PA, Benz EJ: Relative stability of α and β-globin messenger RNAs in homozygous β^+ thalassemia. Proc Natl Acad Sci USA, 74:3960, 1977.
7. Comi P, Giglioni B, Barbarano, C, Ottolenghi S, Williamson R, Novakova M, Masera G: Transcriptional and post-transcriptional defects in β^0 thalassemia. Eur J Biochem 79:617, 1977.
8. Benz EJ, Glass J, Tsistrakis GA, Hillman DG, Cavallesco C, Coupal E, Forget BG, Turner PA, Kantor JA, Nienhuis AW: Heterogeneity of m RNA defects in the thalassemia syndromes. Ann NY Acad Sci 344:101–112, 1980.
9. Benz EJ, Scarpa AL, Tonkonow B, Ritchey AK, deRiel JK, Forget BG: Abnormal messenger RNA processing in β^+ thalassemia. Cell (submitted).
10. Lowenhaupt K, Lingrel JB: A change in the stability of globin mRNA during the induction of mouse erythroleukemia cells. Cell 14:337, 1978.
11. Kinniburgh AJ, Ross J: Processing of the mouse β-globin mRNA precursor: At least two cleavage-ligation reactions are necessary to excise the larger intervening sequence. Cell 17:915, 1979.
12. Benz EJ, Forget BG, Hillman DG, Cohen-Solal M, Pritchard J, Cavallesco C: Variability in the amount of β-globin mRNA in β^0 thalassemia. Cell 14:299, 1978.
13. Lawn RM, Fritsch EF, Parker RC, Blake G, Maniatis T: The isolation and characterization of linked δ- and β-globin genes from a cloned library of human DNA. Cell 15:1157, 1978.
14. Wilson JT, Wilson LB, deRiel JK, Villa-Komaroff L, Efstratiadis A, Forget BG, Weissman SM: Insertion of synthetic copies of human globin genes into bacterial plasmids. Nucleic Acids Res 5:563, 1978.

Relative Stability of Globin and Non-Globin mRNA in Terminally Differentiating Friend Cells

Vladimir Volloch and David Housman
Center for Cancer Research and Department of Biology, Massachusetts Institute of Technology, Cambridge, Massachusetts

INTRODUCTION

The ability to analyze the molecular events which control levels of globin gene expression depends critically on the availability of model systems in which this process can be studied. A great deal of interest and experimental effort has been focussed on the Friend cell system in this connection, but certain aspects of the behavior of the system have been puzzling because the behavior of normal erythroid cells undergoing terminal erythroid differentiation is not reproduced by the supposed model system. In particular, synthesis of the α and β chains of globin in the reticulocyte reach levels of greater than 90% of total protein synthesis. The nature of the molecular mechanisms which control this specialization of the protein synthetic apparatus has been a major reason that particular interest has focussed on the control of globin gene expression. Friend cells, however, do not reproduce this specialization when induced to differentiate under conditions which are normally used by most laboratories [1, 2]. It is unusual to observe levels of globin synthesis above 20% of total protein synthesis in published studies using this system [2], and values equivalent to the 90% of total protein synthesis routinely observed in reticulocytes have never been reported for Friend cells.

Over the past two years we have investigated the Friend cell system from this perspective starting from the hypothesis that physical destruction of end stage cells was an important factor in the failure of the Friend cell system to correctly reproduce terminal erythroid differentiation in vitro. The basic conclusions of our studies support the concept that cell lysis is an important factor in Friend cell cultures induced to differentiate under standard conditions. We have found that by the inclusion of a high molecular weight component in the medium such as BSA or FICOLL, end stage Friend cells can be physically stabilized. When Friend cells are stabilized in this manner a population of cells can be produced in vitro which closely resembles normal mouse reticulocytes in both physical and biochemical characteristics. In particular, the levels of globin synthesis and globin mRNA in such physically stabilized cells approach the levels observed in reticulocytes [3]. The ability to physically stabilize the end stage cells has led us to reevaluate some puzzling aspects of globin mRNA metabolism which have been observed in the

Friend cell system. In particular it has been reported that when Friend cells are exposed to an inducer such as DMSO for three days or more globin mRNA is reduced in stability, compared to other cellular mRNAs [4]. This view of globin mRNA stability in terminally differentiating Friend cells suggests that the control of globin mRNA stability in Friend cells must be radically different from the control of globin mRNA stability in normal erythroid cells differentiating in vivo since there is no obvious way that globin mRNA can become the predominant mRNA in the cell if it is less stable than other cellular mRNAs. The insight gained by the demonstration that cell lysis is an important factor in differentiating Friend cell cultures has led us to the conclusion that the apparent degradation of globin mRNA observed in differentiating Friend cell cultures is actually an artifact of cell lysis. When Friend cells are physically stabilized, we have found that globin mRNA is considerably more stable than other cellular mRNAs. These results imply that the relative stability of globin mRNA in relation to other mRNAs plays an important role in determining the ultimate levels of expression of the globin genes in Friend cells.

PHYSICAL STABILIZATION OF DIFFERENTIATING FRIEND CELLS

Friend cells are normally induced to differentiate by addition of between 1% and 2% DMSO to Friend cells grown in tissue culture medium such as MEM, DME, α, or RPMI supplemented with 10% to 15% fetal calf serum. Under these circumstances, within four to five days of exposure to DMSO the number of benzidine-positive cells in the culture reaches a value between 70% and 95% of the cell population. However, after four days of exposure to inducer, while the number of cells may remain constant, the situation in the culture actually represents a dynamic one in which some cells are lysing while other less-differentiated cells are dividing. To demonstrate this point clearly we performed an experiment in which Friend cells were pulse-labeled ^{14}C TdR for six hours prior to the addition of DMSO. The ^{14}C TdR is incorporated into cellular DNA which should remain stable regardless of the number of cell divisions which occur after the initial incorporation. As long as the progeny of the pulse-labeled cells are physically stable then no loss of incorporated ^{14}C TdR from the cells should be observed. The results shown in Table I demonstrate that in a normally induced Friend cell culture such stability is not observed. On the contrary, between four and six days of exposure to DMSO the proportion of ^{14}C TdR which remains TCA-precipitable drops from 100% to 31%. This result is almost certainly due to cell lysis since no cells which have lost their nuclei are observed in these cultures. The addition of 4% BSA to the medium appears to physically stabilize the cells, however. As shown in Table I, a decrease of less than 10% in TCA-precipitable ^{14}C TdR is observed in the BSA-treated cells during the period when the normally induced culture exhibits a loss of almost 70% of TCA-precipitable ^{14}C TdR. Further characterization of BSA-treated cells by standard morphological criteria leads to the conclusion that a highly differentiated population of cells has been produced which morphologically resemble late orthochromatophilic erythroblasts just prior to enucleation. Examination of cell volume profiles also suggests that the BSA-treated culture is a physically uniform one, exhibiting a mean cell volume of approximately 100 μ^3, close to the value observed for normal reticulocytes, while the untreated population has a much broader cell volume distribution with a mean value of approximately 500 μ^3.

TABLE I. Stability of MEL Cells Prelabeled with ^{14}C Thymidine Grown Under Various Culture Conditions[1]

	CPM (%)								
	Ohr	10	32	60	82	94	108	120	144
Uninduced MEL cells in α^- medium	100	97	96	93	94	95	94	94	93
Induced MEL cells in α^- medium	100	96	95	91	76	65	37	34	31
Induced MEL cells in α^- medium with 4% BSA and 2×10^{-3} M Iron-Dextran	100	96	94	93	94	97	91	94	93

[1] Cells were labeled for 6 hours with 0.5 μCi/ml (methyl - ^{14}C) thymidine (Amersham, 55 mCi/mmol), then concentrated by centrifugation, washed and resuspended in fresh medium, and divided into three subcultures. One culture was treated with 1.8% DMSO, a second culture was treated with 1.8% DMSO in medium supplemented with 4% BSA and 2×10^{-3} M iron in the form of an iron-dextran complex, while a third culture was resuspended in α medium alone. At each time indicated in the table, an equal aliquot of the culture was harvested by centrifugation, nuclei were isolated, lysed, sonicated and ^{14}C thymidine remaining in acid-precipitable form was determined.

Levels of globin synthesis in Friend cells treated with DMSO + 4% BSA are comparable to the levels observed in mouse reticulocytes. In Figure 1, the levels of globin synthesis in BSA-treated cells have been compared to normal cells by SDS-polyacrylamide gel electrophoresis. It is clear that at least 85% of cellular protein synthesis is in the form of polypeptides which comigrate with mouse globin in this gel system. Comparable experiments in which globin mRNA levels have been determined by hybridization to a ^3H cDNA probe lead to similar conclusions—the globin mRNA levels in Friend cells treated with BSA + DMSO are equivalent to the level observed in mouse reticulocytes.

GLOBIN mRNA STABILITY IN BSA-TREATED FRIEND CELLS

Given the ability to produce populations of terminally differentiated Friend cells, we chose to reexamine the question of globin mRNA stability in Friend cells. Our concerns with previous data were centered on the following issues. Treatment of Friend cells with DMSO leads to asynchronous commitment of Friend cells to the terminal differentiation program because of the stochastic nature of the commitment process [5]. A consequence of this asynchrony in commitment is to produce a population of cells which are heterogeneous with respect to the stage of the differentiation program. As cells differentiate the rate of incorporation of ^3H uridine into ribosomal RNA decreases at least tenfold [6, 7], while the rate of incorporation of ^3H uridine into cytoplasmic globin mRNA increases dramatically. If the end stage cells are physically unstable but earlier stages are physically stable then an experiment involving a pulse and chase with ^3H uridine would yield results in which rRNA would appear to be stable compared to globin mRNA. Results of this type were observed by several groups [4, 8]. In general a half-life greater than 60 hours was observed for rRNA while a half-life of approximately 17 hours was observed for

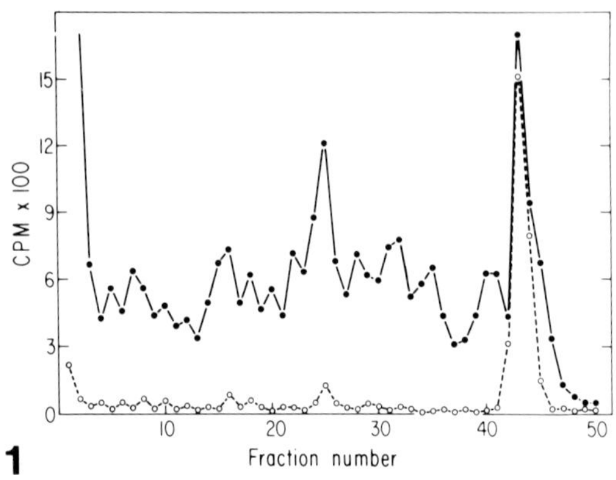

1

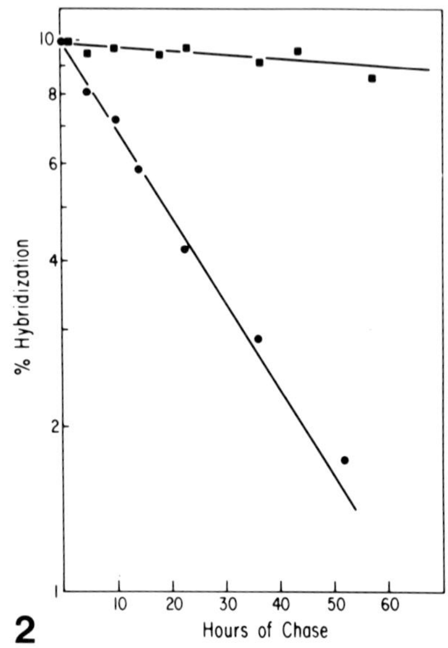

2

Fig. 1. Analysis of polypeptides synthesized in therminally differentiated MEL Cells. MEL cells, clone PC-4-10 of the line 745, growing exponentially in α^- medium with 13% FCS were treated with DMSO in the same medium or in medium supplemented with 5% BSA and 2×10^{-3} M iron-dextran. After 5.5 days approximately 6×10^5 cells were labeled for two hours with ^{35}S methionine, 50 μCi/ml (Amersham, 1,200 Ci/mm). Following incubation, cells were washed with the same medium, concentrated by centrifugation, and cytoplasmic lysates were prepared. The lysates were subjected to PAGE on a 7.5%–20% polyacrylamide gradient slab gel with SDS [12]. Lanes containing labeled material were sliced into 0.4 cm pieces, dissolved in Protosol (New England Nuclear), and radioactivity was counted in Biofluor. Bgr was subtracted. ●———● Cytoplasmic lysate from cells induced in regular medium; ○----------○ cytoplasmic lysate from cells induced in medium supplemented with 5% BSA and 2×10^{-3} M iron-dextran.

Fig. 2. Stability of ^3H globin mRNA under various culture conditions. Cells were treated with 1.8% DMSO for four days either in α^- medium with 13% FCS or in the same medium but supplemented with 4% BSA and 2×10^{-3} M iron-dextran, and then labeled for two hours with ^3H uridine. The cells were then concentrated by centrifugation, washed and resuspended in the same medium, supplemented with 5 mM uridine and 2 mM cytidine. At each time point shown in the figure, an equal aliquot of the culture was harvested by centrifugation and total cytoplasmic RNA was isolated by phenol extraction and hybridized to globin cDNA cellulose as described [8]. From a second aliquot the proportion of ^3H uridine remaining in acid-precipitable form was determined. At the start of the chase the values for the total acid-precipitable ^3H uridine per aliquot was 1.7×10^6 cpm for cells induced in regular medium and 1.45×10^6 cpm for cells induced in the medium supplemented with BSA. Respectively, 6,460 cpm and 8,297 cpm hybridized to globin cDNA-cellulose. Data on proportion of globin mRNA during the chase period are calculated by dividing the number of cpm hybridized to globin cDNA cellulose at each time point by the total number of cpm present in the aliquot at this same time. This value is then divided by the 0 time value to obtain a relative proportion of globin mRNA molecules which have survived the chase period. ●———● Stability of globin mRNA in cells induced in regular medium; ■———■ stability of globin mRNA in cells induced in medium supplemented with 4% BSA and 2×10^{-3} M iron-dextran.

globin mRNA. Using Friend cells treated with 1.5% DMSO without physical stabilization we obtained results similar to those obtained by other groups (Fig. 2). However, if 3% BSA was added to the culture medium globin mRNA was observed to be as stable as rRNA. We conclude that globin mRNA has a half-life of at least 60 hours in terminally differentiating Friend cells.

DESTABILIZATION OF NON-GLOBIN mRNA

The data presented in the previous section show that globin mRNA in terminally differentiating Friend cells has a half-life considerably greater than suggested in previous studies. To fully explain the accumulation of globin mRNA to the high proportion of mRNA observed in reticulocytes, further analysis is necessary. The stability of non-globin mRNA in uninduced Friend cells has been measured previously by binding to oligo(dT) cellulose. Using this criteria for mRNA, two stability classes termed M1 and M2 have been observed. In uninduced Friend cells class M1 has an apparent stability of three hours while class M2 has an apparent stability of 40 hours. If these parameters were to remain constant during terminal erythroid differentiation, then the accumulation of globin mRNA to a proportion of 90% of cellular mRNA would require at least ten days. However, this accumulation clearly occurs in a much shorter period of time. It therefore seems reasonable to presume that non-globin mRNA must be destabilized during terminal erythroid differentiation. Aviv et al have proposed that this is the case [8, 9]. To determine whether the predicted destabilization of non-globin mRNA in fact occurred the experiment shown in Figure 3 was performed. In this experiment BSA was used to physically stabilize terminally differentiating cells as in previous studies. In this experiment Friend cells were pulse-labeled and chased at different times of induction (0, 48, 96, 144 hours) and stability of non-globin poly(A)-containing RNA was measured by oligo(dT) cellulose chromatography after removal of globin mRNA by hybridization to globin cDNA cellulose. The data clearly shows that as differentiation proceeds, non-globin poly(A)-containing RNA is destabilized. For the M2 population in uninduced cells a half-life of 40 hours is observed. After two days of DMSO treatment a half-life of less than 20 hours is observed. By six days a half-life of approximately five to six hours is observed for non-globin mRNA. It is important to note that globin mRNA is as stable as rRNA under these conditions; hence its accumulation as a relative proportion of cellular mRNA is in fact favored as the differentiation of MEL cells proceeds.

DISCUSSION

Over the course of our studies on physical stabilization of end stage Friend cells a number of factors have been critical in achieving a uniform population of end stage cells. As mentioned previously, the stochastic nature of the commitment process leads to an inherent asynchrony in the entry of cells into the terminal differentiation pathway. We have found that the use of a recently cloned cell population is helpful to achieving the most uniform population of terminally differentiating cells. A cell population which has not been cloned prior to the start of an experiment may give rise to a significant number of uncommitted cells which can continue to replicate and overgrow the differentiated populations. An expedient which has served to avoid the necessity of recloning prior to each experiment is to freeze multiple

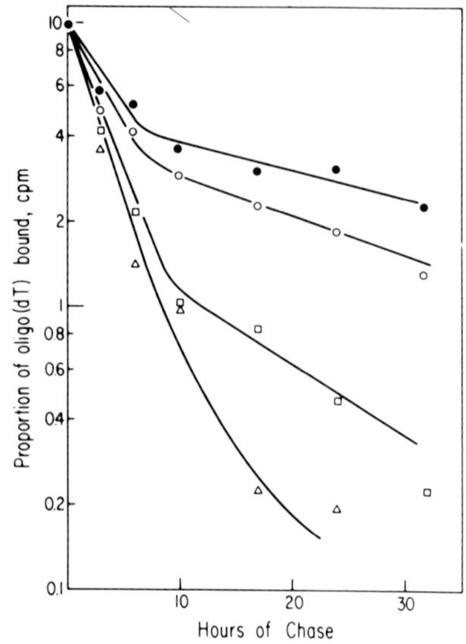

Fig. 3. Decay kinetics of ^{3}H uridine-labeled cytoplasmic poly(A) + RNA from MEL cells exposed to DMSO for various lengths of time. Freshly cloned MEL cells, clone PC-4-10-4, growing in α^{-} medium with 13% FCS supplemented with 3% BSA and 2×10^{-3} M iron-dextran and treated with 1.8% DMSO for different times were labeled for three hours (0, 48, 96 hours induced cells) or 20 hours (144-hour induced cells) with 30 μCi/ml ^{3}H uridine. The cells were then concentrated, washed and resuspended in the same medium, supplemented with 5 mM uridine, 2 mM cytidine. At each time point shown in the figure, an equal aliquot of culture was harvested and total cytoplasmic RNA was isolated and hybridized to globin cDNA cellulose as described [8]. Nonhybridized RNA was subjected to oligo(dT) cellulose affinity chromatography. The proportion of poly(A) + RNA was calculated by dividing the number of cpm eluted from oligo(dT) cellulose by the total number of acid-precipitable cpm present in the aliquot of total cytoplasmic RNA at this same time. This value was then divided by the 0 time value to obtain a relative proportion of ^{3}H poly(A) + RNA remaining during the chase period. ●———● Uninduced cells; ○———○ 48-hour induced cells; □———□ 96-hour induced cells; △———△ 144-hour induced cells.

aliquots of a single clone which shows little or no overgrowth as soon as possible after the clone has been expanded to a sufficient number of cells for freezing. Overgrowth can also be prevented by the addition of hydroxyurea at four days after addition of DMSO.

In studies presented elsewhere in this volume we suggest that the two apparent stability classes observed for non-globin mRNA are actually the result of the removal and readdition of poly(A) from the 3′ end of the mRNA. We suggest that there may be an intimate connection between poly(A) metabolism and mRNA stability. It is of interest to note that our data [15] as well as the data of others [10] indicate that the proportion of nonglobin mRNA in Friend cells which will not bind to oligo(dT) cellulose is approximately 40%. The globin mRNA binds to oligo(dT) cellulose with an efficiency of 95% or more [11]. These results suggest that a higher proportion of globin mRNA may have long stretches of poly(A) at the 3′ end than other cellular mRNAs. We would like to suggest that the relatively high extent of polyadenylation of globin mRNA could be a possible reason for its relatively high stability. Data consistent with this view have been reported [13, 14] but a definitive test of this hypothesis remains to the future.

SUMMARY

We demonstrate that Friend cells treated with DMSO give rise to a population of terminally differentiated cells which are physically unstable under standard culture conditions. The addition of bovine serum albumin or FICOLL physically stabilizes such cells. We have investigated the stability of globin mRNA in terminally differentiating Friend cells and show that globin mRNA is as stable as ribosomal RNA. By contrast, we show that non-globin mRNA becomes less stable as the differentiation program proceeds. These results support the view that differential mRNA stability plays an important role in the generation of the highly specialized translational repertoire of the reticulocyte. We suggest that the Friend cell system is an appropriate experimental system to analyze the molecular mechanisms which control this specialization.

REFERENCES

1. Harrison, PR: Analysis of erythropoiesis at the molecular level. Nature 262:353–356, 1976.
2. Marks, PA, Rifkind RA: Erythroleukemic differentiation. Ann Rev Biochem 47:419–448, 1978.
3. Volloch V, Housman D: Manuscript in preparation.
4. Lowenhaupt K, Lingrel J: A change in the stability of globin mRNA during the induction of murine erythroleukemia cells. Cell 14:337–344, 1978.
5. Gusella J, Geller R, Clarke B, Weeks V, Housman D: Commitment to erythroid differentiation by Friend erythroleukemia cells: A stochastic analysis. Cell 9:221–229, 1976.
6. Sherton C, Kabat D: Changes in RNA and protein metabolism preceeding onset of hemoglobin synthesis in cultured Friend leukemia cells. Dev Biol 48:118–131, 1976.
7. Volloch V: Unpublished results.
8. Aviv H, Volloch V, Bastos R, Levy S: Biosynthesis and stability of globin mRNA in cultured erythroleukemia Friend cells. Cell 8:495–503, 1976.
9. Bastos R, Aviv H: Theoretical analysis of a model for globin mRNA accumulation during erythropoiesis. J. Mol Biol 110:205–218, 1977.
10. Minty A, Gros F: Coding potential of non polyadenylated mRNA in mouse Friend cells. J Mol Biol 139:61–83, 1980.
11. Ross J, Gielen J, Packman S, Ikawa Y, Leder P: Globin gene expression in cultured erythroleukemia cells. J. Mol Biol 87:697–714, 1974.
12. Lieberman D, Volloch V, Aviv H, Nudel U, Revel M: Effects of interferon on hemoglobin synthesis and leukemia virus production in Friend cells. Mol Biol Rep 2:447–458, 1974.
13. Gorski J, Morrison M, Merkel C, Lingrel J: Size heterogeneity of polyadenylate sequences in mouse globin messenger RNA. J Mol Biol 86:363–371, 1974.
14. Sheiness D, Darnell J: Polyadenylic acid segment in mRNA becomes shorter with age. Nature 241:265–268, 1973.
15. Volloch V, Housman D: This volume.

Organization and Expression of Globin Genes, pages 259–266

Regulation of Stability of Non-Globin mRNA in Friend Cells

Vladimir Volloch and David Housman

Center for Cancer Research and Department of Biology, Massachusetts Institute of Technology, Cambridge, Massachusetts

INTRODUCTION

The analysis of mRNA stability in eukaryotic cells is an area of obvious import for many questions involving the control of gene expression. Early work measuring mRNA stability by pulse labeling with ^{3}H uridine followed by a chase with actinomycin D led to the conclusion that cytoplasmic mRNA decayed with a three-hour half-life. Subsequent work using binding to oligo(dT) cellulose as a criterion for mRNA led to a revised view of mRNA stability. These results indicated that polyadenylated cytoplasmic mRNA exhibited biphasic decay kinetics. In a variety of different eukaryotic cell types one class of mRNA molecules decayed with a rather short half-life of between ten minutes and three hours, while the second class decays with a much longer half-life of 20 to 50 hours [1–5]. In MEL cells, these two kinetic classes termed M1 and M2 have been reported to exhibit half-lives of three hours and 30 hours respectively [6]. In the present study we have reexamined the question of mRNA stability in MEL cells using an alternative approach to the measurement of mRNA half-life. We have synthesized cDNA cellulose complementary to total cellular mRNA in quantities sufficient to drive the hybridization of as much as 70% of the polyadenylated mRNA. Hybridization to this cDNA cellulose has given us an alternative method for the measurement of mRNA stability. In pulse-chase experiments the stability of all cellular RNA sequences hybridizing in our cDNA cellulose proved to be equal to the stability of the M2 class of cytoplasmic mRNA measured by the oligo(dT) binding method. Direct analysis of the hybridization characteristics of nonpolyadenylated RNA fraction revealed that the appearance of label in these sequences during a pulse-chase protocol was consistent with conversion of polyadenylated mRNA to the nonpolyadenylated form with kinetics equal to the apparent stability of the M1 mRNA class and a subsequent-steady state equilibrium between polyadenylated and nonpolyadenylated forms for each mRNA. Using our recently developed method for stabilizing MEL cells in the terminal stage of differentiation (elsewhere in this volume) we have also applied the above method of measurement of total mRNA stability to examine the stability of non-globin mRNA during MEL cell differentiation.

MEASUREMENT OF STABILITY OF NON-GLOBIN mRNA IN UNINDUCED FRIEND CELLS

To measure the stability of non-globin mRNA in Friend cells a technique of hybridization of labeled RNA to an excess of cDNA bound to a solid phase was developed. cDNA was hybridized by reverse transcribing cytoplasmic poly(A)$^+$ RNA using oligo(dT) as primer. This cDNA was then covalently bound to ABM cellulose [7], and used for hybridizations with labeled cellular RNA as described by Aviv and Levy [8].

Initial experiments were designed to determine the proportion of cellular mRNA from uninduced Friend cells which could be hybridized by the cDNA cellulose. Exponentially growing Friend cells were labeled with tritiated uridine for either four or 30 hours. Cytoplasmic RNA was isolated as described [6] and subjected to oligo(dT) cellulose chromatography. A constant amount of poly(A)$^+$ RNA was then hybridized to increasing concentrations of cDNA cellulose (Table I).

As one can see, more than 70% of both ^{3}H uridine-labeled RNA preparations can be hybridized to the cDNA cellulose under optimal conditions. In all further experiments we used concentrations of cDNA cellulose on the plateau of this curve, ie, more than 70% of input mRNA bound. These conditions are appropriate for the hybridization of mRNA species in a high or middle class abundance in uninduced cells. To determine stability of cytoplasmic mRNA in uninduced cells exponentially growing Friend cells were labeled with ^{3}H uridine for either two or 30 hours. Unincorporated ^{3}H uridine was removed by several successive centrifugation steps and a chase was performed by addition of cold uridine (5 mM) and cytidine (2 mM) to the culture media.

At different chase times total cytoplasmic RNA was isolated and an aliquot hybridized directly to cDNA cellulose. An equal aliquot was subjected to oligo(dT) cellulose chromatography and thereby separated into poly(A)$^+$ and poly(A)$^-$ fractions. These fractions were also hybridized to cDNA cellulose (Fig. 1). As one can see in Figure 2b, the behavior of poly(A)$^+$ RNA gives the previously observed pattern [6, 9]—two kinetic populations are apparently observed. M1 with a half-life of four hours and M2 with a half life of 40 hours.

However, the results of hybridization with both total RNA and with poly(A)$^-$ RNA are very different. The result with total RNA indicates that there exists only one stability class of mRNA. Also, the proportion of poly(A)$^-$ cytoplasmic mRNA (RNA which hybridizes to cDNA-cellulose) increases for several hours after the chase begins, reaches a maximum, and then declines with the same rate as total mRNA and M2 populations of poly(A)$^+$ RNA.

These results are consistent with the view that high and middle abundance classes of mRNA in Friend cells have a single average rate of decay of approximately 40 hours. According to this view the apparent existence of M1 and M2 classes for mRNA stability is the result of removal of poly(A) from the 3′ end of the mRNA, rather than physical destruction of the mRNA. A model of this type predicts that at any given chase time the sum of poly(A)$^+$ and poly(A)$^-$ RNA will be equal to total mRNA. Our results verify this prediction. The increase in proportion of poly(A)$^-$ mRNA suggests that some poly(A)$^+$ mRNA molecules have a sufficient amount of poly(A) removed so that they no longer bind to oligo(dT) cellulose and are thus assigned to the poly(A)$^-$ fraction.

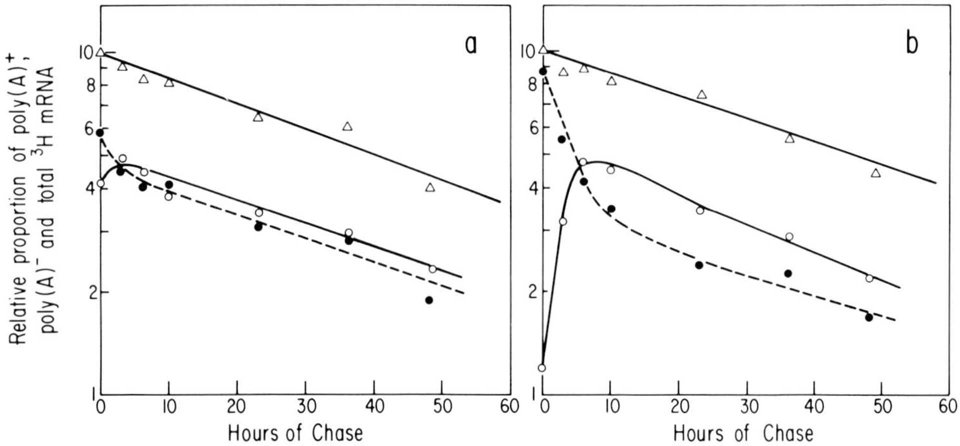

Fig. 1. Decay kinetics of ^{3}H-uridine-labeled cytoplasmic mRNA from MEL cells. MEL cells growing in α^{-} medium with 13% FCS were labeled with ^{3}H uridine for two or 30 hours. The cells were then concentrated, washed, and resuspended in fresh medium supplemented with 5 mM uridine and 2 mM cytidine. At each time point shown in the figure, an equal aliquot of the culture was harvested and total cytoplasmic RNA was isolated [6]. A sample of this RNA was subjected to oligo(dT) cellulose affinity chromatography [11]. Total ^{3}H RNA, poly(A)$^{+}$, and poly(A)$^{-}$ RNA corresponding to the same number of cells were hybridized to total cDNA cellulose [8]. The "0" hour value of hybridization with total ^{3}H RNA was taken as 100% for the calculation of the relative proportions of total, poly(A)$^{+}$ and poly(A)$^{-}$ ^{3}H mRNA remaining after a chase. Calculations are based on acid-precipitable ^{3}H uridine in each total RNA preparation. a. Cells were labeled for 30 hours. b. Cells were labeled for two hours. △——△ hybridization with total ^{3}H RNA. ●——● Hybridization with total poly(A)$^{+}$ ^{3}H RNA. ○——○ Hybridization with total poly(A)$^{-}$ ^{3}H RNA.

The proportion of both labeled poly(A)$^{+}$ and poly(A)$^{-}$ RNA reach steady-stage values different from both zero and the total mRNA level. This observation suggests that the reverse process, the addition of poly(A) tails to the poly(A)$^{-}$ RNA, also operates in Friend cells. This model is presented in the Appendix.

As the length of the labeling period increases, one would predict that the proportion of labeled poly(A)$^{+}$ and poly(A)$^{-}$ RNA will come closer to this steady-state value. One also expects that the sum of the proportions of poly(A)$^{+}$ and poly(A)$^{-}$ RNA will at all chase times equal the proportions of total mRNA. Finally, one also expects that the degradation kinetics of all three populations will be the same as in Figure 1b.

The results presented in Figure 1a fit these predictions very well. As the labeling period increases from two to 30 hours, the initial proportions of poly(A)$^{-}$ mRNA increase from 10% to 40%, and the proportion of poly(A)$^{+}$ RNA decreases from 90% to 60%. The kinetic behavior of all three populations is the same in both experiments.

To determine the actual proportion of cytoplasmic mRNA which is poly(A)$^{-}$ the following experiment was performed. Cells were labeled for different periods of time, and cytoplasmic RNA was prepared. An aliquot of each of these samples was hybridized to cDNA cellulose. Poly(A)$^{-}$ RNA derived from an equal aliquot was also hybridized to cDNA cellulose (Table II). The results clearly show that the

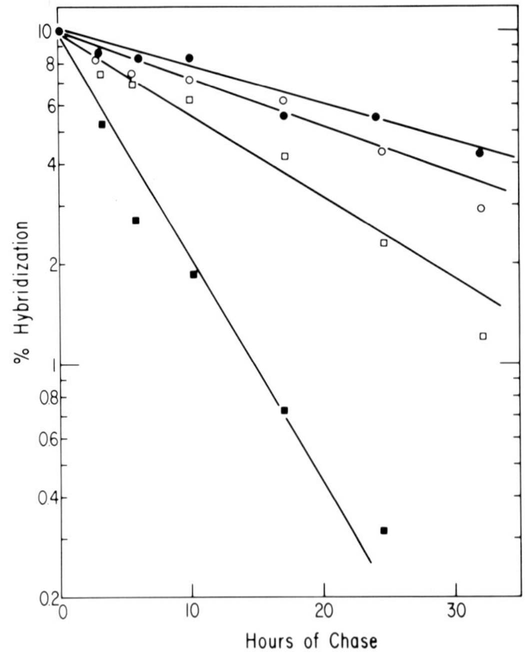

Fig. 2. Destabilization of non-globin mRNA in differentiating Friend cells. MEL cells, clone PC4-7, growing in α^- medium with 13% FCS, supplemented with 4% BSA and 5×10^{-3} M iron-dextran were treated with 1.8% DMSO for different periods of time (0, 48, 96, 120 hours). The cells were then labeled with ^{3}H uridine for two hours (0-, 48-, 96-hour DMSO treated cells) or for 24 hours (120-hour treated cells). After labeling cells were concentrated, washed, and resuspended in 50% fresh-50% conditioned medium of the same composition as described above, supplemented with 5 mM uridine and 2 mM cytidine. At each time point shown in the figure, an equal aliquot of culture was harvested by centrifugation and total cytoplasmic RNA was isolated [6]. These RNA were then hybridized to total cDNA cellulose [8]. The "0" hour of chase value of hybridization was taken as 100% for the calculation of the relative proportion of ^{3}H RNA remaining after a chase. ●——● Hybridization with ^{3}H RNA from cells treated with DMSO for 0 hours. ○——○ Hybridization with ^{3}H RNA from cells treated with DMSO for 48 hours. □——□ Hybridization with ^{3}H RNA from cells treated with DMSO for 96 hours. ■——■ Hybridization with ^{3}H RNA from cells treated with DMSO for 120 hours.

TABLE I. Hybridization of ^{3}H-Uridine-Labeled Poly(A)$^+$ Cytoplasmic RNA With Different Concentrations of Total cDNA-Cellulose[1]

cDNA cellulose/μl	10	30	50	100	150	200	250
4 hour-labeled RNA, % hybridized	9	29	44	57	64	73	76
30 hour-labeled RNA, % hybridized	8	26	40	52	66	70	72

[1]MEL cells were exponentially growing in α^- medium with 13% FCS were labeled with ^{3}H uridine (30 μCi/ml) for four or 30 hours. Then cells were collected by centrifugation, lysed, and total cytoplasmic and poly(A)$^+$ RNA were isolated as described [6]. Hybridization was performed as detailed by Levy and Aviv [8]. Aliquots of four and 30 hour-labeled poly(A)$^+$ RNA (corresponding to the same number of cells) were hybridized for 30 hours with different dilutions of total cDNA cellulose.

TABLE II. Hybridization with Total cDNA Cellulose of ^3H Cytoplasmic RNA Labeled for Different Times[1]

	Time of labeling (hours)						
	1	3	6	10	18	24	36
Total cpm hybridized	12,711	18,837	21,256	14,523	12,112	12,414	10,024
Poly(A)$^-$ RNA, cpm hybridized	767	1,648	3,381	3,324	3,732	4,719	3,867
Percent of total hybridized RNA in poly(A)$^-$ form	6	9	16	23	31	38	39

[1]MEL cells growing in α^- medium with 13% FCS were labeled with ^3H uridine for different times from one to 36 hours. For the 10-, 18-, 24-, and 36-hour time points specific activity of ^3H-uridine (30 μCi/ml) was decreased two, three, four, and six times respectively. By the end of labeling period cells were harvested by centrifugation, total cytoplasmic RNA was isolated [6] and separated into poly(A)$^+$ and poly(A)$^-$ fractions by oligo(dT) cellulose affinity chromatography [11]. Total ^3H RNA and poly(A)$^-$ ^3H RNA, derived from an aliquot of total RNA equal to those used in hybridization, were hybridized to total cDNA cellulose. Percent of total hybridized RNA in poly(A)$^-$ form is calculated:

$$\frac{\text{poly(A)}^- \text{ mRNA, cpm hybridized}}{\text{total mRNA, cpm hybridized}} \times 100.$$

proportion of mRNA found in the poly(A)$^-$ form increases dramatically as a function of the length of the pulse-labeling period. The data suggest that in the steady state approximately 40% of cytoplasmic mRNA is in the poly(A)$^-$ form.

DESTABILIZATION OF NON-GLOBIN mRNA DURING TERMINAL DIFFERENTIATION OF FRIEND CELLS

During normal erythroid differentiation globin mRNA accumulates to a value of over 95% of total reticulocyte mRNA. Under normal culture conditions terminally differentiated Friend cells are physically unstable [Volloch and Housman, this volume]. We have developed a method of stabilization of Friend cells in the terminal stages of the differentiation process. These terminally differentiated cells eventually become enucleated and accumulate both hemoglobin and globin mRNA to levels similar to those observed in reticulocytes.

If the stability of non-globin mRNAs remains the same during erythroid differentiation, then the accumulation of globin mRNA to these levels would require at least ten days. However, this accumulation occurs in a much shorter period of time.

Aviv et al [6, 10] proposed that in order for globin mRNA to accumulate at its observed rate, the M2 population of non-globin poly(A)$^+$ mRNA must be destabilized during differentiation. The lack of availability of a method for stabilization of Friend cells has prevented the testing of this hypothesis, however. Using our method of stabilization of Friend cells in the terminal stage of differentiation we have measured the stability of non-globin mRNA during differentiation. Given the considerations stated in the first section of this paper, it seems appropriate to measure the average stability of total non-globin mRNA as opposed to the stability of M2 population during differentiation. To make such measurements, Friend cells, physically stabilized by addition of BSA to culture media, were pulse-labeled and chased at different times as indicated (0, 48, 96, 144 hours). Total cytoplasmic RNA was prepared at different chase times, and globin mRNA was removed by hybridization to globin cDNA-cellulose. The remaining RNA was then hybridized to total cDNA cellulose (Fig. 2). Clearly, as differentiation proceeds, non-globin mRNA is destabilized from a half-life of 40 hours in uninduced cells to a half-life of approximately five to six hours in six-day induced cells. However, globin mRNA remains very stable (as stable as rRNA) during differentiation [Volloch and Housman, this volume]. We suggest that the destabilization of non-globin mRNAs plays an important role in the emergence of globin mRNA as the predominant mRNA species in the reticulocyte during erythroid differentiation. We also suggest that the Friend cell system appears to be an excellent model system in which to study the molecular mechanisms which control this process.

SUMMARY

The metabolism of non-globin mRNA in Friend cells has been investigated. In previous studies two apparent classes of stability for mRNA molecules, M1 and M2, have been observed using oligo(dT) cellulose binding a criteria for mRNA

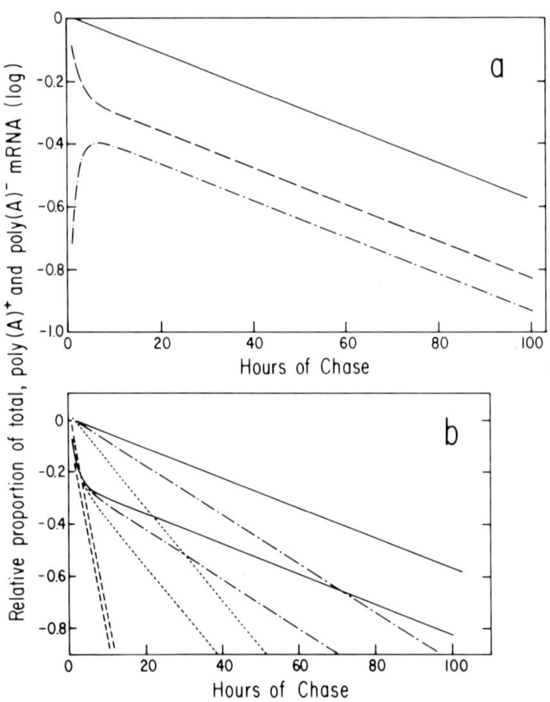

Fig. 3. Computer analysis of the model presented in Appendix. a. Parameters used: $R_1 = 0.25$, $R_2 = 0.3$, $R_3 = 0.03$, $R_4 = 0$. ———— F + G — — — F —— • —— G. b. Destabilization of mRNA as a result of R_3 increase. Parameters used: $R_1 = 0.25$, $R_2 = 0.3$, (————) $R_3 = 0.03$, (—— • ——) 0.1, (• • • • •) 0.15, $R_4 = 0.02$, (— — — —) 0.3.

fractionation. In the present study, we have devised an alternative procedure to directly fractionate high and middle abundant mRNA species by hybridization to cDNA cellulose. Using this approach we have obtained data which suggest that the two stability classes, M1 and M2, observed in previous studies are the result of the removal of poly(A) from mRNA molecules. The data obtained by hybridization to cDNA cellulose suggests that mRNA molecules which can be purified by this technique have a single average half-life of approximately 40 hours. When the same methods are used to measure mRNA half-life during Friend cell differentiation a significant decrease in the average mRNA half-life from 40 hours to approximately six hours is observed. These results are consistent with the view that globin mRNA becomes the predominant mRNA species in the late erythroblast by a process which involves destabilization of non-globin cellular mRNAs.

APPENDIX

Metabolism of poly(A)$^+$ and poly(A)$^-$ mRNA during the chase of pulse-labelled RNA (assuming for simplicity that at time zero of chase all cytoplasmic RNA is polyadenylated and neglecting residual export of labeled mRNA from nuclei) can be described by the following equations:

$$\frac{dF}{dt} = -(k_1 + k_4)F + k_2G$$

$$\frac{dG}{dt} = k_1F - (k_2 + k_3)G$$

where: F(t)—function describing current amount of the labeled poly(A)$^+$ RNA, G(t)—function describing current amount of the labeled poly(A)$^-$ RNA, k_1—degradation rate constant of poly(A) from poly(A)$^+$ RNA, k_2—addition rate constant of poly(A) to poly(A)$^-$ RNA, k_3—degradation rate constant of poly(A)$^-$ mRNA, k_4—degradation rate constant of poly(A)$^+$ mRNA. The solution of this system of equations is as follows:

$$\frac{d_2F}{dt^2} + (k_1 + k_2 + k_3 + k_4)\frac{dF}{dt} + (k_1k_3 + k_2k_4 + k_3k_4)\,F = 0$$

$$\frac{d_2G}{dt^2} + (k_1 + k_2 + k_3 + k_4)\frac{dG}{dt} + (k_1k_3 + k_2k_4 + k_3k_4)\,G = 0$$

and:

$$F = A_1e^{\alpha t} + B_1e^{\beta t} \qquad G = A_2e^{\alpha t} + B_2e^{\beta t}$$

where

$$\alpha,\beta = (-(k_1+k_2+k_3+k_4) \pm [(k_1+k_2+k_3+k_4)^2 - 4\,(k_1k_3+k_2k_4+k_3k_4)]^{1/2})/2$$

and:

$$\begin{aligned} A_1 &= -(k_1 + k_4 + \beta) / (\alpha - \beta) \\ A_2 &= k_1 / (\alpha - \beta) \\ B_1 &= (k_1 + k_4 + \alpha) / (\alpha - \beta) \\ B_2 &= -k_1 / (\alpha - \beta) \end{aligned}$$

A mathematical analysis demonstrates that the stability of total mRNA ($M = F + G$) follows an exponential function which, when plotted on a semi-logarithmic scale, assumes the shape of a straight line. The behavior of poly(A)$^+$ RNA can be described by two exponential functions, implying that it consists of two kinetic populations ($F = M_1 + M_2$).

The analysis implies that the stability of M_2 always equals that of total mRNA and thus M_2 destabilization can be achieved only through a decrease in the stability of the total mRNA. An increase in the parameter k_4 also results in M destabilization, but its main effect is the decrease of the proportion of M_2 in M. In both cases there is also a strong destabilization of M_1.

Another two possible ways of mRNA destabilization that do not imply substantial destabilization of M_1 are: an increase of k_3 and/or a decrease of k_2.

An example of the behavior of function of F_1G, and $(F + G)$ is presented in Figure 3a.

An example of the effect of changing the constant is presented in Figure 3b.

REFERENCES

1. Lenk R, Herman R, Penman S: mRNA abundance and lifetime: A correlation in Drosophila cells but not in HeLa. Nucleic Acids Res 5: 3057–3070, 1978.
2. Berger S, Cooper M: Very short-lived and stable mRNAs from resting lymphocytes. Proc Natl Acad Sci USA 72: 3873–3877, 1975.
3. Singer R, Penman S: mRNA in HeLa cells: Kinetics of formation and decay. J Mol Biol 78: 321–334, 1973.
4. Puckett L, Cambers S, Darnell J: Short lived mRNA in HeLa cells and its impact on the kinetics of accumulation of cytoplasmic polyadenylate. Proc Natl Acad Sci USA 72: 389–393, 1975.
5. Perry R, Bard E, Hames B, Kelley D, Schibler U: The relationship between hn RNA and mRNA. Prog Nucleic Acids Res Mol Biol 19: 275–292, 1976.
6. Aviv H, Volloch V, Bastos R, Levy S: Biosynthesis of globin mRNA in Friend cells. Cell 8: 495–503, 1976.
7. Noyes B, Stark A: Nucleic acid hybridization using DNA covalently coupled to cellulose. Cell 5: 301–310, 1975.
8. Levy S, Aviv H: Quantitation of labeled globin mRNA by hybridization with excess complementary DNA covalently bound to cellulose. Biochem 15: 1844–1847, 1975.
9. Lowenhaupt R, Lingrel J: A change in the stability of globin mRNA during the induction of MEL cells. Cell 14: 337–344, 1978.
10. Bastos R, Aviv H: Theoretical analysis of a model for globin mRNA accumulation during erythropoiesis. J Mol Biol 110: 205–218, 1977.
11. Aviv H, Leder P: Purification of biologically active globin mRNA by chromatography on oligo (dT) cellulose. Proc Natl Acad Sci USA 69: 1408–1413, 1972.

Molecular Pathology

Organization and Expression of Globin Genes, pages 269–278

Globin Gene Pathology: Clues to Gene Function and Hemoglobin Switching

Arthur Bank, A. Lee Burns, Michael Baird, and Robert Pergolizzi

Departments of Medicine, and Human Genetics and Development, Columbia University, New York

INTRODUCTION

The thalassemia syndromes and related disorders of man have provided insights into the relationship between globin gene structure and function. Restriction endonuclease analysis provided the initial impetus for defining the physical map of the globin genes in normal cells and using mutants of human globin synthesis [1–3]. In the case of the α-globin genes, restriction mapping analysis has indicated that there are two linked α-globin genes, 2.5 kilobases (kb) apart, per haploid chromosome set [4, 5]. The major defects in α-globin expression in the α thalassemias are due to the deletion of one to four of these α-globin genes [4,6]. There is an embryonic α-like gene, the ζ gene, as well. The non-α-globin gene organization is more complicated. There are at least five non-α-globin genes linked together on a single fragment of DNA approximately 50 kb long, and reading from 5′ to 3′ they include one ϵ, two γ, one δ, and one β gene [6–8]. The hemoglobins (Hbs) found include the embryonic Hbs Gower I ($\epsilon_2\zeta_2$) and Gower II ($\alpha_2\epsilon_2$); fetal Hb ($\alpha_2\gamma_2$) and the adult Hbs; HbA ($\alpha_2\beta_2$) and HbA$_2$ ($\alpha_2\delta_2$).

Quantitative defects in expression of the non-α-globin genes are most commonly seen in disorders of β-globin biosynthesis known as the β thalassemias [9]. In β^+ thalassemia, there is decreased β-globin, and in β^0 thalassemia, absence of β-globin. These disorders are accompanied by marked anemia due to the deficit in β-globin production, and the inability to compensate for this decrease by an increase in γ-globin (Table I). By contrast, in two other disorders of β-globin biosynthesis, $\delta\beta$ thalassemia and HPFH, there is complete absence of δ and β-globin biosynthesis, yet in $\delta\beta$ thalassemia the anemia is mild, while in HPFH, no anemia is found (Table I). This is a result of the continued high level expression of γ-globin in these disorders in contrast to the β^+ and β^0 thalassemias. Restriction endonuclease analysis of β^+/β^0 thalassemias and $\delta\beta$ and HPFH DNAs have provided insights into the relationships between globin gene structure and function in these disorders.

DELETIONS IN $\delta\beta$ THALASSEMIA AND HPFH

There is deletion of δ and β-globin structural sequences in different patients with $\delta\beta$ thalassemia [1, 2, 10–13]. In one type of homozygous $\delta\beta$ thalassemia studied in

TABLE I. Quantitative Defects in Expression of Non-α Globin Genes

	Anemia	HbA	HbA_2	HbF	β and δ genes
Normal	None	Present	NL	NL[1]	NL[1]
β^+ thal	Severe	Decreased	NL	↑[2]	Not deleted
β^0 thal	Severe	Absent	NL	↑[2]	Not deleted
$\delta\beta$ thal	Mild	Absent	Absent	↑↑[2]	Deleted
HPFH	None	Absent	Absent	↑↑↑[2]	More deleted

[1]NL = Normal.
[2]↑-↑↑↑ = Mild to greatly increased.

our laboratory [1, 2, 13], all of the β-globin gene fragments were absent, and only a single δ-globin gene fragment representing the 5′ end of the δ-globin gene was detectable using a δ-globin cDNA probe (Fig. 1). The location of the residual δ-globin gene fragment as 5′ δ was established by the presence of a 15 kb Bam HI fragment which hybridized to both β and γ cDNA, indicating that the 5′ end of the δ gene was present, since it is known that this 15 kb fragment is generated as the linking fragment between the $^A\gamma$ gene and the δ-globin gene. Further analysis by us and others has confirmed this and clarified the extent of deletion in this $\delta\beta$ thalassemia DNA at its 5′ end [10–14]. The 3′ end of the deletion in $\delta\beta$ thalassemia has also been mapped and is about 1 kb 3′ to the δ-globin gene [11].

At least three different types of HPFH have been defined by gene mapping [1, 2, 10–15]. In two types of Black homozygotes, no normal δ or β-globin gene fragments are detectable, indicating complete deletion of these genes. The extent of the deletion 5′ to the δ gene has been mapped in one of these patients, and is approximately 3 kb 5′ to the δ gene, while in the other it is about 7 kb 5′ to the δ gene [14] (Fig. 1). In both of these cases, there has been maximal γ-globin gene expression and no anemia. Thus, the extent of deletion in this region between the γ and δ genes does not appear to affect γ gene expression. The results of these studies have also suggested that the deletion of sequences in the region of the δ gene and 5′ to it may permit optimal reactivation of γ-globin biosynthesis. In the so-called Greek type of HPFH, no detectable deletion has been found [12, 15].

In β^+ and β^0 thalassemia, in which there is no detectable deletion of sequences in the δ region, there is no significant compensation by increased γ. By contrast, in $\delta\beta$ thalassemia, in which most of the δ gene is deleted although the 5′ end remains, there is more suppression of γ-globin biosynthesis than in HPFH. A patient homozygous for $\delta\beta^A\gamma$ thalassemia, in whom the δ, β, and $^A\gamma$ genes are all deleted, has also been described and appears at first glance to be an exception to the concept that absence of sequences in the δ region permits maximal γ-globin gene biosynthesis, since these individuals have anemia [10, 16]. The most likely explanation for this apparent paradox is that the full expression of γ-globin genes requires a full complement of these genes. In $\delta\beta^A\gamma$ thalassemia, the absence of two of the four γ genes limits the total potential for γ-globin gene biosynthesis, and, thus, results in a phenotype similar to that of $\delta\beta$ thalassemia, although the genotype is more similar to that in HPFH.

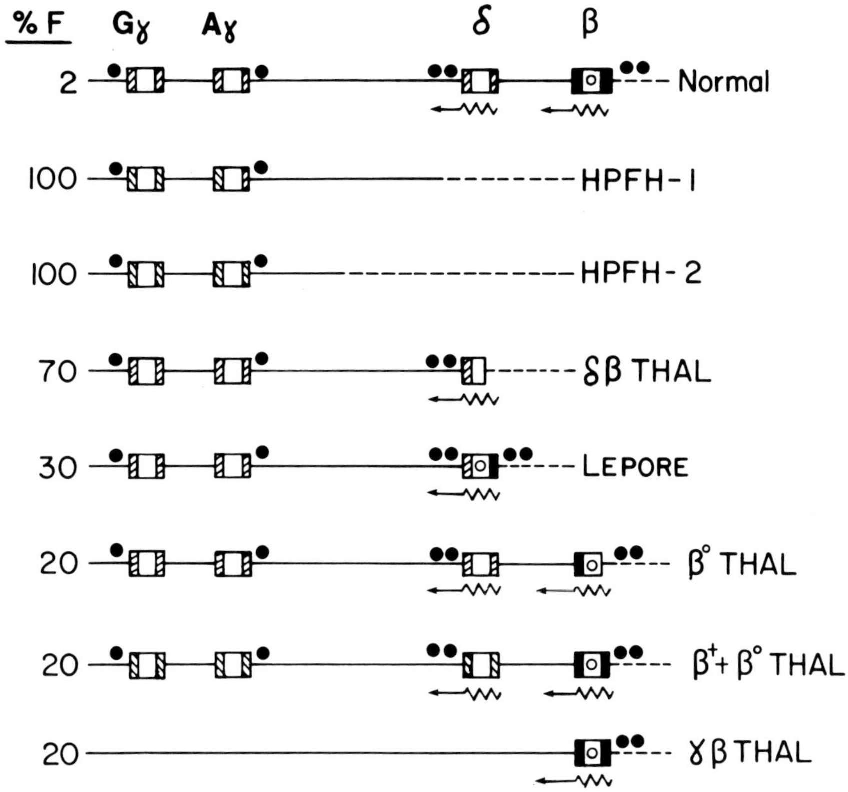

Fig. 1. The physical map of the globin genes in normal and abnormal human cells. The $^{G}\gamma$, $^{A}\gamma$, δ, and β genes are diagrammed in rectangles. The globin structural gene regions are shown by the hatched or black areas while the intervening sequences are shown by the clear areas with or without circles. Relative percentage of HbF in homozygotes for the different disorders is shown to the left. The solid black circles surrounding the globin genes indicate those regions in which repetitive sequences have been identified; the wavy solid lines with arrows (underneath genes) indicate potential IVS regions from which products are produced which may affect genes more 5′. It is of interest that the reiterated sequences 5′ to the δ gene as well as IVS II in the δ gene are deleted in HPFH.

β^{+} AND β^{0} THALASSEMIA

In β^{0} thalassemia, there is significant heterogeneity in genotypes resulting in similar phenotypes (Fig. 1). In rare individuals, restriction endonuclease analysis demonstrates deletion of the 3′ end of the β-globin gene [17, 18]. When it exists, this deletion is only present in one of two β^{0} thalassemia genes in individuals homozygous for β^{0} thalassemia. It is unclear whether the deletion is a primary defect or results from secondary nucleotide changes occurring on the background of an inactivated β-globin gene. The vast majority of patients with β^{0} thalassemia do not have detectable deletions in the β-globin gene by restriction analysis [2, 13, 17, 18]. Several types of lesions in β^{0} thalassemia might result in the β^{0} thalassemia phenotype without detectable deletion. Small deletions or point mutations could give normal restriction patterns unless the specific restriction enzyme used fortuitously identifies the specific nucleotide change. In one patient with β^{0} thalassemia, a single

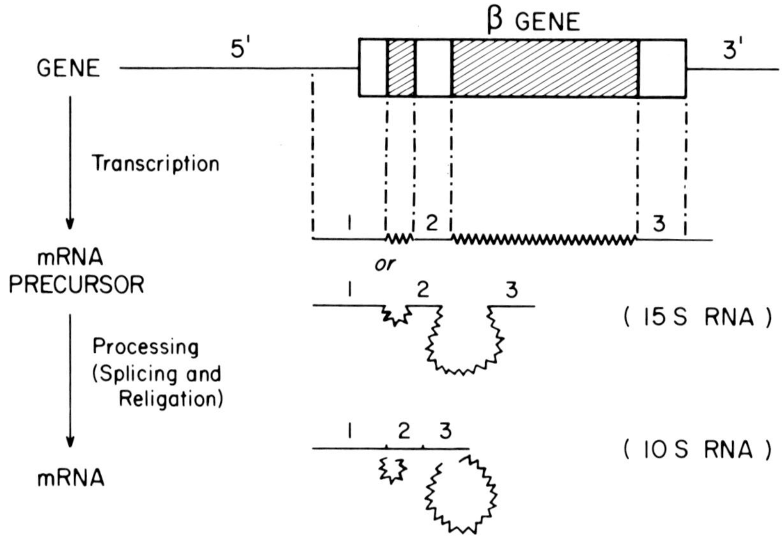

Fig. 2. β-Globin gene transcription. The β-globin structural gene in the rectangle is interrupted by two intervening sequences—"hatched" areas. Transcription of the gene begins 5′ to it, and includes transcription of the intervening sequences (wavy lines) contiguous to that of globin structural sequences (solid lines 1, 2, and 3). The mRNA precursor is subsequently processed by splicing and religation to release the two intervening sequences from the 15S RNA to form the final 10S RNA product in which structural gene components 1, 2, and 3 are joined.

nucleotide defect in codon 17, resulting in a nonsense mutation, has been localized by sequence analysis of cDNA synthesized using the patient's mRNA as template [19]. Even in this case, restriction analysis was normal. In other individuals with β^0 thalassemia, β-globin mRNA biosynthesis is either markedly decreased or absent. In these individuals, it is presumed that there is a defect either in transcription of the β-globin gene or in the processing of β-globin mRNA (Fig. 2). Cloning and sequencing of β-globin genes from patients with these disorders is currently underway and should provide information on the specific lesions in these individuals.

The β^+ thalassemias are also heterogeneous both at the clinical level with varying severity of anemia, and in the amount of HbA and HbF biosynthesis. Restriction endonuclease analysis of over 100 individuals with β^+ thalassemia, to date, has not, as yet, revealed detectable deletions or other changes in the vicinity of the β gene [2, 13, 17]. Since there is continued β-globin biosynthesis in β^+ thalassemia, the defect must be either in DNA sequences required for the normal initiation of β-globin gene transcription, or in β-globin mRNA processing (Fig. 2). Recent reports [20, 21] indicate that β-globin mRNA processing may be defective in certain individuals with β^+ thalassemia as evidenced by decreased turnover, and abnormal accumulation of β-globin mRNA precursors compared to normal (Fig. 2).

In order to investigate the DNA defect in β^+ thalassemia in greater detail, we have cloned a β-globin gene from a patient with β^+ thalassemia, and have analyzed its nucleotide content [22]. To compare the β-globin gene from the β^+ thalassemia patient to a similar gene from a normal individual (HβG1 provided by Dr. Tom Maniatis) [23], specific regions of the β-globin genes were subcloned; these include a 4.4 kb Pst I fragment containing the entire β-globin gene, and a 0.95 kb fragment containing the entire large intervening sequence (IVS II) of the β gene. Since the possibility of an RNA processing defect in this patient with β thalassemia was raised by the finding of abnormal β-globin mRNA turnover, we initially analyzed the β IVS II subclone of this patient, and compared it to the normal IVS II subclone of HβG1. Restriction endonuclease analysis, using 24 different enzymes, shows no differences in the size or number of DNA fragments obtained, using normal and β^+ thalassemia IVS subclones. Since the complete sequences of the IVS II and pBR 322 containing the IVS II insert are known, the size of the restriction fragments could be verified, and are consistent with those found. In this analysis, approximately 11% of the nucleotide sequence in the IVS II have been evaluated.

One of the enzymes used, Hph I, recognizes the 5′ end of IVS II, while another enzyme, Alu I, recognizes the 3′ end of IVS II; since neither of these enzymes leads to a change in fragment pattern, the 5′ and 3′ borders of IVS II in the β-globin gene of this patient with β^+ thalassemia is unchanged despite a possible defect in β-globin mRNA processing. These results do not exclude a defect either in the small IVS (IVS I) of this patient, or a change in a sequence within IVS II, which is required from the normal processing of IVS II.

We are currently determining the entire nucleotide sequence of both IVS I and II of this individual with β^+ thalassemia, and also intend to examine the 5′ flanking region for nucleotide changes which may be responsible for the defect in β^+ thalassemia. When such changes are found, they will have to be distinguished from polymorphisms which may be present in the particular populations under analysis. We intend to do this by comparing DNA from several different patients of similar ethnic groups with and without thalassemia by both restriction endonuclease analysis and cloning.

It is also possible that nucleotide sequences more distant from the β structural gene may influence the processing of β globin RNA, and may be defective in this patient. Recent reports of a patient with so-called $\gamma\beta$ thalassemia, in whom there is deletion in the γ and δ genes associated with a β thalassemia phenotype [24], are consistent with this possibility; however, available genetic data indicate that the β^+ thalassemia gene is closely linked to the β structural locus [25].

REGULATION OF NON-α GENES IN MAN

The β thalassemias and related disorders provide a set of mutants from which information has been obtained on the relationship between non-α-globin gene structure and function. Several generalizations can be made: (1) Deletion of DNA sequences in the δ-globin gene region is associated with increased activation of γ-globin genes; this is the case in HPFH and $\delta\beta$ thalassemia as compared with β^+ and β^0 thalassemia; (2) greater activation of γ-globin biosynthesis is present in HPFH in which there is the complete absence of δ-globin components than in $\delta\beta$ thalassemia in which a part of the δ gene is preserved; and (3) in homozygotes for hemoglobin Lepore there is less activation of γ-globin expression than in $\delta\beta$ thalassemia or HPFH.

A variety of interactions between DNA, RNA, and protein components in the nucleus may be responsible for the regulation of globin biosynthesis, and no specific site of control can be defined at the molecular level at present. The recent finding of reiterated sequences 5′ to the ϵ gene and to the $^G\gamma$ gene, 3′ to the $^A\gamma$ gene, 5′ to the δ gene, and 3′ to the β gene, are of great interest [8]. These regions have been shown to be preferentially transcribed by RNA polymerase III in a cell-free system [26]. The relevance of this in vitro transcription to in vivo transcription is not yet known. However, these regions may define the limits of specific globin gene transcription domains, with the fetal (γ) and adult ($\delta\beta$) domains being discreet entities in equilibrium [15]. It has been suggested that in HPFH and $\delta\beta$ thalassemia the loss of the $\delta\beta$ flanking domains (Fig. 1) leads to a change in the equilibrium, and favors γ biosynthesis [15]. In $\gamma\beta$ thalassemia, it is proposed that the loss of δ sequences leads to a relative decrease in activity of the $\delta\beta$ domain and decreased β biosynthesis [24].

Another possible level of control may involve the RNA produced as the result of processing of globin RNA precursors. Control may be exercised by interaction of RNA processing intermediates of different genes with each other. Normally, the processing of δ-globin RNA precursors may lead to RNA products which feedback inhibit the processing of γ-globin mRNA precursors and thus decrease γ-globin gene mRNA. Alternatively, products of the δ and β-globin mRNA processing may interact with the γ-globin gene DNA itself, and inhibit the transcription of γ-globin genes. In HPFH, the lack of any δ or β-globin gene RNA processing intermediates may permit the full expression of γ-globin genes either by the lack of inhibition of γ-globin gene transcription or of γ-globin mRNA processing. In $\delta\beta$ thalassemia, the presence of the small IVS may result in a low level of inhibition of γ-globin gene expression, either at the DNA or RNA level. In hemoglobin Lepore, the presence of an intact small and large IVS in the Lepore gene may be sufficient to completely suppress γ-globin gene expression, as it does normally. Similarly, the normal intermediates in β^0 and β^+ thalassemia may also lead to the suppression of γ-globin biosynthesis. In $\gamma\beta$ thalassemia, there is a defect in β-globin biosynthesis, which may be explained by an additional component of this model. Delta gene IVS sequences, absent in $\gamma\beta$ thalassemia, may be required for optimal β-globin gene expression. This model implies that upstream genes produce RNA intermediates which stimulate the transcription and/or processing of downstream or 3′ genes in the same way that downstream or 3′ genes produce RNA intermediates which inhibit upstream or 5′ genes. RNA processing regulation would provide a role for the δ-globin gene at the RNA level and not at the polypeptide level. It would also explain why δ-globin activation is more pronounced in bone marrow cells in which RNA synthesis occurs than in reticulocytes where it does not.

One mechanism by which RNA processing regulation could occur is shown in Figure 3. Here, RNA processing products of the δ gene act on the γ mRNA or γ gene to inhibit γ mRNA processing; at the same time, other products generated by processing of the δ gene, stimulate either β-globin mRNA transcription or β mRNA processing. The end result is decreased γ mRNA and increased β mRNA. In order to explore the possibility that this model is feasible, it will be necessary to examine the RNA intermediates of the δ, β, and γ-globin IVS, as well as flanking and structural gene sequences. It may be possible to study these events by a careful analysis of the size and specific transcriptional origins of the RNAs present in human bone marrow cells at different stages of development.

We have made a preliminary assessment of the feasibility of this model of the potential interactions of RNA products of the γ, δ, and β IVS with each other, by computer analysis of the sequences provided by others [27–30]. In this analysis, we have assumed that regions of greater secondary structure, within IVS I and II, are

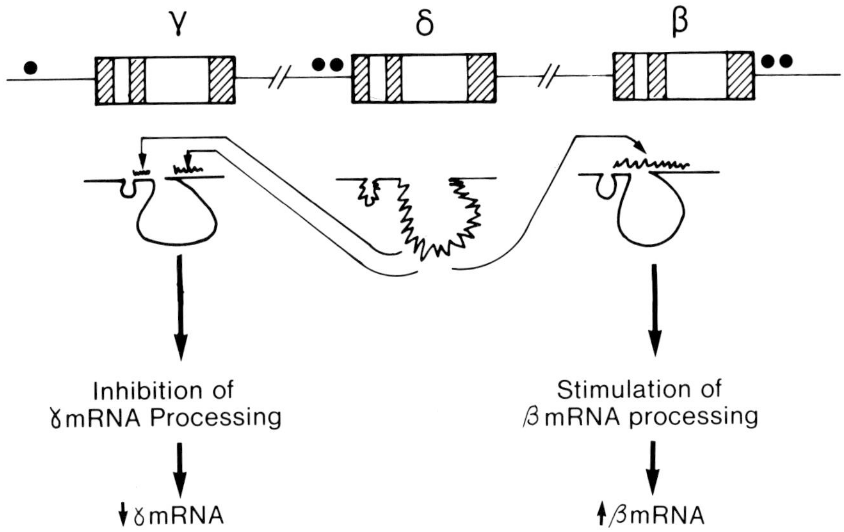

Fig. 3. The mechanism of RNA processing regulation. In the proposed scheme, RNA sequences from within the δ intervening sequences act both 5′ and 3′ to either inhibit γ mRNA processing or stimulate β mRNA processing. These events occur by the interaction of δ IVS RNA products with the IVS RNA precursors for γ and β-globin genes. In the example shown, δ IVS intermediates can prevent the formation of splice sites at the edges of the γ IVS II, and stimulate β mRNA processing by acting on the β IVS II.

the best potential areas for secondary splice sites within these IVS, and have focused on those in which stems, of relatively high stability, and loops, circumscribed within the IVS, are present (Fig. 4). Six such regions in the β-globin gene have been defined: one within IVS I and five within IVS II.

It has recently been suggested that a possible mechanism for facilitating RNA splicing is the ability of eukaryotic small RNAs to stabilize potential splice sites [31–33]. The presence of stems, as well as these RNAs complementary to IVS RNA might accentuate the availability of certain specific sites within the RNAs for splicing. Thus, we have looked to see if there are unique sequences in the γ, δ, or β genes, which are complementary for these specific regions. This analysis has revealed that there are unique complements to the flanking sequences of some of these regions in the β gene. For example, the most stable complement to the loop-stem structure in Figure 4 in IVS I is located only within the δ-globin IVS; thus, the possibility exists that δ-globin IVS fragments may facilitate β-globin IVS processing. Other δ IVS fragments may inhibit γ mRNA processing by binding to only one side of a splice junction, and thus interfering with RNAs spanning such junctions (Fig. 3).

Clearly, there are many levels at which the biosynthesis of globin genes can be controlled. In fetal life, the initial switch from ϵ to γ-globin biosynthesis may be regulated by different mechanisms than the reactivation of γ-globin genes in adult life. The primary activation of γ-globin biosynthesis appears to be normal in the β thalassemias, while γ reactivation is extremely limited in these disorders. A physical change in the DNA may signal γ-globin biosynthesis; for example, by transposition of inverted repeats 5′ to the ϵ or γ genes. RNA controls might further modulate

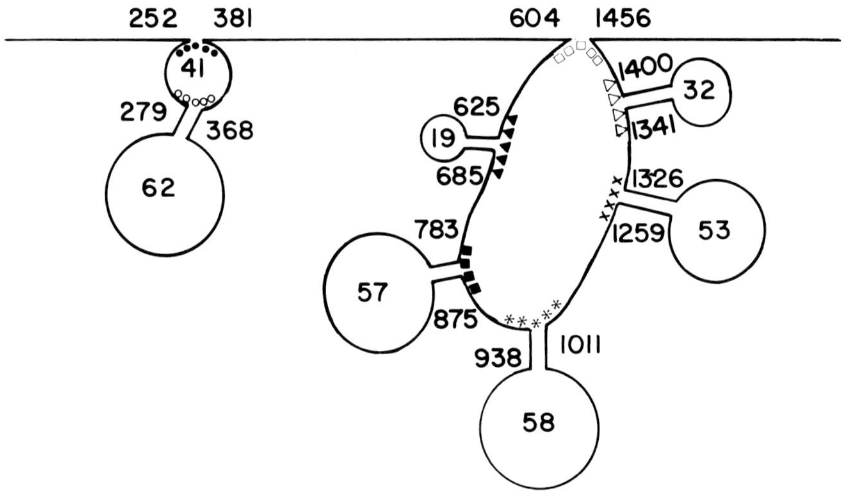

Fig. 4. Secondary structure within β IVS I and II in the β mRNA precursor. The numbers outside the circles indicate the relative nucleotide numbers derived from reference 28. IVS I extends from nucleotide 252 to 381; IVS II from 604 to 1456. The loops and stems drawn were derived from computer analysis of the nucleotide sequences of these genes. The stems drawn indicate regions in which the average energy is –5 kcal or greater. The numbers within the loops indicate the number of nucleotides within those loops. The stems vary between eight and 20 nucleotides in length. In the small β IVS, there is a stem and loop between nucleotides 279 and 368. Nucleotides 283 and 365 form the respective 5′ and 3′ ends of a perfect 9-base pair stem in this region. A computer search for sequences complementary to the nucleotides immediately flanking this stem (5′ to 283 and 3′ to 365) reveals that only one sequence is found in any of the other IVSs which is complementary to 7 consecutive bases—4 on one side and 3 on the other side of the stem. This sequence (AATCCTT) is between nucleotides 920 and 960 in the δ IVS. The uniqueness of this complement lends some credence to the notion that there is an interaction between RNA intermediates of the δ and β-globin genes, although clearly this must be viewed as speculative at present. Other potential complementary regions are diagrammed by the opened-closed triangles, closed circles, xs, and asterisks.

the progressive change from ϵ to γ expression. Environmental influences may determine which cells are selected at a given stage. Late in fetal life, another signal, perhaps again a physical change in DNA, may activate the δ and β genes; again, RNA intermediates produced by this activation may serve a regulatory function. Again, the switch from γ to adult hemoglobin production may simply involve selection of those cells which produce hemoglobin A.

Globin gene deletion mutants and the β thalassemias hold the promise of being useful in identifying changes in DNA structure associated with changes in function which may more precisely define the mechanisms of globin gene regulation. It is hoped that the study of globin gene pathology will also have relevance to the general problem of the mechanisms regulating the normal switch from γ to $\delta\beta$ gene expression in man.

SUMMARY

The β thalassemias and related disorders provide human mutants in which to explore the relationship of globin gene structure to function. Deletions of nucleotide sequences in the δ and β region in $\delta\beta$ thalassemia and hereditary persistence of fetal hemoglobin (HPFH) are associated with increased γ-globin gene expression, suggesting that sequences in this region regulate γ-globin gene expression. In most cases of β^+ and β^0 thalassemias, no deletion of δ or β genes are detectable to explain decreased or absent β-globin biosynthesis. Analyses of cloned β genes in β^+ thalassemia have not, as yet, revealed significant DNA changes. Globin gene expression may be regulated by differential activation of DNA sequences flanking the γ or δ-β complex. Alternatively, RNA processing may be a site of control of globin gene expression.

ACKNOWLEDGMENTS

This work was supported by NIH grants AM-25274 and CA-18314, National Science Foundation 24120, Cooley's Anemia Foundation, 312-2581, and March of Dimes Defects Foundation NF-1-664.

REFERENCES

1. Mears JG, Ramirez F, Leibowitz D, Nakamura F, Bloom A, Konotey-Ahulu F, Bank A: Changes in restricted human cellular DNA fragments containing globin gene sequences in the thalassemias and related disorders. Proc Natl Acad Sci USA 75:1222–1226, 1978.
2. Mears JG, Ramirez F, Leibowitz D, Bank A: Organization of human δ and β globin genes in cellular DNA and the presence of intragenic inserts. Cell 15:15–23, 1978.
3. Flavell RA, Kooter JM, de Boer E, Little PFR, Williamson R: Analysis of the $\beta\delta$ globin gene loci in normal and Hb Lepore DNA: Direct determination of gene linkage and intergene distance. Cell 15:25–41, 1978.
4. Orkin SH, Old JM, Weatherall DJ, Nathan DG: Partial deletion of β globin gene DNA in certain patients with β^0 thalassemia. Proc Natl Acad Sci USA 76:2400–2404, 1979.
5. Lauer J, Shen C-K, Maniatis T: The chromosomal arrangement of human α-like globin genes: Sequence homology and α globin gene deletions. Cell 20:119–130, 1980.
6. Embury S, Lebo R, Dozy A, Kan YW: Organization of the α globin genes in the Chinese α thalassemia syndromes. J Clin Invest 63:1307–1310, 1979.
7. Maniatis T, Hardison RC, Lacy E, Lauer J, O'Connell C, Quon D: The isolation of structural genes from libraries of eucaryotic DNA. Cell 15:687–701, 1978.
8. Fritsch EF, Lawn RM, Maniatis T: Molecular cloning and characterization of the human β-like globin gene cluster. Cell 19:959–972, 1980.
9. Bank A, Mears JG, Ramirez F: Disorders of human hemoglobin. Science 207:486–493, 1980.
10. Fritsch EF, Lawn RM, Maniatis T: Characterization of deletions which affect the expression of fetal globin genes in man. Nature 279:598–603, 1979.
11. Bernards R, Kooter JM, Flavell RA: Physical mapping of the globin gene deletion in $\delta\beta^0$ thalassemia. Gene 265–280, 1979.
12. Tuan D, Biro PA, de Riel JK, Lazarus H, Forget BG: Restriction endonuclease mapping of the human γ globin gene loci. Nucleic Acids Res 6:2519–2544, 1979.
13. Bank A, Mears JG, Ramirez F, Burns AL, Feldenzer J: Organization of human globin genes in normal and thalassemic cells. In Axel R, Maniatis T, Fox FC (eds): "Eucaryotic Gene Regulation," New York: Academic Press, 1979, pp 355–366.
14. Ottolenghi S, Comi P, Giglioni B, Williamson R, Vullo G, Conconi F: Direct demonstration of β mRNA in homozygous Ferrara β^0 thalassemia patients. Nature 266:231–240, 1977.
15. Bernards R, Flavell RA: Physical mapping of the globin gene deletion in hereditary persistence of fetal hemoglobin (HPFH). Nucleic Acids Res 8:1521–1534, 1980.
16. Orkin SH, Alter BP, Altay C: Deletion of the $^{A}\gamma$ globin gene in $^{G}\gamma\delta\beta$ thalassemia. J Clin Invest 64:866–869, 1979.
17. Flavell RA, Bernards R, Kooter JM, de Boer R: The structure of the human β globin gene in β thalassemia. Nucleic Acids Res 6:2749–2760, 1979.
18. Orkin SH, Old JM, Weatherall DJ, Nathan DG: Partial deletion of β globin gene DNA in certain patients with β^0 thalassemia. Proc Natl Acad Sci USA 76:2400–2404, 1979.

19. Chang JC, Kan YW: β^0 thalassemia, a nonsense mutation in man. Proc Natl Acad Sci USA 76:2886–2889, 1979.
20. Nienhuis AW, Turner P, Benz EJ: Relative stability of α and β globin messenger RNAs in homozygous β^+ thalassemia. Proc Natl Acad Sci USA 74:3960–3964, 1977.
21. Maquat LE, Kinniburgh AJ, Beach LR, Honig GR, Lazerson J, Ershler WB, Ross J: Processing of the human β globin mRNA precursor to mRNA is defective in three patients with β^+ thalassemia. Proc Natl Acad Sci USA 77:4287–91, 1980.
22. Burns AL, Spence S, Mears JG, Ramirez F, Kosche K, Feldenzer J, Bank A: Isolation and characterization of the β globin gene in β^+ thalassemia. Blood 54:51a, 1979 (Abstract).
23. Lawn RM, Fritsch EF, Parker RC, Blake G, Maniatis T: The isolation and characterization of linked δ and β globin genes from a cloned library of human DNA. Cell 15:1157–1174, 1978.
24. Van Der Ploeg LHT, Konings A, Oort M, Roos D, Bernini L, Flavell RA: $\gamma\beta$ thalassemia studies showing that deletion of the γ and δ genes influences β globin gene expression in man. Nature 283:637–642, 1980.
25. Weatherall DJ, Clegg JB: "The Thalassemia Syndromes," 3rd ed. Oxford: Blackwell, 1979.
26. Duncan C, Biro PA, Chowdary PV, Elder JT, Wang RRC, Forget BG, De Riel JK, Weissman SM: RNA polymerase III transcriptional units are interspersed among human non-α globin genes. Proc Natl Acad Sci USA 76:5095–5099, 1980.
27. Fritsch EF, Lawn RM, Maniatis T: Molecular cloning and characterization of the human β-like globin gene cluster. Cell 19:959–972, 1980.
28. Efstratiadis A et al: DNA sequence comparison of the human β-like globin genes. Cell 21:653–668, 1980.
29. Spritz R, De Riel JK, Forget BG, Weissman SM: Nucleotide sequence of the human δ globin gene. Cell, 21:639–646, 1980.
30. Slightom J, Blechl AE, Smithies O: Human fetal $^G\gamma$ and $^A\gamma$ globin genes: Complete nucleotide sequences suggest that DNA can be exchanged between these duplicated genes. Cell 21:627–638, 1980.
31. Lerner MR, Boyle JA, Mount MS, Wolin LS, Steitz JA: Are mRNPs involved in splicing? Nature 283:220–224, 1980.
32. Rogers J, Wall R: A mechanism for RNA splicing. Proc Natl Acad Sci USA 77:1877–1879, 1980.
33. Avvedimento VE, Vogeli G, Yamade Y, Maizel JV, Pastan I, de Crombrugghe B: Correlation between splicing sites within an intron and their sequence complementarity with U_1 RNA. Cell, 21:689–696, 1980.

Organization and Expression of Globin Genes, pages 279–286

Structural and Functional Characterization of Cloned β^+ Thalassemic Globin Gene Fragments

R.A. Spritz, P. Jagadeeswaran, P.A. Biro, J.T. Elder, P.V. Choudary, J. K. deRiel, J. L. Manley, M.L. Gefter, S.M. Weissman, and B.G. Forget

Departments of Medicine and Human Genetics, Yale University School of Medicine, New Haven, Connecticut (R.A.S., P.J., P.A.B., J.T.E., P.V.C., J.K.dR., S.M.W., B.G.F.) and Department of Biology, Massachusetts Institute of Technology, Cambridge (J.L.M., M.L.G.)

INTRODUCTION

The thalassemia syndromes are characterized by quantitatively deficient synthesis of either α or β-globin chains of Hb A ($\alpha_2\beta_2$) in the erythroid cells of affected individuals [1, 2]. Our current understanding of the molecular basis of the thalassemias has been recently reviewed [3–6]. Two major types of β thalassemia have been described, β^0 thalassemia and β^+ thalassemia. In β^0 thalassemia there is complete absence of β-globin chain synthesis, associated with either the complete absence of β mRNA or with the presence of small amounts of apparently nonfunctional β-globin mRNA. In β^+ thalassemia, the most common type of β thalassemia, decreased amounts of β-globin chains are produced. These β-globin chains appear normal by peptide mapping analysis, and the small amounts of β-globin produced are associated with correspondingly low levels of β-globin mRNA in the erythroid cells of affected individuals. This mRNA appears to be qualitatively normal as assessed by its ability to direct globin synthesis in cell-free protein synthesizing systems, and by its ability to hybridize completely to a cDNA copy of normal β-globin mRNA. In all likelihood, β^+ thalassemia results from a defect at the level of globin gene transcription, globin mRNA processing, mRNA transport, mRNA stability, or some combination of these processes.

Family studies have demonstrated that the β thalassemic defect is allelic with or very tightly linked to the β-globin structural locus [1]. These findings suggest that cloning of the β-globin gene from a β^+ thalassemic patient, and subsequent structural and functional analyses, might indeed yield information on the molecular basis of this disorder, and thus provide significant insights into the fundamental mechanisms of gene expression. In this paper, we report the isolation by recombinant DNA technology, as well as the structural and functional characterization of β-globin gene DNA fragments from an individual with homozygous β^+ thalassemia.

J.K. deRiel's current address is Fels Research Institute, Temple University School of Medicine, Philadelphia, Pennsylvania 19140.

CLONING OF DNA FRAGMENTS OF A HUMAN β^+ THALASSEMIC GLOBIN GENE

Total DNA was prepared from the spleen of a Greek Cypriot patient with typical transfusion dependent homozygous β^+ thalassemia. The parents of this patient were shown to have elevated levels of Hb A_2, characteristic of the heterozygous state. The patient's DNA was digested to completion with the restriction endonuclease Eco RI, which cleaves the human β-globin gene at codon 122, to the 3′-side of the large intervening sequence, and produces a 5.2 kilobase (kb) DNA fragment containing the 5′-portion of the gene, and a 3.6 kb DNA fragment containing the 3′-portion of the gene (Fig. 1). The DNA fragments were fractionated by electrophoresis in a large horizontal slab gel of agarose, and the positions in the gel of the β-globin gene fragments were identified by blotting of a strip of the gel according to the method of Southern [7] followed by hybridization of the nitrocellulose paper strip containing the transferred DNA to ^{32}P-labeled nick-translated β-globin cDNA from the plasmid clone pJW102 [8]. DNA from the appropriate regions of the master gel was eluted and cloned in the bacteriophage vector λ gt-WES. Recombinant clones were identified by hybridization to radioactively labeled β-globin cDNA from plasmid pJW102 [8]. The recombinant clone containing the 5′-portion of the β-globin gene was designated C6, and that containing the 3′-portion of the gene was designated D47. In order to facilitate structural and nucleotide sequence analyses, the cloned β-globin gene fragments were subcloned in the plasmid pBR322 and the filamentous bacteriophage M13mp2. All procedures were performed in accordance with the NIH guidelines for recombinant DNA research.

RESTRICTION ENDONUCLEASE CLEAVAGE ANALYSIS

In order to investigate the possibility that deficient β-globin synthesis in β^+ thalassemia might result from a small rearrangement in the β-globin gene or its flanking sequences, DNA from the β^+ thalassemic gene clone C6 and the corresponding 5′-Eco RI fragment from a normal β-globin gene [9] were digested with the restriction endonucleases Hinf I, Alu I, and Hae III, and fractionated by electrophoresis in an agarose gel. Figure 2 illustrates side-by-side digestion patterns from the thalassemic and normal β-globin gene DNA fragments. It is evident that there are no significant differences between the digestion patterns of the thalassemic and normal DNA, indicating that there are, in fact, no detectable deletions, duplications, or other rearrangements in the cloned thalassemic β-globin gene fragment.

A detailed restriction map of the thalassemic β-globin gene fragments was developed (Fig. 1), and was used to guide nucleotide sequence analysis. Comparison of the restriction maps of the normal and β^+ thalassemic β-globin genes demonstrated no differences.

NUCLEOTIDE SEQUENCE ANALYSIS

Nucleotide sequence analysis of the β^+ thalassemic β-globin gene DNA fragments was carried out by the method of Sanger, Nicklen, and Coulson [10]. on single-stranded DNA templates prepared either from M13-cloned material or by Exonuclease III degradation of double-stranded plasmid DNA. Additional nucleotide sequencing procedures included modifications of the methods of Maxam and

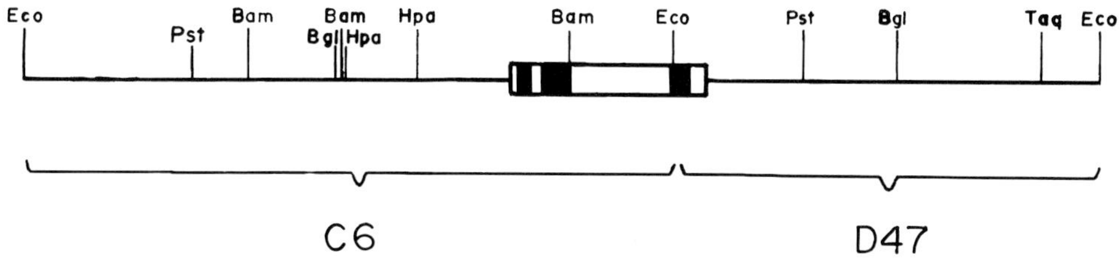

Fig. 1. Partial restriction endonuclease map of cloned β^{+} thalassemic β-globin gene fragments, C6 and D47. The solid blocks represent coding sequences of the gene whereas the open blocks represent the two intervening sequences (introns) of the gene, and, at either extremity, the sequences corresponding to the 5′- and 3′-untranslated sequences of the globin mRNA. Eco, Eco RI; Hpa, Hpa I; Pst, Pst I; Bgl, Bgl II; Bam, Bam HI; Taq, Taq I.

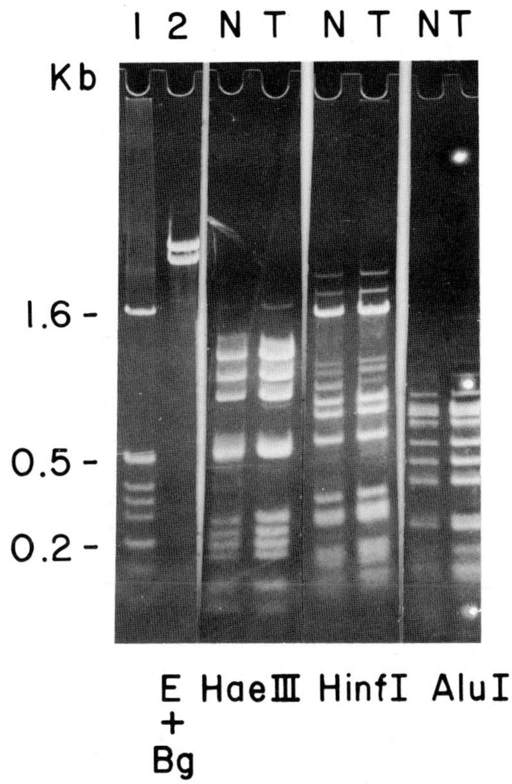

Fig. 2. Restriction endonuclease digests of normal and β^{+} thalassemic globin gene DNA fragments. The genomic DNA inserts corresponding to the 5′-Eco RI fragment of β^{+} thalassemic clone C6 and the corresponding fragment of a normal β-globin gene subclone were excised from their plasmid vectors by digestion with Eco RI + Bgl II and fractionated by agarose gel electrophoresis. The eluted DNA fragments (lane 2) containing the 5′-portion of the globin gene and its 5′-flanking DNA were subsequently redigested with the indicated enzymes (Hinf I, Alu I, and Hae III), fractionated by electrophoresis in a 1% agarose gel, and stained with ethidium bromide. No differences were detected between the normal (N) and thalassemic (T) DNAs. The size markers in lane 1 were obtained by Hinf I digestion of plasmid pBR322 [25].

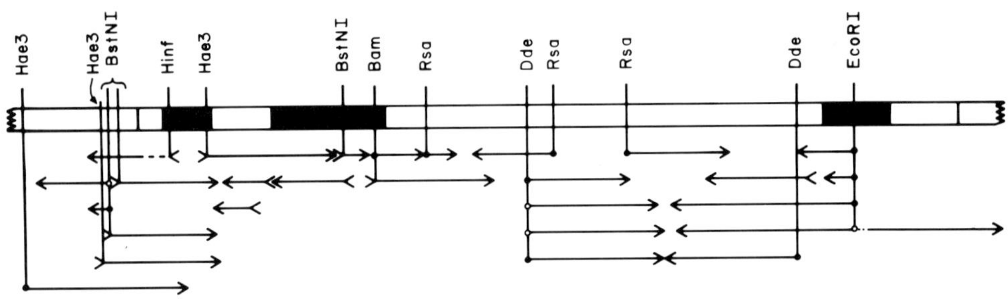

Fig. 3. DNA sequencing strategy of the β^+ thalassemic β-globin gene fragments. Tailed arrows indicate regions sequenced by the method of Sanger, Nicklen, and Coulson [10], closed circles regions sequenced by the method of Maxam and Gilbert [11], and open circles regions sequenced by the method of Maat and Smith [12]. Arrows pointing to the right indicate regions sequenced on the nontranscribed strand, and those to the left, regions sequenced on the transcribed DNA strand.

Gilbert [11] and of Maat and Smith [12]. Figure 3 depicts the general strategy utilized in the DNA sequence analysis of the thalassemic gene.

The 1,904-nucleotide sequence determined extends from a point 216 nucleotides 5′ to the cap site of the β-globin mRNA to a point 82 nucleotides 3′ to the polyadenylation site. The general structure of the human β-globin gene is similar to that of other known mammalian non-α globin genes. The protein encoding region is interrupted by two intervening sequences, confirming previous data obtained by gel blotting [13, 14] and DNA sequence analysis of a cloned normal β-globin gene [9]. The small intervening sequence occurs between codons 30 and 31, and is 130 nucleotides long. The large intervening sequence is 850 nucleotides in length, and lies between codons 104 and 105.

Comparison of the nucleotide sequence of the β^+ thalassemic gene with that of a normal human β-globin gene [15] reveals only one difference. There are no sequence differences in the 5′- or 3′-flanking regions, in the 5′- or 3′-untranslated regions, in the large IVS, or in the protein encoding regions of the β globin gene itself. It should be noted however that the coding sequence of both the normal [15] and thalassemic β-globin genes differ, at codon position 53, from that previously reported for normal β-globin mRNA [16]. The sequence GCA previously assigned for this codon [16] was not determined by direct sequence analysis of cDNA or mRNA, but was deduced indirectly on the basis of short oligonucleotides present or absent in cRNA of one individual. It is possible that polymorphism exists at this codon position in different individuals.

The only difference of DNA sequence between the normal and thalassemic β globin genes occurs in the internal portion of the small IVS, well away from the junctions of the IVS with the adjacent protein encoding sequences. Of particular note is that all four of the coding-intervening sequence junctions are normal in the β^+ thalassemic gene.

NUCLEOTIDE DIFFERENCE IN THE SMALL INTERVENING SEQUENCE

The small intervening sequence of the β^+ thalassemic gene differs from that of the normal human β-globin gene at only one site. Twenty-two nucleotides before the 3′-end junction of the small IVS with the adjacent protein encoding sequence, an

```
                5'                                      3'

                                                    LeuLeu
β-Thal    CCTATT[A]GTCTATTTTCCCACCCTTAGGCTGCTG

Normal    CCTATT[G]GTCTATTTTCCCACCCTTAGGCTGCTG
```

Fig. 4. Nucleotide sequences near the 3′-junction of the small IVS with the adjacent protein coding sequence. The authentic and possible internal splice sites are indicated by the solid line. The single nucleotide change in the β^{+} thalassemic IVS-1 is indicated by the box.

adenine nucleotide replaces a guanine. It is noteworthy that this substitution creates a sequence that is homologous to that of the authentic 3′-splice junction in six of seven nucleotides (Fig. 4). Furthermore, the one divergent nucleotide at the normal splice junction may not be rigidly specified, as the sequences of the human δ and ϵ-globin genes differ from those of the γ and β-globin genes at this base residue. If the atypical sequence in the small intron of the β^{+} thalassemic gene is a functional splice sequence, it might compete or interfere with RNA splicing at the authentic splice site, resulting in improper splicing of some of the β-globin precursor mRNA molecules [17, 18]. Such abnormal precursor mRNAs would probably be unstable, and might not be efficiently transported to the cytoplasm, or the peptide encoded by this RNA might be unstable, since no abnormal peptide appears on carboxymethyl-cellulose column chromatography of proteins synthesized by reticulocytes of β^{+} thalassemic patients [1, 2].

TRANSCRIPTION OF THE THALASSEMIC β-GLOBIN GENE IN VITRO

The low level of β globin mRNA in erythroid cells from patients with β^{+} thalassemia may reflect a defect in the DNA sequences that direct transcription. The sequences involved in transcription by RNA polymerase II have not yet been fully defined. However, Ziff and Evans [19], Benoist et al [20], and Efstratiadis et al [21] have identified highly conserved sequences in specific regions upstream from a number of eukaryotic structural genes. The nucleotide sequences upstream from both the β^{+} thalassemic and normal β-globin gene are identical, however, suggesting that these regions are not involved in the defective function of the thalassemic gene. Brown and colleagues [22, 23] have identified a region in an internal portion of the Xenopus 5S ribosomal RNA gene, the integrity of which is necessary for transcription of this gene in vitro by RNA polymerase III. The small intervening sequences of the human β-like globin genes are among the most highly conserved portions of these genes [21], suggesting that these sequences may have some as yet undefined role in gene regulation, perhaps analogous to the internal "control" region of the Xenopus 5S gene. If so, the single nucleotide substitution in the small intervening sequence of the thalassemic gene could result in decreased transcription of this gene.

In order to investigate the transcription of the cloned β^{+} thalassemic globin gene, we have transcribed the gene in vitro using the cell-free system of Manley et al [24]. This system utilizes a whole-cell extract to transcribe DNA templates by RNA polymerase II. Figure 5 illustrates the transcripts produced by the normal and thalassemic β-globin gene templates, and by a $^{G}\gamma$ globin gene template.

As shown in Figure 5, both the normal β globin and $^{G}\gamma$ globin gene templates transcribed faithfully in this system; the slightly longer length of the $^{G}\gamma$ gene transcript reflects the fact that the large intervening sequence of this $^{G}\gamma$ globin gene is 35 nucleotides longer than that of the β-globin gene [P. Jagadeeswaran, unpublished observations]. Moreover, the β^+ thalassemic transcript is qualitatively and quantitatively identical to that of the normal β-globin gene, suggesting that transcription of this thalassemic gene is normal, at least in this in vitro system.

CONCLUSIONS

We have cloned fragments of a human β-globin gene from a patient with homozygous β^+ thalassemia, whose erythroid cells accumulate decreased amounts of cytoplasmic β-globin mRNA and, as a result, synthesize reduced amounts of

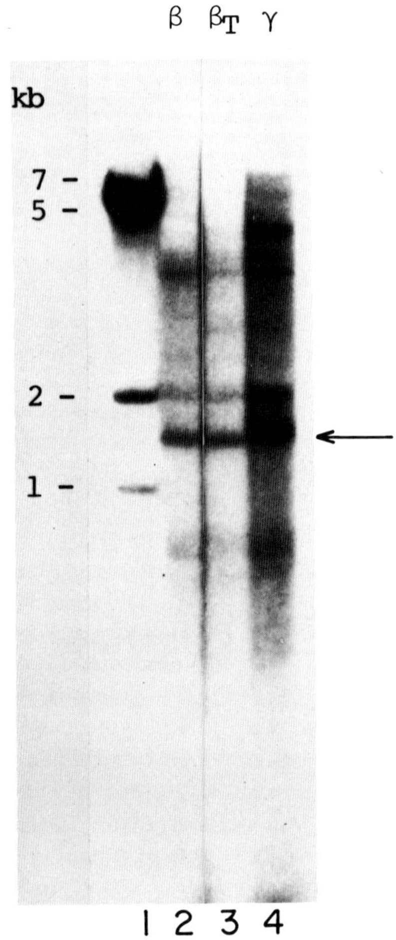

Fig. 5. In vitro transcription of 5′-Eco RI fragments of human globin genes: normal β (lane 2); β^+ thalassemic (lane 3), and normal $^{G}\gamma$ (lane 4). Of each plasmid subclone 2.5 μg were digested with restriction endonuclease Eco RI, and transcribed in a whole cell extract according to Manley et al [24]. The ^{32}P-labeled transcripts were then fractionated by electrophoresis in a gel of 1.4% agarose and the gel subjected to autoradiography. Arrow indicates globin-specific transcripts. The labeled 2 kb fragment consists of end-labeled 18S RNA and was present in control reactions lacking DNA templates. Lane 1 contains size markers derived from SV40 virus DNA.

β-globin chains. Detailed structural and nucleotide sequence analyses indicate that this gene differs from the normal human β-globin gene only in the internal portion of the small intervening sequence. Furthermore, transcription of the cloned thalassemic β-globin gene indicates that this gene is transcribed normally in vitro. These data suggest that the basic molecular defect in this patient may be an abnormality of RNA processing, due to the mutation in the small intervening sequence, or to a mutation distant from the β-globin gene itself. Testing of these hypotheses will most likely require analysis of globin gene expression using in vitro systems that have the ability to process transcribed precursor mRNA molecules into mature mRNA.

SUMMARY

β-globin gene fragments from an individual with homozygous β^+ thalassemia have been cloned by use of recombinant DNA technology and subjected to nucleotide sequence analysis and to in vitro transcription in a cell-free system with RNA polymerase II activity. The entire nucleotide sequence of the cloned β^+ thalassemic globin gene has been determined including the entire coding sequence, the entire small and large intervening sequences (IVSs), the 5′- and 3′-untranslated sequences, 216 nucleotides of the 5′-flanking DNA, and 82 nucleotides of the 3′-flanking DNA. No nucleotide differences, compared to the sequence reported for a normal β-globin gene, have been detected in the coding sequence, 5′- and 3′-untranslated sequences, 3′-flanking sequence, large IVS, and the first 110 nucleotides of the 5′-flanking DNA that could be compared. The only difference from the reported normal sequence was found in the small IVS and consisted of a single nucleotide base substitution within the body of the IVS; as a result of this base change, a heptanucleotide sequence is generated within the IVS that is virtually identical to the sequence at the 3′-junction of the small IVS with the coding sequence. The nucleotide sequences at all four junctions between IVS and coding sequence were normal. The β^+ thalassemic clone containing the 5′-portion of the gene was actively transcribed in a cell-free transcription system. The RNA transcript was qualitatively and quantitatively similar to that obtained from a similar clone of normal β-globin gene DNA. The cloned β^+ thalassemic globin gene that we have isolated therefore does not appear to be defective in its ability to be transcribed. Any posttranscriptional defects that may be associated with the gene product cannot be attributed to abnormalities of the nucleotide sequence in untranslated sequences or at the splice junctions of the intervening sequences. Either the single nucleotide difference within the body of the small IVS is related to a posttranscriptional defect in processing of β mRNA precursor molecules, or the β^+ thalassemic gene is defective at a more distant site than has been analyzed.

ACKNOWLEDGMENTS

We thank E. Coupal, R. Ezekiel, and V. Vellucci for skilled technical assistance; S.J. Baserga for helpful discussions; and Dr. T. Maniatis and colleagues for providing us with the subclones of the normal β-globin gene and the normal β globin gene sequence prior to its publication.

This work was supported in part by grants AM-19482, GM20124, GM07439, CA90255, and AI13357 of the National Institutes of Health, and grant NP6H of the American Cancer Society. R.A. Spritz is the recipient of fellowships from the Cooley's Anemia Foundation, Inc. and the Charles E. Culpepper Foundation, Inc.

REFERENCES

1. Weatherall DJ, Clegg JB: "The Thalassemia Syndromes," 2nd ed. Oxford: Blackwell Scientific Publications, 1972, pp 1–374.
2. Bunn HF, Forget BG, Ranney HM: "Human Hemoglobins," Philadelphia: W.B. Saunders, 1977, pp 140–192.
3. Forget BG: Molecular lesions in thalassemia. Trends Biochem Sci 3:86–90, 1978.
4. Bank A: The thalassemia syndromes. Blood 51:369–384, 1978.
5. Forget BG: Molecular genetics of human hemoglobin synthesis. Ann Int Med 91:605–616, 1979.
6. Bank A, Mears JG, Ramirez F: Disorders of human hemoglobin. Science 207:486–493, 1980.
7. Southern EM: Detection of specific sequences among DNA fragments separated by gel electrophoresis. J Mol Biol 98:503–517, 1975.
8. Wilson JT, Wilson LB, deRiel JK, Villa-Komaroff L, Efstratiadis A, Forget BG, Weissman SM: Insertion of synthetic copies of human globin genes into bacterial plasmids. Nucleic Acids Res 5:563–581, 1978.
9. Lawn RM, Fritsch EF, Parker RC, Blake G, Maniatis T: Isolation and characterization of linked δ and β globin genes from a cloned library of human DNA. Cell 15:1157–1174, 1978.
10. Sanger F, Nicklen S, Coulson AR: DNA sequencing with chain-terminating inhibitors. Proc Natl Acad Sci USA 74:5463–5467, 1977.
11. Maxam AM, Gilbert W: A new method for sequencing DNA. Proc Natl Acad Sci USA 74:560–564, 1977.
12. Maat J, Smith AJH: A method for sequencing restriction fragments with dideoxynucleoside triphosphates. Nucleic Acids Res 5:4537–4545, 1978.
13. Flavell RA, Kooter JM, DeBoer E, Little PFR, Williamson R: Analysis of the $\beta\delta$ globin gene loci in normal and Hb Lepore DNA: Direct determination of gene linkage and intergene distance. Cell 15:25–41, 1978.
14. Mears JG, Ramirez F, Leibowitz D, Bank A: Organization of human δ and β globin genes in cellular DNA and the presence of intragenic inserts. Cell 15:15–23, 1978.
15. Lawn RM, Efstratiadis A, O'Connell C, Maniatis T: The nucleotide sequence of the human β globin gene. Cell 21:647–651, 1980.
16. Marotta CA, Wilson JT, Forget BG, Weissman SM: Human β globin messenger RNA. III. Nucleotide sequences derived from complementary DNA. J Biol Chem 252:5040–5053, 1977.
17. Kantor JA, Turner PH, Nienhuis AW: Beta thalassemia: Mutations which affect processing of the β-globin mRNA precursor. Cell 21:149–157, 1980.
17:915–921, 1979.
18. Maquat LE, Kinniburgh AJ, Beach LR, Honig GR, Lazerson J, Ershler WB, Ross J: Processing of the human β globin mRNA precursor to mRNA is defective in three patients with β^+ thalassemia. Proc Natl Acad Sci USA 77:4287–4291, 1980.
19. Ziff EB, Evans RM: Coincidence of the promoter and capped 5′-terminus of RNA from the adenovirus 2 major late transcription unit. Cell 15:1463–1475, 1978.
20. Benoist C, O'Hare K, Breathnach R, Chambon P: The ovalbumin gene-sequence of putative control regions. Nucleic Acids Res 8:127–142, 1980.
21. Efstratiadis A, Posakony JW, Maniatis T, Lawn RM, O'Connell C, Spritz R, deRiel JK, Forget BG, Weissman SM, Slightom J, Blechl AE, Smithies O, Baralle FE, Shoulders CC, Proudfoot NJ: The structure and evolution of the human β globin gene family. Cell 21:653–668, 1980.
22. Sakonju S, Bogenhagen DF, Brown DD: A control region in the center of the 5S RNA gene directs specific initiation of transcription: I. The 5′-border of the region. Cell 19:13–25, 1980.
23. Bogenhagen DF, Sakonju S, Brown DD: A control region in the center of the 5S RNA gene directs specific initiation of transcription: II. The 3′-border of the region. Cell 19:27–35, 1980.
24. Manley JL, Fire A, Cano A, Sharp PA, Gefter ML: DNA-dependent transcription of adenovirus genes in a soluble whole-cell extract. Proc Natl Acad Sci USA 77:3855–3859, 1980.
25. Sutcliffe JG: pBR322 restriction map derived from the DNA sequence: Accurate DNA size markers up to 4361 nucleotide pairs long. Nucleic Acids Res 5:2721–2728, 1978.

Organization and Expression of Globin Genes, pages 287–298

Heterogeneity of $\delta\beta$ Thalassemia and HPFH in Mediterranean Populations: Gene Mapping Studies

Sergio Ottolenghi, Barbara Giglioni, Roberto Taramelli, Paola Comi, Umberto Mazza, Giuseppe Saglio, Clara Camaschella, Pietro Izzo, Antonio Cao, Renzo Galanello, Enric Gimferrer, Montserrat Baiget, and Alessandro M. Gianni

Ist. Patologia Generale, Centro Studio Patologia Cellulare CNR (S.O., B.G., R.T., P.C.), Ist. Clin. Medica I (A.M.G.), University of Milano, Ist. Medicina Interna, Clin. Medica II (U.M., G.S., C.C.), University of Torino, Ist. Pathologia Medica, University of Bari (P.I.), Ist. Clin. Pediatrica II, University of Cagliari (A.C., E.G.), Italy; and Hospital de la Santa Cruz y San Pablo, Barcelona, Spain (E.G., M.B.)

INTRODUCTION

Hereditary disorders involving the human non-α-globin gene cluster ($^{G}\gamma$-$^{A}\gamma$-δ-β) provide a useful model for studying the regulation of gene activity in eukaryotes; in particular, Hereditary Persistence of Fetal Hb (HPFH) and $\delta^0\beta^0$ thalassemia—which are characterized by persistent expression of the γ genes in adults and reduced or abolished HbA synthesis—have been extensively studied in recent years.

Following the discovery of the Lepore (δ-β) and Kenya ($^{A}\gamma$-β) fusion globin chains, and the relationship of the latter with $^{G}\gamma$-HPFH, a model was proposed suggesting that deletions within the δ-β-globin gene cluster and adjacent DNA areas (3′ to the $^{A}\gamma$-globin gene) might be responsible for the failure of the γ-globin genes (cis to the deletion) to undergo the normal postnatal repression [1, 2]. Several aspects of this model have been confirmed, by either liquid hybridization experiments (showing deletions of δ-β-globin structural sequences [2]), or gene mapping studies demonstrating extensive deletions of adjacent DNA regions beside coding areas [3–12]. However, the precise definition of a specific regulatory area (if any), was not possible on the basis of these data; while the $^{A}\gamma$-δ bridging area was partially or completely deleted in some cases of hereditary persistence of HbF [6, 8, 9, 11], the 5′ end of the deleted area could be mapped to the δ-globin gene intron in a type of $\delta^0\beta^0$ thalassemia [5–7]. A comparison of the deletion in $\delta^0\beta^0$ thalassemia with that observed in the Lepore $\delta\beta$ gene (that is not associated to HPFH) indicates that DNA areas 3′ to the β-globin gene are missing in the former syndrome, but are present in the second one, a fact that suggests [7] that the deletion of these sequences may have some importance in determining the HbF persistence; yet these sequences are presumably present in the DNA from the $^{G}\gamma$-HPFH-associated Hb Kenya. As an alternative hypothesis, it was also suggested that structural δ-globin gene sequences may play a role in the repression of γ-globin genes [5–12]; the low level of HbF production in Lepore patients as compared to $\delta^0\beta^0$ thalassemia ones might be due to the presence of a "normal" δ-globin gene structure (the δ-β fusion gene) in the latter case. A further difficulty in the definition of the relationship between DNA deletions and gene activity arises as the consequence of the remarkable heterogeneity of the globin synthetic pattern (β, $^G\gamma$, $^A\gamma$) in these syndromes, and in particular of the expression of the $^G\gamma$ and $^A\gamma$ genes, which probably reflects, at least to a certain extent, a genetic effect [2]. Although in most instances both $^G\gamma$ and $^A\gamma$ genes are expressed at a comparable level, in other cases either the $^G\gamma$ or the $^A\gamma$ appears to be the predominant or even the only chain synthesized; since most cases so far reported are in the heterozygous state, uncertainties as to the relative proportion of the minor chain and the contribution of the normal allele trans to the HPFH gene make it unclear whether both γ genes or only one are expressed in a given case. An example of such a situation is provided by the $^A\gamma$-HPFH, the so-called Greek type [13]; initially $^A\gamma$ (but not $^G\gamma$) and probably β-globin were reported to be synthesized cis to the mutation at levels high enough to give a balanced synthetic pattern in affected individuals [14, 15]; later, cases apparently of the same ethnic background, were reported, which synthesized some $^G\gamma$ in addition to $^A\gamma$ chains, and gave a rather unbalanced synthetic pattern in the heterozygote [16]. To what extent genetic rather than analytical factors affect the reported data, is unclear.

TYPES OF DEFECTS OBSERVED

We have recently exploited our isoelectricfocusing technique [17] (which allows us to clearly separate and quantitate the $^G\gamma$ from the $^A\gamma$ chains) to characterize the $^G\gamma{}^A\gamma$ pattern of heterozygous, and when possible, of homozygous Mediterranean patients with various types of HPFH and $\delta^0\beta^0$ thalassemia. Here we wish to report initial restriction enzyme analysis of HPFH or $\delta^0\beta^0$ thalassemic patients from five distinct groups, and to attempt to correlate their globin pattern with the molecular lesions.

Table I summarizes the main hematological and globin data from the patients studied; three of the families might be considered to fall within the limits of classical $\delta^0\beta^0$ thalassemia, while two families should better be regarded as HPFH (although a clear-cut distinction between these two entities is probably artificial).

Type 1 (HPFH From Puglia)

Heterozygotes produce 15% HbF, of which 95% is of the $^A\gamma$ type. From balanced biosynthetic data and the observation that HPFH and HPFH/β thalassemic patients in the family have a per cell Hb content similar to their normal or β-thal counterparts, respectively, it can be deduced that the β-globin gene cis to the mutation is active, and the combined γ + β activity is similar to the expected activity of a normal β-globin gene. Indeed, no abnormality was found in their DNA, when probing β-globin gene with Eco RI, PstI, BglII, Bam HI, XbaI; however, BglII digestion reveals, in addition to the normal 13 kb γ band, a smaller (≃ 8 kb) γ band (see Fig. 1A). Since BglII cuts outside the γ gene cluster, a deletion might be occurring either between the two γ genes, or at one side. Digestion with PstI (Fig.

TABLE I. Main Haematological and Globin Data for δβ Thalassemia and HPFH Heterozygotes[1]

	Hb g/dl	MCH pg	HbF %	HbA_2 %	β/α synth. ratio	γ/α synth. ratio	$^G\gamma/^A\gamma + {}^G\gamma$	cis β-globin gene activ.	$^G\gamma/^A\gamma + {}^G\gamma$ in homozygotes
1) Puglia HPFH	14.7 (13–16)	30 (29–31)	14 (12–16)	2.2 (1.9–2.7)	0.82	0.12	0.05–0.1	+	—
2) Southern Italian HPFH	14 (12–16)	29 (27–31)	30 (29–32)	1.9 (1.7–2.0)	0.66	0.27	0.15–0.20	—	—
3) Sardinian $\delta\beta^0$	13.7 (13–15)	26 (21–29)	15 (10–25)	2.4 (1.4–3.2)	0.61	0.11	0.05–0.1	—*	—
4) Spanish $\delta^0\beta^0$	13.1 (12–15)	23 (20–26)	8 (3–21)	2.3 (1.3–2.65)	—	—	0.3–0.35	—**	0.45
5) Southern Italian $\delta^0\beta^0$	12.5 (12–13)	22 (21–25)	8 (7–10)	2.0 (1.8–2.0)	0.65	0.08	0.1–0.15	—**	0.45

[1] Data from Refs. 18, 19, 20, and unpublished results.
*Double heterozygous $\delta\beta/\beta^0$ thalassemic patients have no Hb A.
**Homozygotes have HbF only.

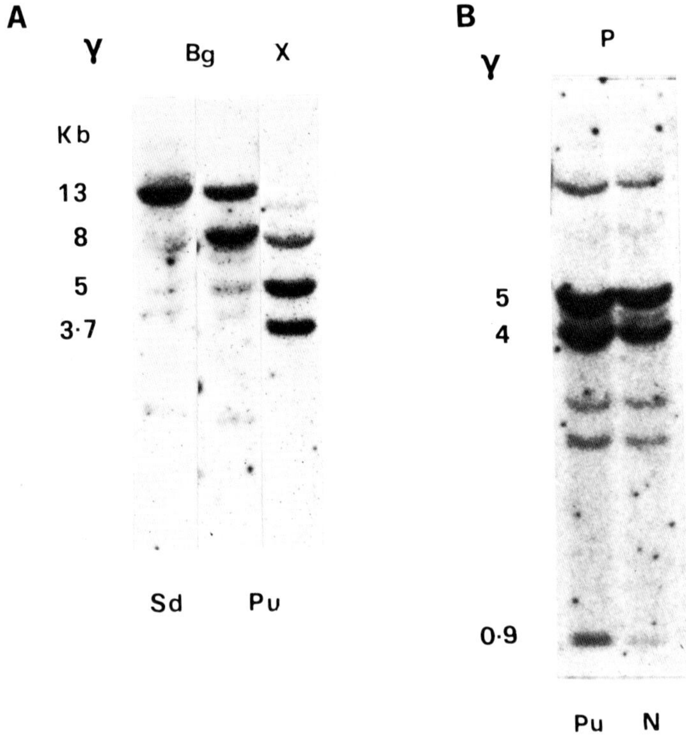

Fig. 1. Comparison of γ-globin gene fragments in HPFH from Puglia (Type 1, Pu), Sardinian $\delta\beta^0$ thalassemia (Type 3, Sd), and Normal (N) DNA. DNA was prepared from blood white cells, digested with a restriction enzyme, electrophoresed in the native state on an agarose gel, blotted on to nitrocellulose filters, and hybridized to a γ-globin plasmid as described previously [5]. The fragments lengths were calculated using as size markers the normal γ bands obtained with several additional enzymes. A). Bg: BglII; X: XbaI. B). P: PstI.

1B), XbaI, and BamHI (Fig. 4A) (each of which gives a fragment extending in the 5′ → 3′ direction from within the $^{A}\gamma$ to adjacent areas just before -PstI- or past -Bam HI, XbaI- the 3′ BglII site) fails to show any abnormal band, thus ruling out a deletion on the 3′ side of the γ gene cluster. On the other hand, following redigestion of the separated BglII 13 kb and 8 kb bands with EcoRI, the abnormal fragment appears to contain the normal expected 3 kb fragment, corresponding to the 5′ side of the $^{G}\gamma$ gene. This result leaves only the possibility of a deletion in the $^{G}\gamma$-$^{A}\gamma$ intergene area. Indeed, the expected 2.5 kb and 1.65 kb BglII-EcoRI fragments are missing from the ≃ 8 kb BglII band (Fig. 2A). In addition, a new faint fragment (2.3 kb) appears (Fig. 2A,B) that is more clearly observed under conditions of low stringency; whether it is related to the deletion causing HPFH, or is a polymorphism of some non-α gene in these patients (see for example Fig. 4A) is unclear. However, these data, although not yet defining with precision the deletion, clearly locate it to the $^{G}\gamma$-$^{A}\gamma$ intergene area.

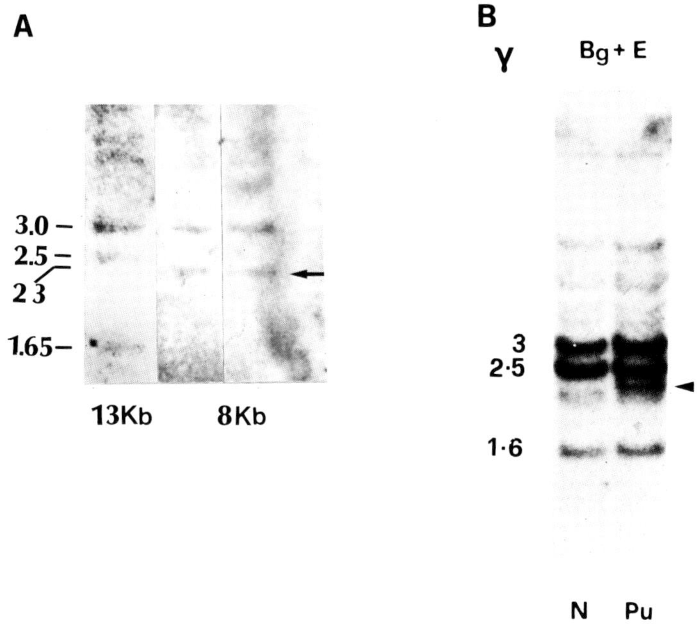

Fig. 2. A. Eco RI digestion of the BglII 13 and 8 kb bands from HPFH from Puglia. Following BglII digestion and electrophoresis, the gel was sliced, the DNA was eluted on to hydroxylapatite, recovered by washing with 0.4M sodium phosphate pH 6.8, desalted, and redigested with EcoRI. Three slices were interposed between the 13 kb and 8 kb fragments. B. Double digestion of Normal (N) and HPFH DNA with Eco RI and BglII. Arrow points to the abnormal band.

Type 2 (Southern Italian HPFH)

Heterozygotes from this family show a balanced globin synthesis with a high proportion of γ-globin synthesis compensating for the severe reduction of β-globin synthesis. This pattern is exceptional in Mediterranean areas and strongly recalls the Negro HPFH syndrome, although our patients synthesize predominantly $^{A}\gamma$ chains (80%–85% of γ chains) in contrast to the $^{G}\gamma$ excess more commonly observed in African patients. Digestion of their DNA with a number of restriction enzymes, followed by hybridization analysis with cloned β and γ-globin plasmids of the agarose gel fractionated restriction fragments, fails to demonstrate any abnormality (Fig. 3); however, β bands in the HPFH DNA are consistently fainter than the normal corresponding ones. Although this pattern is consistent with normality, it is also compatible with total deletion of the relevant band, because the normal bands from the non-HPFH chromosome would appear. However, the $^{G}\gamma$ and $^{A}\gamma$ genes, being the ones certainly expressed, must be present; we therefore analyzed this DNA using an enzyme (BamHI) that gives a fragment containing part of the $^{A}\gamma$-globin gene and part of the δ-globin gene. This experiment demonstrates an approximately 14 kb fragment that contains γ-globin (Fig. 4A) but no β-like (Fig. 4B) sequences, and seems to be analogous to a fragment previously described [6] in

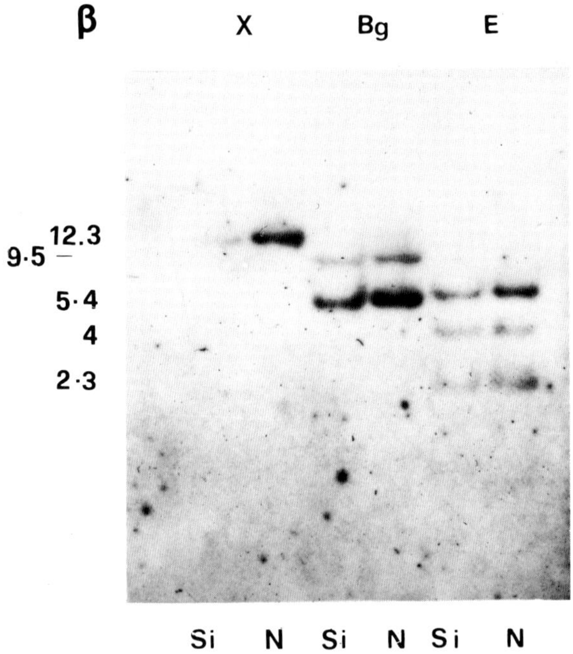

Fig. 3. β-Globin fragments in Normal (N) and Southern Italian HPFH (Type 2, Si). X: Xbal; E: Eco RI; Bg: BglII.

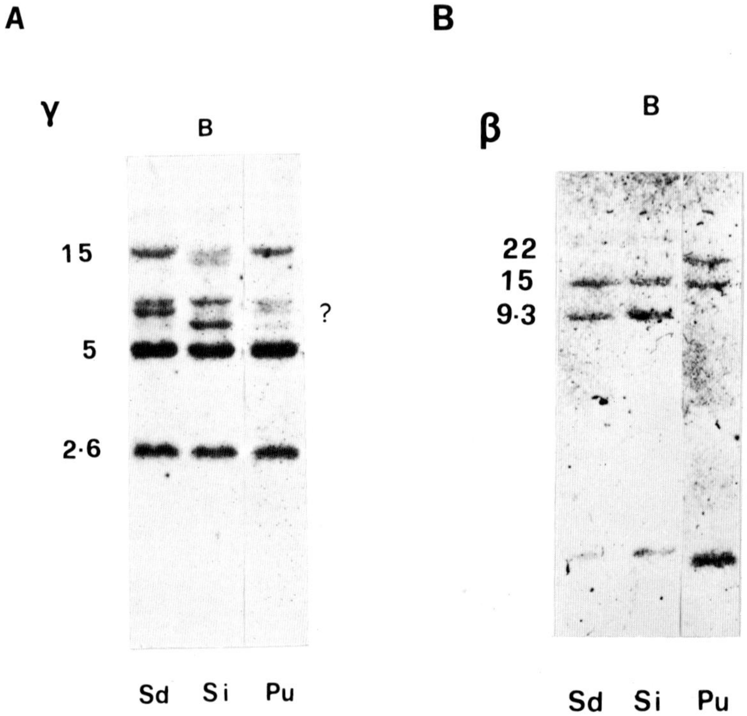

Fig. 4. Bam HI γ and β-globin fragments in HPFH Types 1 (Pu) and 2 (Si), and $\delta\beta^0$ thalassemia Type 3 (Sd). A. γ probe. B. β probe. The 22 kb BamHI β fragment is a known polymorphism [23]. The faint bands marked ? are polymorphic unidentified β-like fragments.

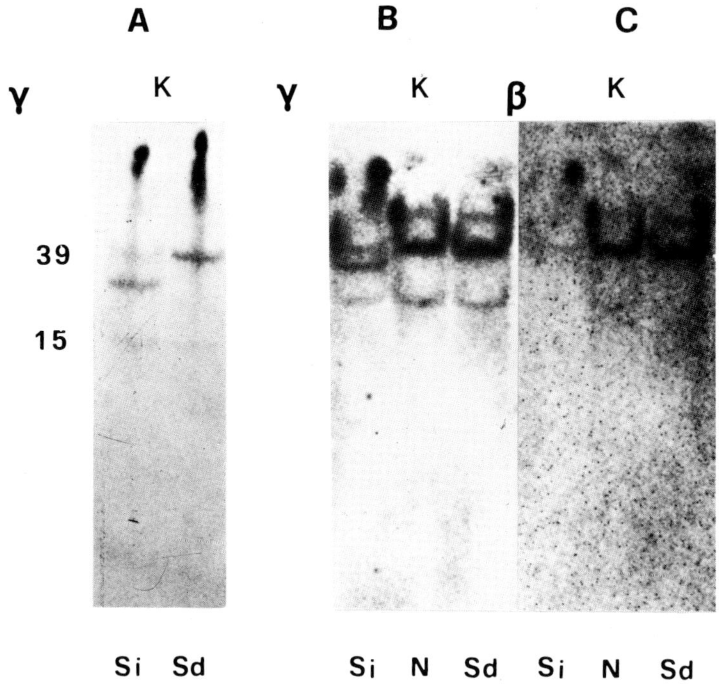

Fig. 5. KpnI γ and β fragments. A and B. γ probe, comparison of Normal (N) HPFH 2 (Si) and Sardinian $\delta\beta^0$ DNA (Sd, cases 1 and 2). C. Same as in B, rehybridized with β probe.

Negro-type HPFH; on this basis and because of the normality of the γ bands, we thought that a deletion within the δ-β cluster might have occurred. This is borne out by digestion with KpnI, which usually gives a single band containing the entire $^G\gamma$-$^A\gamma$-δ-β cluster; in this case a shorter abnormal band is apparent, which hybridizes to the γ, but not to the β probe (Fig. 5A,B), indicating that a deletion of the whole δ-β-globin cluster has occurred.

Type 3 (Sardinian $\delta\beta^0$ Thalassemia)

Several $\delta\beta$ thalassemic families of Sardinian origin, independently discovered in Cagliari and Turin, appear to share common hematological parameters and globin synthetic pattern, thus allowing the definition of a homogeneous entity. In heterozygotes the increased synthesis of γ-globin chains leads to only slightly unbalanced α/non-α ratios; however, the β-globin gene appears to be inactive, as shown by the absence of any HbA in double heterozygous $\delta\beta^0/\beta^0$ patients in two of the families studied. Most of the HbF (95%) is of the $^A\gamma$ type; the $^G\gamma$ gene, although not significantly expressed in heterozygotes, is also in a potentially active state, as shown by the high ($\simeq$20%) proportion of $^G\gamma$ chains in a $\delta\beta^0$ thalassemic heterozygote with chronic hemolysis due to G6PD deficiency. Digestion of Sardinian $\delta\beta^0$ thal DNA with Eco RI (Fig. 6) does not give any abnormal band, with both β and γ probes; this is still consistent with the total deletion of structural gene sequences corresponding to a band; since EcoRI cuts each globin gene into a 5′ and 3′ piece,

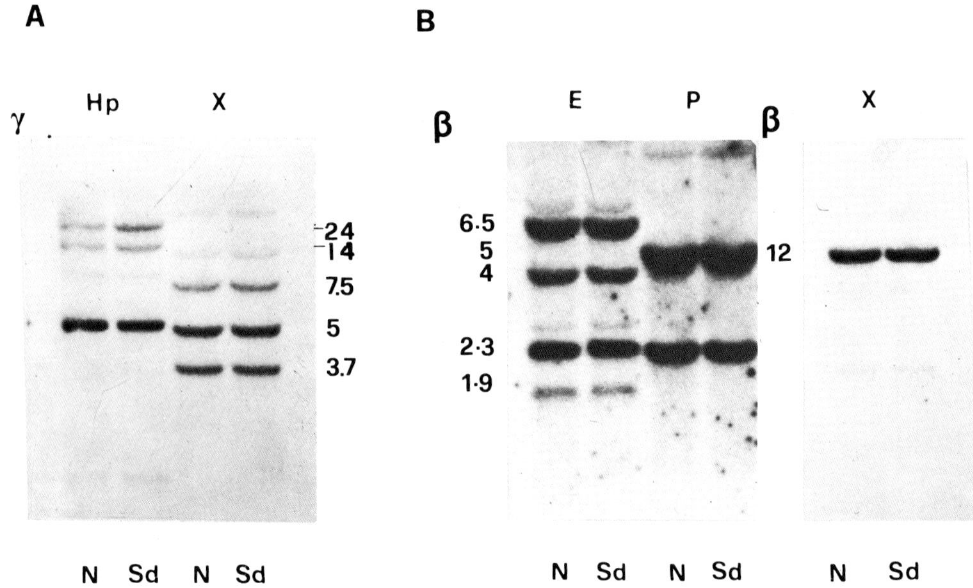

Fig. 6. γ and β fragments in Sardinian $\delta\beta^0$ (Sd) and Normal (N) DNA. A. Hp: HpaI; X: XbaI, γ probe. B. P: PstI; E: Eco RI; X: XbaI, β probe.

deletion of either of these pieces could remain unnoticed in heterozygotes. However, such a partial deletion should result in the appearance of an abnormal band when tested with an enzyme cutting at both sides of the gene. Digestion with several additional enzymes (PstI, BglII, XbaI, HpaI) does not, however, provide any evidence for abnormal β or γ bands; rather it consistently demonstrates very similar intensities of the bands from the $\delta\beta^0$ thal DNA to those shown in normal DNA. This is particularly significant when XbaI (Fig. 7A) (which gives a single δ-β-globin band), BglII (which gives a single $^G\gamma$-$^A\gamma$ band; Fig. 1), BamHI (which gives a fragment bridging $^A\gamma$ to δ gene; Fig. 4), and KpnI (which reveals the $^G\gamma$-$^A\gamma$-δ-β cluster; Fig. 7) are used. Thus large deletions within the non-α-globin gene cluster do not appear to occur in these patients.

Type 4 (Spanish $\delta^0\beta^0$ thalassemia) and Type 5 (Southern Italian $\delta^0\beta^0$ Thalassemia)

These Spanish and Italian $\delta^0\beta^0$ thalassemic families correspond to the classical hematological description of this syndrome. However, Spanish heterozygotes have a higher $^G\gamma$-$^A\gamma$ ratio, and their $^A\gamma$ chain is totally of the $\gamma^{Sardinia}$ variety, demonstrating an independent genetic origin of the mutation. Indeed, restriction enzyme analysis of a Spanish $\delta^0\beta^0$ homozygote DNA shows the complete absence of the δ-β cluster (Fig. 7), in contrast to the Italian cases in which the 5′ δ-globin exon is preserved [5–7, 12]. In the Spanish $\delta^0\beta^0$ thal the 14 kb HpaI γ fragment is absent, while the XbaI and the BglII fragments are present, indicating that the deletion extends at least 1.2 kb 5′ to the δ-globin gene (and possibly up to ≃8 kb).

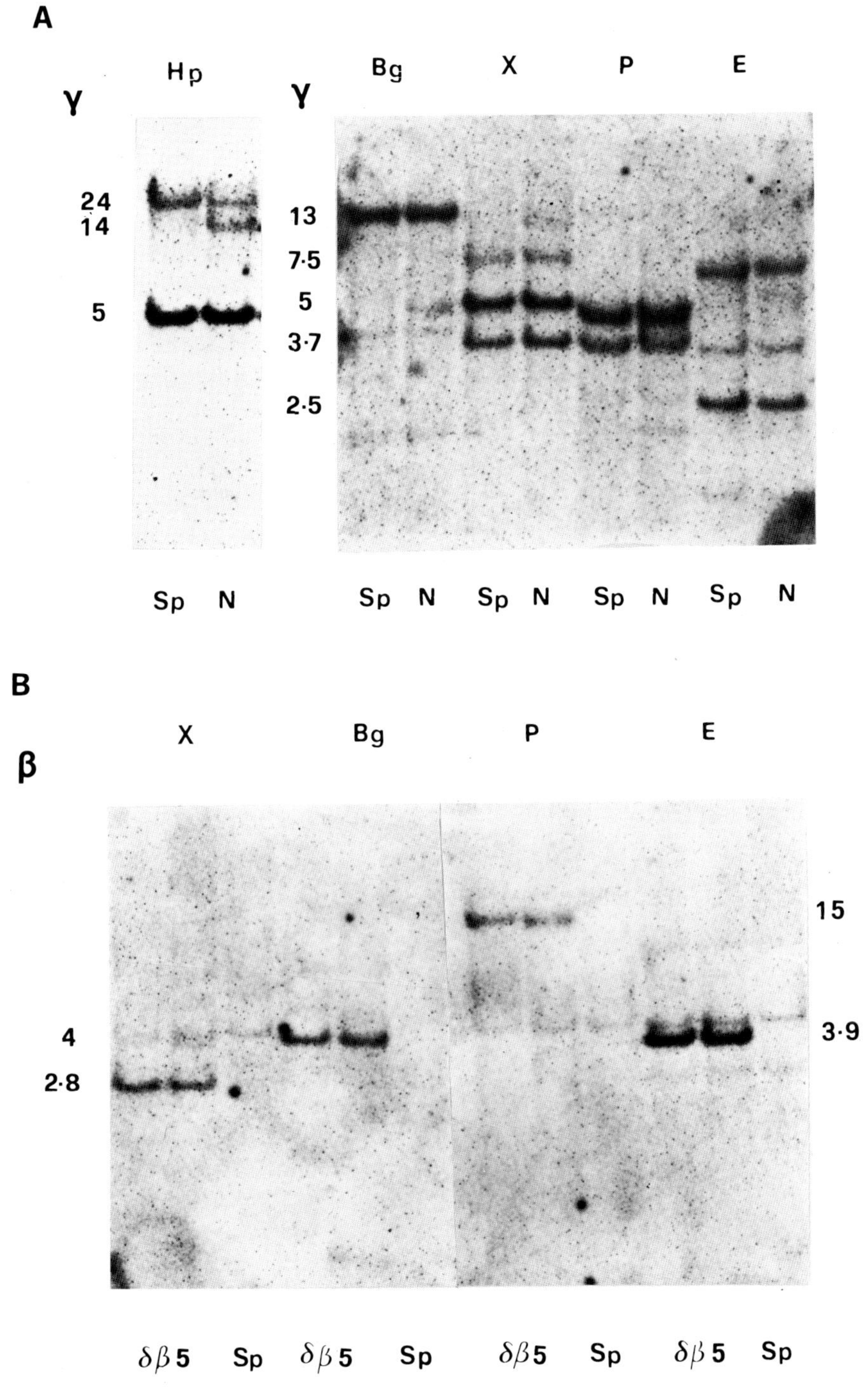

Fig. 7. γ and β fragments in two $\delta^0\beta^0$ Southern Italian homozygotes ($\delta^0\beta^0 5$) (from two families) and in the Spanish $\delta^0\beta^0$ homozygote (Sp). A. Hp:HpaI; X: XbaI; Bg: BglII; E: Eco RI; P: PstI γ probe. B. Same, β probe.

DISCUSSION

The present data, although not defining exactly the extent of the deletion or the site of the mutation, provide evidence for a very wide degree of molecular heterogeneity of HPFH and $\delta^0\beta^0$ thalassemia. In three different families, partial (in the classical $\delta^0\beta^0$ thal) or complete (Spanish and Negro-type) deletions of the δ-β-globin cluster occur; these cases are therefore in agreement with the notion that DNA areas within and around the δ-β cluster are necessary for the postnatal repression of γ-globin gene activity. However, no simple correlation emerges so far from the comparison of the globin synthetic data and the extent of the deletion; thus the Spanish type and the classical $\delta^0\beta^0$ thalassemia share many common properties (similar HbF level and hematological parameters), although in the former case the deletion additionally removes the 5′ half of δ-globin gene, extending for at least 1.3 kb 5′ to the δ sequences into the region usually thought to be involved in the repression of the γ-globin genes.

Interestingly, all three types of deletions show a great predominance of $^A\gamma$ expression over $^G\gamma$ in the heterozygotes; in the classical $\delta^0\beta^0$ thalassemic heterozygote (type 5) it can be calculated that the amount of $^G\gamma$ chains per cell is around 0.1–0.2 pg, a value that is similar or only slightly higher than that we find in control β thal heterozygotes, which are not thought to have derepressed γ-globin genes; yet the $^G\gamma$ gene is in a potentially activated state, as shown by the high level of $^G\gamma$ chains in two homozygotes from two different families studied ($^G\gamma/^A\gamma + {}^G\gamma = 0.45$). In the Spanish family the $^G\gamma/^A\gamma + {}^G\gamma$ ratio is higher ($\simeq 0.3$) than in the Italian $\delta^0\beta^0$ thalassemia; it is possible that, in order to make the $^G\gamma$ gene active, the deletion must extend 5′ to the δ gene, as it appears to be in the Spanish and Negro type.

The Sardinian $\delta\beta^0$ thal patients provide still another type of $\delta\beta$ thalassemia in which persistent HbF production (almost exclusively of the $^A\gamma$ type) appears to be associated with total suppression of the β-globin gene on the same chromosome. Yet restriction enzyme analysis clearly differentiates these patients from the typical $\delta^0\beta^0$ thal usually found in Italy (type 5), and provides evidence for the presence of an apparently intact δ-β gene cluster. Although the unlikely possibility that two independent mutations coexist on the same chromosome cannot be ruled out formally, extensive genetic studies of several families fail to show segregation of HPFH and β thal [20]. Thus it seems likely that a single, so far unidentified mutation, probably mapping close to the γ-β gene cluster, completely suppresses β-globin gene activity while simultaneously acting to prevent repression of the γ-globin gene region. An apparently intact γ-β gene cluster has recently been reported [9, 11] in Greek-type HPFH; in this condition, however, the β-globin gene seems to be active [13–16] and hematological characteristics are different from those observed in our patients [13, 15].

Finally, in the family from Puglia (type 1), hematological and globin synthetic data strongly suggest the presence of an active β-globin gene cis to the HPFH mutation. In these patients our gene mapping studies locate a small DNA rearrangement to the $^G\gamma$-$^A\gamma$ intergene area; previous data [10] indicated that a large deletion spanning the ϵ-$^G\gamma$-$^A\gamma$-δ-globin genes causes the partial (or complete?) inactivation of the apparently intact β-globin gene. Within the frame of a general model for HPFH, it was suggested [11] that the gross gene rearrangement might alter the boundary of an "adult domain," thus preventing the establishment of the normal chromatin configuration necessary for β-globin gene expression, by some long range direct effect. However, in our case the DNA rearrangement lies in the γ-globin area, remote from the proposed adult boundary; since β-globin gene activity in cis appears to be reduced roughly in proportion to the increased expression of the ${}^{A}\gamma$ gene, it is alternatively possible that this stems from an indirect effect on the activity of the β-globin gene region of the active transcription abnormally occurring on the γ globin area; in vivo evidence for such an inverse relationship is provided by the group of heterocellular HPFH [21] and by studies with cultured erythroid cells showing a rather precise negative correlation of γ versus β activity and ${}^{G}\gamma/{}^{A}\gamma$ versus $\beta/\beta + \gamma$ synthetic ratios in cord blood erythroid cells [22, 23].

SUMMARY

We studied five different types of Hereditary Persistence of Fetal Hemoglobin (HPFH) or $\delta\beta$ thalassemia of Mediterranean origin. ${}^{G}\gamma/{}^{G}\gamma + {}^{A}\gamma$-globin chain ratios were analyzed by isoelectricfocusing of globin chains. Gene mapping studies using cloned β and γ-globin probes were carried out in an attempt to correlate globin synthetic patterns with molecular defects at the DNA level. In a family with HPFH (from Puglia) a small deletion in the ${}^{G}\gamma$-${}^{A}\gamma$ intergene area appears to be responsible for persistent expression of the ${}^{A}\gamma$ gene (${}^{G}\gamma/{}^{A}\gamma + {}^{G}\gamma$ in heterozygotes is $\simeq 0.05-0.1$) and a corresponding decrease of the β-globin gene activity. In the Sardinian type of $\delta\beta^0$ thalassemia (${}^{G}\gamma/{}^{A}\gamma + {}^{G}\gamma \simeq 0.05$), no abnormality is found by restriction enzyme analysis. In a Southern Italian type showing a deletion extending from within the δ-globin intron to a region 3′ to the β-globin gene [5–7], the ${}^{G}\gamma/{}^{A}\gamma + {}^{G}\gamma$ ratio is low in heterozygotes [$\simeq 0.1$], although it is much higher [$\simeq 0.45$] in two homozygotes from the same families, indicating that the ${}^{G}\gamma$ gene is only potentially active in the absence of erythropoietic stress. Finally, in a Spanish type (${}^{G}\gamma/{}^{A}\gamma + {}^{G}\gamma$ $\simeq 0.3$), the $\delta\beta$-globin deletion extends 5′ to the δ gene, removing at least 1.2 kb of nonstructural DNA. Deletion of these sequences may be necessary for persistent expression of the ${}^{G}\gamma$ gene, although not causing significant increase of γ synthesis as compared to other $\delta^0\beta^0$ thalassemias.

ACKNOWLEDGMENTS

This work was partially supported by a CNR grant N.80.01133.83 (Progetto Finalizzato Medicina Preventiva) to S.O.

REFERENCES

1. Huisman THJ, Schroeder WA, Efremov GD, Duma H, Mladenovski B, Hyman CB, Rachmilevitz EA, Bouver N, Miller A, Brodie A, Shelton JB, Apell G: The present status of the heterogeneity of fetal hemoglobin in β thalassemia: An attempt to unify some observations in thalassemia and related conditions. Ann NY Acad Sci 232:107–124, 1974.
2. Weatherall DJ, Clegg JB: Recent development in the molecular genetics of human hemoglobin. Cell 16:467–479, 1979.
3. Mears JG, Ramirez F, Liebowitz D, Nakamura F, Bloom H, Konotey-Ahulu F, Bank A: Changes in restricted human cellular DNA fragments containing globin gene sequences in thalassemia and related disorders. Proc Natl Acad Sci USA 75:1222–1226, 1978.
4. Orkin SH, Alter BP, Altay C, Mahomey MG, Lazarus H, Hobbins JC, Nathan DG: Application of endonuclease mapping to the analysis and prenatal diagnosis of thalassemia caused by globin gene deletion. N Eng J Med 299:166–172, 1978.

5. Ottolenghi S, Giglioni B, Comi P, Gianni AM, Polli E, Acquaye GTA, Oldham JM, Masera G: Globin gene deletions in HPFH, $\delta^0\beta^0$ thalassemia and Hb Lepore disease. Nature 278:654–656, 1979.
6. Fritsch EF, Lawn RM, Maniatis T: Characterization of deletions which affect the expression of fetal globin genes in man. Nature 279:598–603, 1979.
7. Bernards R, Kooter SM, Flavell RA: Physical mapping of the globin gene deletion in $\delta^0\beta^0$ thalassemia. Gene 6:265–280, 1979.
8. Tuan D, Biro A, de Riel JK, Lazarus H, Forget BG: Restriction endonuclease mapping of the human γ globin gene loci. Nucleic Acids Res 6:2519–2544, 1979.
9. Tuan D, Murnane MJ, deRiel JK, Forget BG: Heterogeneity in the molecular basis of Hereditary Persistence of Fetal Hemoglobin. Nature 285:335–337, 1980.
10. Van der Ploeg LHT, Konings A, Oort N, Roos D, Bernini L, Flavell RA: γ-β thalassemia studies showing that deletion of the γ and δ genes influences β-globin gene expression in man. Nature 283:637–642, 1980.
11. Bernards R, Flavell RA: Physical mapping of the globin deletion in hereditary persistence of fetal hemoglobin (HPFH). Nucleic Acids Res 8:1521–1534, 1980.
12. Bank A, Mears G, Ramirez F: Disorders of human hemoglobin. Science 207:486–493, 1980.
13. Fessas P, Stamatoyannopoulos G: Hereditary persistence of fetal hemoglobin in Greece. A study and a comparison. Blood 24:223–240, 1964.
14. Huisman THJ, Schroeder WA, Stamatoyannopoulos G, Bouver N, Shelton JR, Shelton JB, Apell G: Nature of fetal hemoglobin in the Greek type of hereditary persistence of fetal hemoglobin with and without concurrent β thalassemia. J Clin Invest 49:1035–1040, 1970.
15. Sofroniadou K, Wood WG, Nute PE, Stamatoyannopoulos G: Globin chain synthesis in the Greek type of ($^A\gamma$) hereditary persistence of fetal hemoglobin. Br J Haematol 29:137–148, 1975.
16. Clegg JB, Metaxatou-Mavromati A, Kattamis C, Sofroniadou K, Wood WG, Weatherall DJ: Occurence of $^G\gamma$ HbF in Greek HPFH: Analysis of heterozygotes and compound heterozygotes with β thalassemia. Br J Haematol 43:521–536, 1979.
17. Comi P, Giglioni B, Ottolenghi S, Gianni AM, Ricco G, Mazza U, Saglio G, Camaschella C, Pick PG, Gianazza E, Righetti PG: $^G\gamma$ and $^A\gamma$ globin separation and quantitation by isoelectric focusing. Biochem Biophys Res Commun 87:1–8, 1979.
18. Camaschella C, Ciocca-Vasino MA, Guerrasio A, Balegno G, Barberis E, Delponte D, Saglio G: Biosynthetic studies and γ chain composition in the Greek type of hereditary persistence of fetal hemoglobin and in its association with β thalassemia. Acta Haematol 643:1–10, 1978.
19. Gimferrer E, Baiget M: Hematological, biochemical, clinical and genetical data in a case of homozygous $\delta\beta$ thalassemia. Biol Clin Haematol 1:65–73, 1979.
20. Cao A, Melis MA, Galanello R, Angius A, Furbetta M, Giordano P, Bernini LF: $\delta\beta$(F) thalassemia in Sardinia. Submitted for publication.
21. Weatherall DJ, Cartner R, Clegg JB, Wood WG, Macrae IA, Mackenzie A: A form of Hereditary Persistence of Fetal Hemoglobin characterized by uneven cellular distribution of hemoglobin A and A_2 in homozygotes. Br J Haematol 29:205–220, 1975.
22. Comi P, Giglioni B, Ottolenghi S, Gianni AM, Polli E, Barba P, Covelli A, Migliaccio G, Condorelli M, Peschle C: Globin chain synthesis in single erythroid bursts from cord blood: Studies on $\gamma \rightarrow \beta$ and $^G\gamma \rightarrow {}^A\gamma$ switches. Proc Natl Acad Sci USA 1:362–365, 1980.
23. Ottolenghi S, Peschle C, Comi P, Giglioni B, Pozzoli ML, Cappellini MD, Migliaccio AR, Lettieri F, Migliaccio G, Barba P, Covelli A, Gianni AM: In vivo and in vitro erythropoiesis: The Friend System. In GB Rossi (ed): Amsterdam: Elsevier, North Holland Biomedical Press 1980, pp 41–47.
24. Kan YW, Lee KY, Furbetta M, Angius A, Cao A: Polymorphism of DNA sequence in the β globin gene region. Application to prenatal diagnosis of β^0 thalassemia in Sardinia. N Engl J Med 302:185–188, 1980.

Globin Gene Transfer

Organization and Expression of Globin Genes, pages 301–312

Gene Transfer and In Vitro Expression

W.F. Anderson, P.J. Kretschmer, L. Sanders-Haigh, L. Killos, and E.G. Diacumakos

Laboratory of Molecular Hematology, National Heart, Lung, and Blood Institute, National Institutes of Health, Bethesda, Maryland (W.F.A., P.J.K., L.S.-H., L.K.) and Laboratory of Cytobiology, The Rockefeller University, New York (E.G.D.)

INTRODUCTION

The transfer of expressible genes into tissue culture cells has been accomplished in a number of different ways. These include somatic cell hybridization [1, 2], chromosome-mediated gene transfer [3, 4], DNA-mediated gene transfer [5–11], and SV40 hybrid transvection [12, 13]. Our laboratory has had a long-standing interest in the study of globin gene expression [14–16]. Our work in recent years has centered on an attempt to transfer functional globin genes into mammalian cells. Such an endeavor requires the ability to isolate a known gene and a procedure for transferring this specific gene into a living cell. We wanted to transfer a thymidine kinase (TK) and a human β-globin gene into a mouse TK^- L cell to determine if the cell could be genetically corrected for TK deficiency and if the globin gene could be replicated and expressed.

That the TK gene can be transferred to TK^- L cells by DNA-mediated gene transfer has been clearly shown by Axel and his colleagues [5–7] and others [8–11]. In this technique, picogram quantities of TK DNA are mixed with microgram quantities of salmon sperm carrier DNA and then layered onto mouse TK^- cells in the presence of calcium phosphate. The TK gene is incorporated and expressed by approximately one cell in 10^6 (when 20 picograms of TK DNA are used). These transformed cells can be recognized by their ability to grow in selective (HAT) medium. Other DNA fragments (eg, containing a human or rabbit β-globin gene) can be cotransformed along with the TK gene under appropriate conditions. In contrast, we wished to develop a procedure whereby a single copy of a specific gene could be inserted into the nucleus of a single cell under conditions whereby the injected cell could be grown in culture into a cloned population with or without selective pressure. This is the reason we chose the physical microinjection pro-

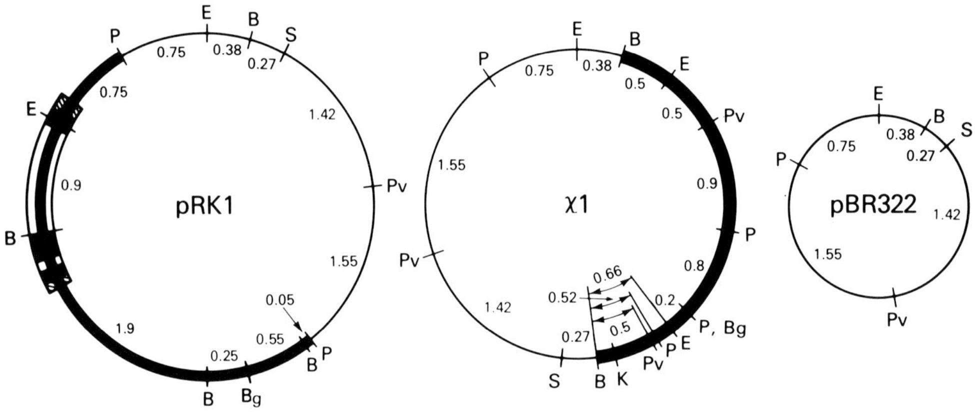

Fig. 1. Restriction endonuclease maps of pRK1, χ1, and pBR322. The restriction endonucleases used were Eco RI (E), Bam HI (B), Sal I (S), Pvu II (Pv), Pst I (P), Bgl II (Bg), and Kpn I (K). The numbers are in kilobases. The thin portion of the circle represents pBR322; the thick portion of the circle is the 4.4 kb Pst I fragment containing the human β-globin gene in (pRK1) or the 3.5 kb Bam HI fragment from herpes simplex type 1 containing the thymidine kinase gene (in χ1). The globin gene itself is represented by the boxed area, from bottom to top—hatched: 5′ nontranslated region; solid: 5′ coding region; clear: first intron; solid: middle coding region; clear: second intron; solid: 3′ coding region; hatched: 3′ nontranslated region.

cedure developed by Diacumakos [17, 18]. A minute volume (10^{-} ml) of solution can be injected into any region of a cell with no apparent interference with cell function.

GENERAL EXPERIMENTAL STRATEGY

Mouse TK⁻ cells were microinjected with a mixture of two plasmids: one containing the Herpes simplex TK gene and the other the human β-globin gene. The restriction endonuclease map of the two recombinant plasmids used for microinjection, together with the vector plasmid pBR322, are shown in Figure 1. On the left is pRK1, which is a pBR322 plasmid [19] with a 4.4 kb Pst I fragment of human DNA (isolated from HβG1 [20]) containing the human β-globin gene insert [21]. In the middle is χ1, another pBR322 plasmid with a 3.5 kb Bam HI fragment of herpes simplex type 1 containing the TK gene. This latter plasmid was developed by Enquist and his colleagues [22]. Our experiment is outlined in Figure 2. One copy (on the average) of pRK1 and one copy of χ1 were comicroinjected into the nucleus of individual mouse TK^- L cells attached to a coverslip. The coverslips were then incubated in HAT selective medium, where transformed cells grew as individual colonies. The individual colonies were isolated, grown to 10^8–10^9 cells, and analyzed. The presence of the injected genes was detected by the technique of restriction endonuclease digestion and Southern blot analysis while the expression of the genes was measured by analysis of herpes simplex TK enzyme activity or by liquid cDNA-mRNA molecular hybridization to detect human β-globin mRNA.

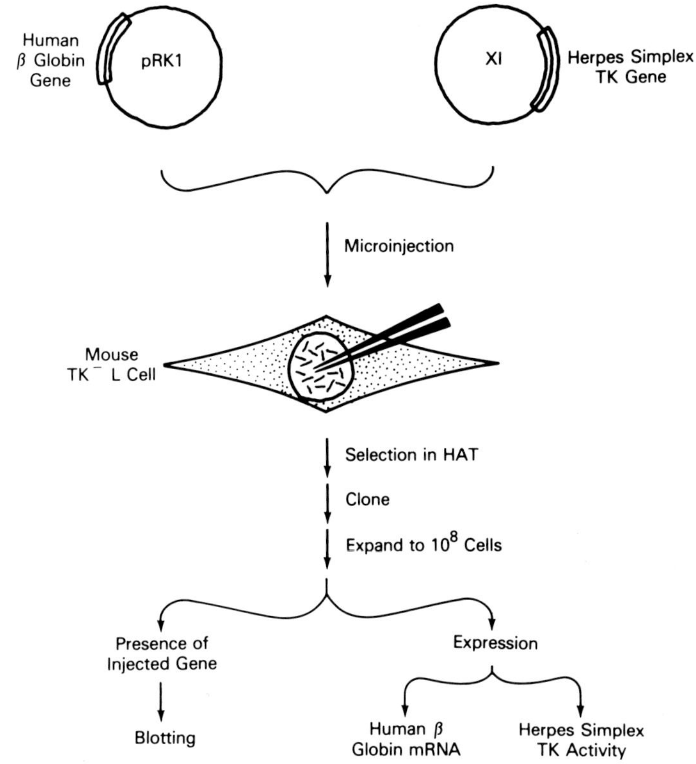

Fig. 2. Flow chart for microinjection of pRK1 and $\chi 1$ into mouse TK^- L cells. See text.

DETECTION OF TK ACTIVITY IN TRANSFORMED CELLS

Seven transformed colonies each of which had been grown in HAT medium over the past ten months were obtained; growth of injected TK^- cells in HAT selective medium indicates that either the microinjected herpes simplex TK gene is functioning or the mouse TK gene itself has been activated. A definitive technique using ^{125}I-dC for detection of herpes simplex TK enzyme activity has been described [23]. The herpes simplex TK enzyme can phosphorylate ^{125}I-dC, which is then incorporated into trichloracetic acid-insoluble DNA, while mouse TK enzyme cannot. However, mouse cells possess an enzyme, dC deaminase, which can convert ^{125}I-dC into a usable substrate. The inhibitor tetrahydrouridine (THU) blocks dC deaminase activity. Therefore, incorporation of ^{125}I-dC into DNA in the presence of THU is specific evidence for herpes simplex TK enzyme activity. Colonies, C1 and C2, were analyzed for the expression of herpes simplex TK enzyme giving the results in Table I. The mouse TK^- L cell ($LMTK^-$) shows no incorporation of ^{125}I-dC; wild-type A9 cells do incorporate ^{125}I-dC in the absence, but not in the presence, of THU; and TK^- mouse fibroblasts transformed with herpes simplex virus were positive for herpes simplex TK enzyme. Both colonies C1 and C2 were positive for herpes simplex TK enzyme activity. Colonies C2B, C2D, and C3 have

TABLE I. Measurement of Herpes Simplex Virus Type I Thymidine Kinase* in Microinjected TK⁻ L Cells

Cell line	Incorporation of ^{125}IdC (CPM)	
	−THU	+THU
Controls		
LMTK⁻	46	54
A9 (TK⁺)	2,576	107
TFC3 (HS TK⁺)	3,477	2,917
Clones		
Parental LTK⁻	48	53
C_1	1,541	1,227
C_2	1,979	2,000

*Assay: Summers WC, Summers WP: J of Virology, 24:314–318 (1977)

also been shown to possess herpes simplex TK activity. Since the assays were done on clones of 5 × 10^4 cells expanded from, presumably, single cells, the injected herpes simplex TK gene is clearly replicating and functioning in these cloned cells.

RETENTION AND EXPRESSION OF GLOBIN GENES IN TRANSFORMED CELLS

Southern blot analyses [24] of DNA extracted from the microinjected cells were performed. The DNA was digested with the enzymes Eco RI, Pst I, Bgl II, Bam HI, Pvu II, Sal I, and Kpn I either singly or in combination. The digested DNA was electrophoresed on 1% agarose gel slabs, transferred to nitrocellulose filters, and then annealed with the ^{32}P-labeled nick-translated Hha I fragment [25, 26] isolated from the human beta-globin cDNA plasmid clone JW-102 [27]. Data for colony C1 are shown in Figure 3 and for colony C2 in Figure 4 (and in tabular form in Table II). As can be seen, the human β-globin gene is present in this colony. Since the nick-translated DNA fragment used as the human β-globin plasmid cDNA probe contains some sequences of pBR322, Southern blots were also carried out using as a probe the gel-purified ^{32}P-labeled 4.4 kb Pst I human β-globin genomic fragment (Fig. 5). Results confirmed the existence of human β-globin gene sequences in the microinjected cell DNA.

Human β-globin messenger RNA was detected by hybridizing the RNA extracted from several of the microinjected colonies with ^{32}P-labeled anti-mRNA single-stranded human β-globin cDNA (isolated from JW-102) as probe. The data are shown in Figure 6. XX-8 is a somatic cell hybrid (2S MEL × human fibroblast) retaining human chromosome 11 [26] and is known to contain roughly 500 molecules of human beta-globin mRNA per induced cell. The data for C2B, C3, and C2C indicate that approximately seven, five, and two molecules of human β-globin mRNA sequences are present per microinjected L cell. Thus, the human β-globin gene is present and expressing at a very low level. These results are similar to those obtained by Wold et al [28] using the DNA-mediated gene transfer technique.

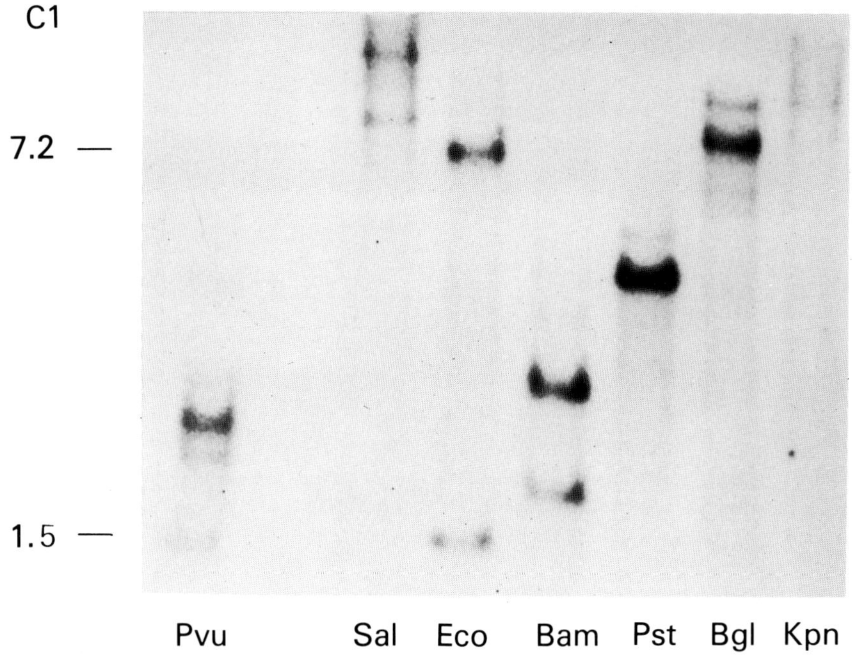

Fig. 3. Southern blot analysis of the DNA isolated from microinjected mouse L cell colony C1 using a ^{32}P-labeled human β-globin cDNA as probe. High molecular weight DNA was isolated from 10^9 cells (60–90 generations), digested with the specified restriction endonuclease, electrophoresed on 1% agarose slab gels, transferred to nitrocellulose filters, and then annealed with the ^{32}P-labeled nick-translated Hha I fragment isolated from the human β-globin cDNA plasmid clone JW-102 [26, 27]. Thirty to forty μg were loaded into each lane. Restriction endonuclease used: Pvu II, Sal I, Eco RI, Bam HI, Pst I, Bgl II, and Kpn I. The unlabeled lane contained lambda DNA marker fragments for use in size determinations.

REARRANGEMENTS OF MICROINJECTED PLASMIDS

In what form is the human β-globin gene in the microinjected cell? In Table II are collected all of the data for the seven microinjected colonies with the seven restriction endonucleases used. Even though the patterns are different for different colonies, there are a surprisingly high number of fragments the same size as those obtained when the originally injected plasmid, pRK1, is digested. The implication from these data is that perhaps covalently closed-circular pRK1 itself is present in the microinjected cells. If this were the case, it should be possible to rescue the pRK1 plasmids by transforming Escherichia coli with undigested microinjected mouse L cell DNA.

DNA from cells grown for 60 generations following microinjection yielded pRK1 with moderate efficiency (approximately one bacterial colony per μg of DNA). Surprisingly, however, not only was pRK1 recovered, but in addition, a number of other unique plasmids were found; these appear to be the result of recombination events occurring within the mouse cell. For example, as shown in Figure 7, pPK578

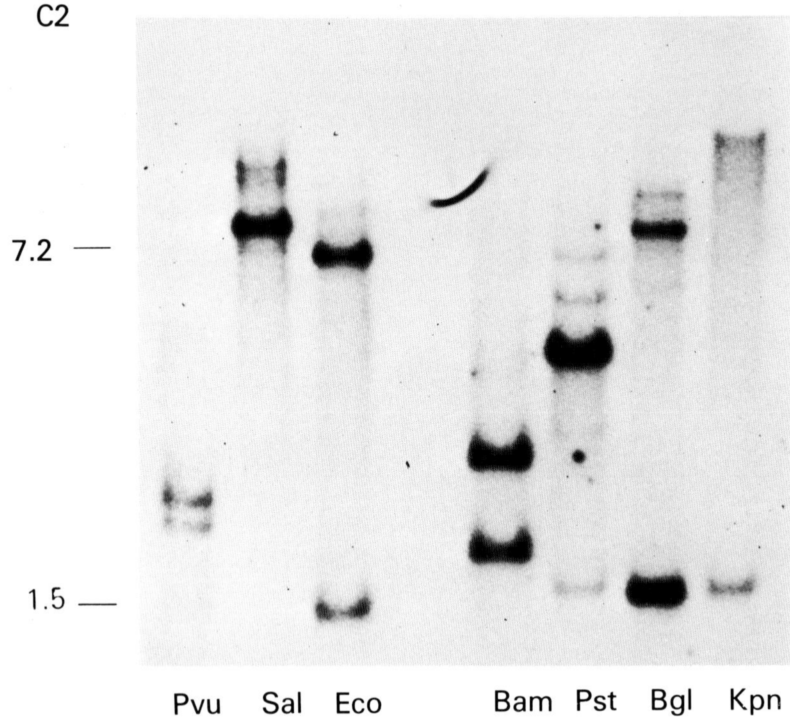

Fig. 4. Southern blot analysis of the DNA isolated from microinjected mouse L cell colony C2 using a ^{32}P-labeled human β-globin cDNA as probe. See legend to Figure 3 for details.

TABLE II. Restriction Bands of DNA From Microinjected Colonies (in KB)

DNA source	Eco RI	Pst I	Bgl II	Bam HI	PvuII	Sal I	Kpn I
C1	7.2	4.4	11	2.8	2.1	~20	8.7
	1.5		8.7	1.9	2.0	8.7	
C2	7.2	7.0	11	2.8	2.1	~20	~20
	1.5	6.0	8.7	1.9	2.0	15	1.5
		4.4	1.5			8.7	
		1.5					
C2A	7.2	4.4	–	2.8	8.7	~20	–
	1.5	3.2		1.9		15	
						8.7	
C2B	9.0	4.4	8.7	2.8	8.7	8.7	~20
	7.2			1.9	6.0		
	4.5				4.3		
	1.5						
C2C	14	4.4	–	3.2	15	–	~20
	7.2			2.8	8.7		
	1.5			1.9			
C2D	7.2	4.4	8.7	3.2	8.7	~20	~20
	4.5		6.0	2.8	3.1	8.7	
	1.5		5.0	1.9	2.1		
C3	7.2	4.4	10	4.2	8.7	~20	~20
	4.5	4.2	8.7	2.8	3.1	8.7	8.7
	1.5		6.0	1.9	2.1		
			4.8				
pRK1	7.2	4.4	8.7	2.8	8.7	8.7	–
	1.5			1.9			

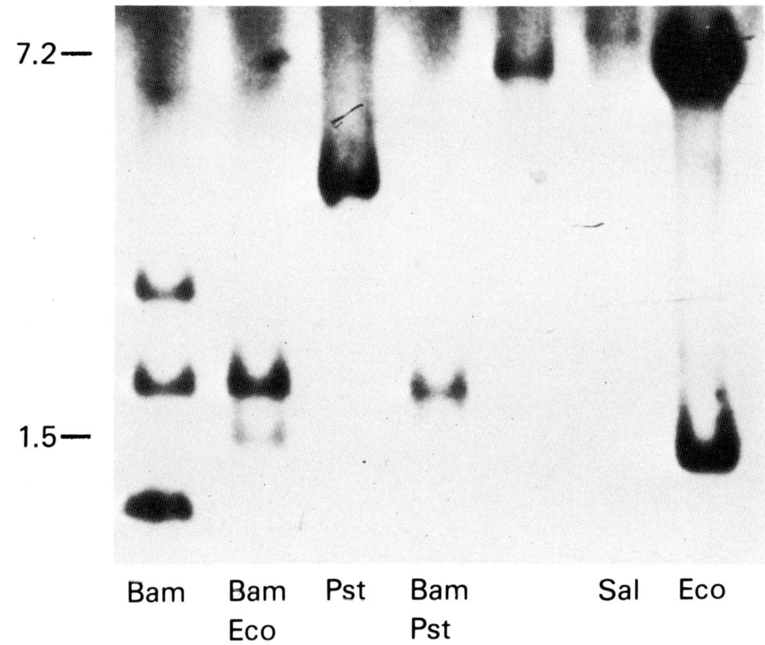

Fig. 5. Southern blot analysis of the DNA isolated from microinjected colony C3 using the ^{32}P-labeled 4.4 kb Pst I fragment from HβG1 [20] (ie, pure human DNA) as probe. These data demonstrate the presence of human DNA in the microinjected L cell DNA and correlate well with the data in Table 2, which were obtained with a human cDNA probe which contained a small amount of pBR322 sequences (thus producing additional bands on the Southern blots). If the human DNA were present as pRK1 in the microinjected DNA, then the following fragment sizes should be seen: Bam HI-2.8, 1.9, 0.8 kb; Pst I-4.4 kb; Sal I-8.7 kb; Eco RI-7.2, 1.5 kb (see Fig. 1). These are the fragments seen. However, double digests of Bam + Eco should give: 1.9, 1.5, 0.95, and 0.8 kb fragments, but only the larger fragments are seen (1.9 and 1.5 kb). Likewise, the Bam + Pst digestion should give 1.9, 1.7, and 0.85 kb fragments, but only the 1.9 (and ? 1.7) kb bands are observed. More refined Southern blot analysis using various portions of the 4.4 kb Pst I fragment as probes is now underway [L. Sanders-Haigh, unpublished observations]. The unlabeled lane contains a low level of Eco RI digested DNA, while the "Eco" labeled lane contains a high level of DNA.

appears to be a plasmid of 20.7 kb containing two intact copies of pRK1 and a third copy of pRK1 which contains a deletion of approximately 5.6 kb. pPK682, 576, and 193 all appear to have arisen from a dimer of pRK1, but containing deletions of different sizes: 5.6 kb, 1.15 kb, and 2.8 kb, respectively. Figure 8 illustrates these plasmids and the suggested deletions. The 5.6 kb deletion in pPK578 and pPK682 appears to be identical.

If the pRK1 molecule is able to combine with itself to form dimers and trimers, followed by a deletion event, it should be possible to isolate a recombinant between pRK1 and χ1 from the DNA of the microinjected cells. pPK579, Figure 9, shows a 13.7 kb plasmid which appears to have a complete copy of pRK1, combined with a copy of χ1 which contains a deletion of 2.9 kb. Southern blot analyses showed that the putative human β-globin gene region in the pRK1 portion of pPK579 hybridizes with the human β-globin probe but not with a 3.5 kb herpes simplex TK probe, whereas the 0.5 kb Bam HI/Eco RI fragment in the putative χ1 area hybridizes only with the TK probe.

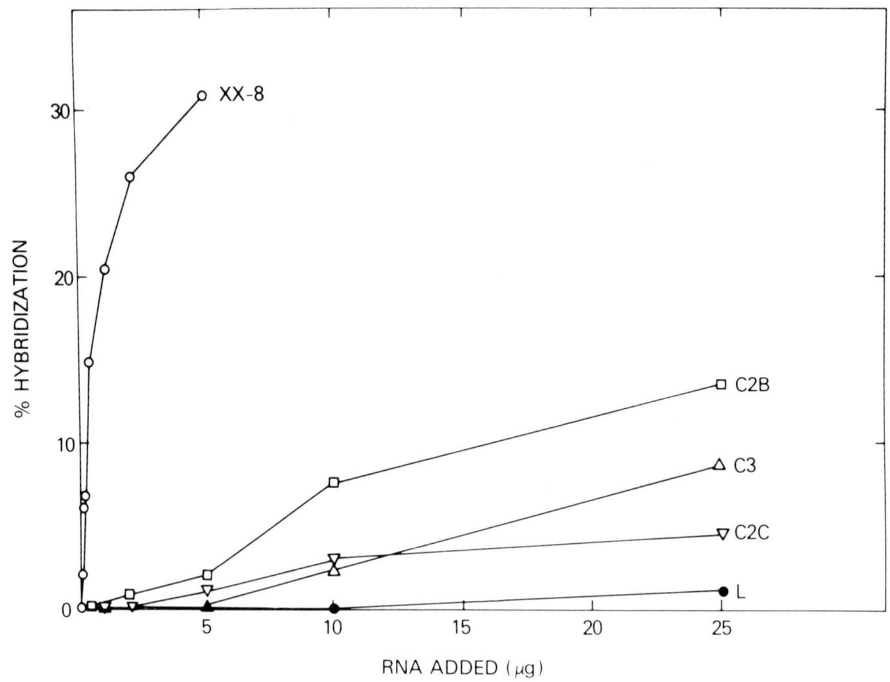

Fig. 6. Hybridization of RNA extracted from microinjected mouse L cells (30–60th generations) with ^{32}P-human β-globin cDNA. RNA from the somatic cell hybrid XX-8 (2S mouse erythroleukemia cell X human fibroblast [26]) was used as a standard since it has been determined [L. Sanders-Haigh, unpublished observations] that induced XX-8 cells contain approximately 500 human β-globin mRNA molecules per cell. Based on estimated slopes, colony C2B contains 7; C3; 5; C2C; 2; and parental L cells 0 copies of human β-globin mRNA.

Thus, five different classes of recombinant molecules were recovered. All five of these recombinant classes appear to contain a common deletion endpoint occurring at a similar region of the pBR322 segment of pRK1. Four of the recombinant classes apparently arose from a deletion of segments of pRK1 trimer or dimer molecules, while the fifth recombinant class resulted from recombination between pRK1 and χ1 followed by a deletion event within this recombinant. Extensive reconstitution experiments suggest that these plasmid recombination and deletion events occurred within the mouse cell. Recovery of intact pRK1 molecules from the mouse DNA (at a level of 0.5 extrachromosomal plasmids per mouse cell) suggests that microinjected bacterial plasmids might exist in the nonintegrated state in mouse L cells.

SUMMARY

Recombinant plasmids containing a specific human or herpes simplex gene inserted into the plasmid pBR322 were microinjected into the nuclei of mouse TK^{-}L cells where they replicated and were expressed. These plasmids were recovered from the DNA isolated from cells grown for many generations. The plasmids appear to undergo recombination and deletion events within the mouse cells.

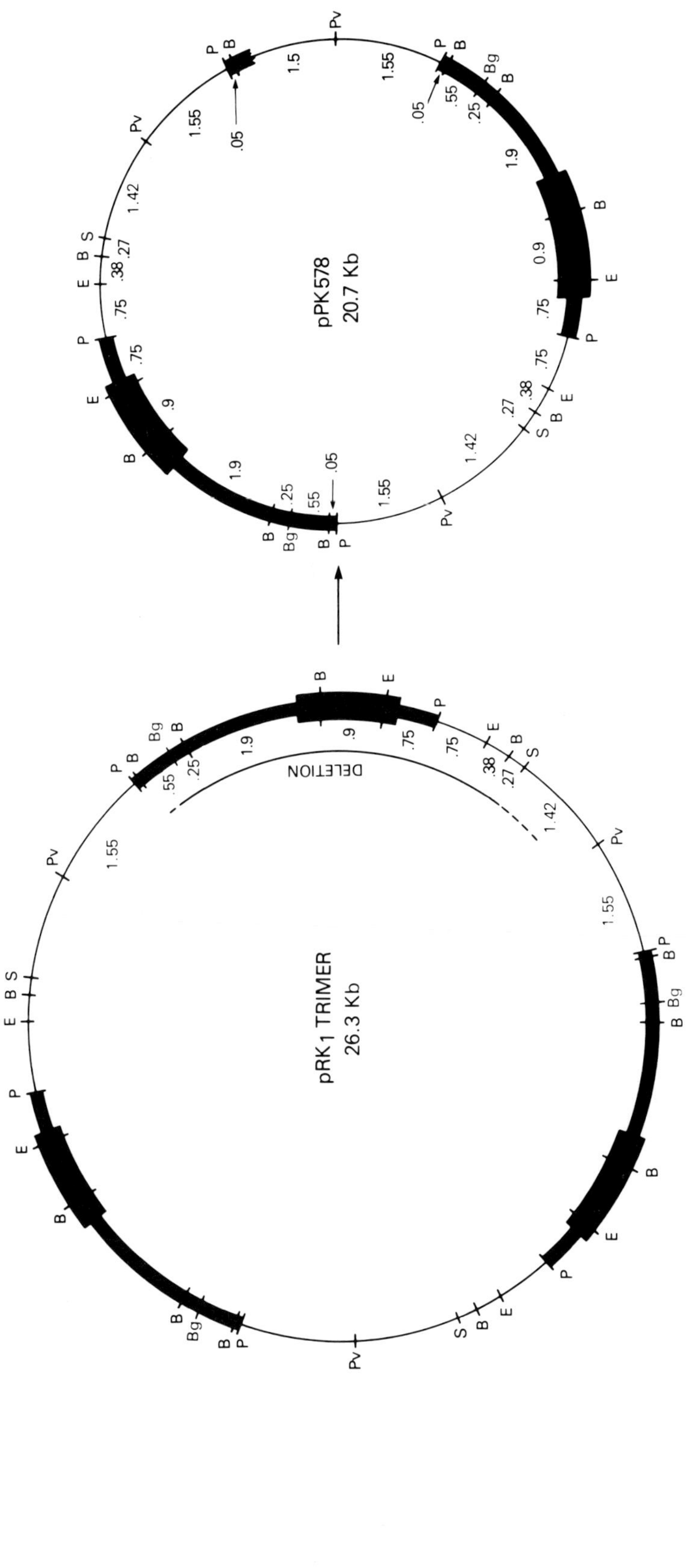

Fig. 7. Suggested origin (from a pRK1 trimer) and restriction endonuclease map of recombinant plasmid pPK578. For symbols, see legend to Figure 1.

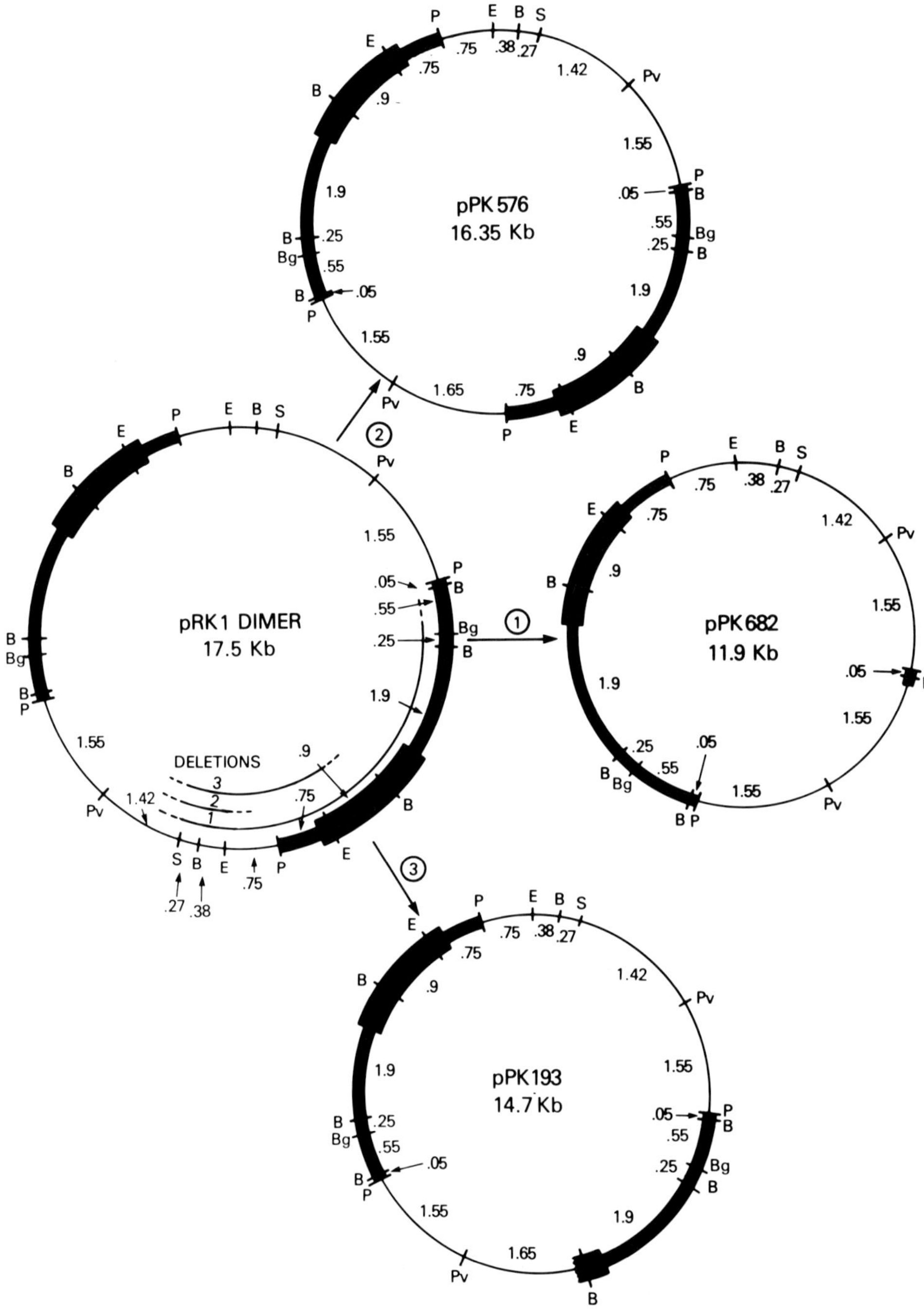

Fig. 8. Suggested origin (from a pRK1 dimer) and restriction endonuclease maps of recombinant plasmids, pPK682, pPK576, and pPK193. For symbols, see legend to Figure 1.

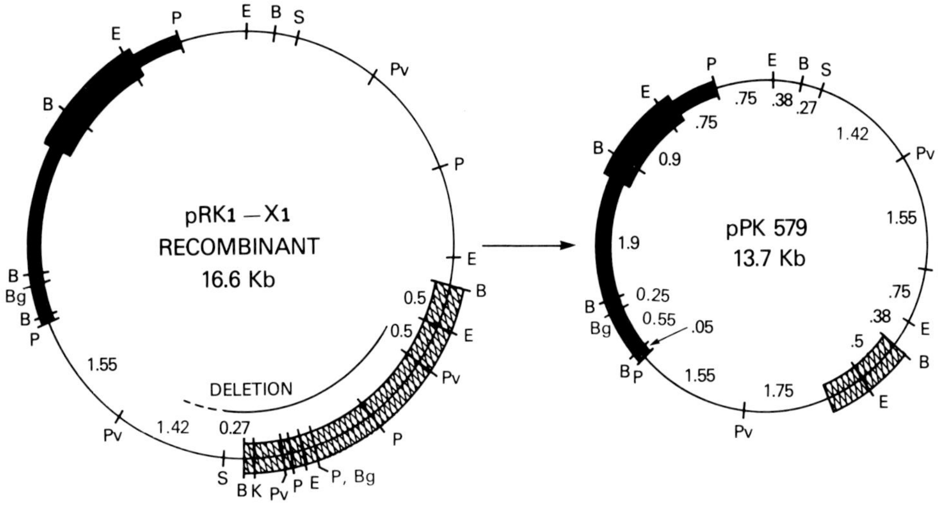

Fig. 9. Suggested origin (from a pRK1/χ1 chimera) and restriction endonuclease map of recombinant plasmid pPK579. For symbols, see legend to Figure 1.

REFERENCES

1. Davis FM, Adelberg EA: Use of somatic cell hybrids for analysis of the differentiated state. Bact Rev 37:197–214, 1973.
2. Davidson RL: Control of expression of differentiated functions in somatic cell hybrids. In Davidson RL, de la Cruz F (eds): "Somatic Cell Hybridization," New York: Raven Press, 1974, pp 131–150.
3. McBride OW, Ozer HL: Transfer of genetic information by purified metaphase chromosomes. Proc Natl Acad Sci USA 70:1258–1262, 1973.
4. Willecke K, Ruddle FH: Transfer of the human gene for hypoxanthine-guanine phosphoribosyltransferase via isolated human metaphase chromosomes into mouse L cells. Proc Natl Acad Sci USA 72:1792–1796, 1975.
5. Wigler M, Silverstein S, Lee L-S, Pellicer A, Cheng Y-C, Axel R: Transfer of purified herpes virus thymidine kinase gene to cultured mouse cells. Cell 11:223–232, 1977.
6. Pellicer A, Wigler M, Axel R, Silverstein S: The transfer and stable integration of the HSV thymidine kinase gene into mouse cells. Cell 14:133–141, 1978.
7. Wigler M, Pellicer A, Silverstein S, Axel R: Biochemical transfer of single-copy eucaryotic genes using total cellular DNA as donor. Cell 14:725–731, 1978.
8. Bacchetti S, Graham FL: Transfer of the gene for thymidine kinase to thymidine kinase-deficient human cells by purified herpes simplex viral DNA. Proc Natl Acad Sci USA 74:1590–1594, 1977.
9. Maitland NJ, McDougall JK: Biochemical transformation of mouse cells by fragments of herpes simplex virus DNA. Cell 11:233–241, 1977.
10. Mantei N, Boll W, Weissman C: Rabbit β-globin mRNA production in mouse L cells transformed with cloned rabbit β-globin chromosomal DNA. Nature 281:40–46, 1979.
11. Peterson JL, McBride OW: Cotransfer of linked eukaryotic genes and efficient transfer of hypoxanthine phosphoribosyltransferase by DNA-mediated gene transfer. Proc Natl Acad Sci USA 77:1583–1587, 1980.
12. Mulligan RC, Howard BH, Berg P: Synthesis of rabbit β-globin in cultured monkey kidney cells following infection with a SV40 β-globin recombinant genome. Nature 277:108–114, 1979.
13. Hamer DH, Smith KD, Boyer SH, Leder P: SV40 recombinants carrying rabbit β-globin gene coding sequences. Cell 17:725–735, 1979.
14. Anderson WF, Barker JE, Elson NA, Merrick WC, Steggles AW, Wilson GN, Kantor JA, Nienhuis AW: Activation and inactivation of genes determining hemoglobin types. J Cell Physiol 85:477–494, 1975.

15. Anderson WF, Willing MC, Axelrod DE, Gopalakrishnan TV, Diacumakos EG: A new approach in the search for globin gene regulatory factors. In Stamatoyannapoulos G, Nienhuis AW (eds): "Cellular and Molecular Regulation of Hemoglobin Switching," New York: Grune & Stratton, 1978, pp 779–792.
16. Anderson WF, Killos L, Sanders-Haigh L, Kretschmer P, Diacumakos E, Nienhuis A, Willing M, Vembu D: Regulation of human globin gene expression after gene transfer. In Sigler PB (ed): "Molecular Basis of Mutant Hemoglobin Dysfunction," Vol. 1 (University of Chicago Symposia on Sickle Cell Anemia), New York: North-Holland Publishing Co., in press, 1981.
17. Diacumakos EG: Methods for micromanipulation of human somatic cells in culture. Meth Cell Biol VII:287–311, 1973.
18. Diacumakos EG, Holland S, Pecora P: A microsurgical methodology for human cells in vitro: Evolution and applications. Proc Natl Acad Sci USA 65:911–918, 1970.
19. Bolivar F, Rodriguez FL, Greene PJ, Betlach MC, Heyneker HL, Boyer HW, Crosa JH, Falkow S: Construction and characterization of new cloning vehicles. II. A multipurpose cloning system. Gene 2:95–113, 1977.
20. Lawn RM, Fritsch EF, Parker RC, Blake G, Maniatis T: The isolation and characterization of linked δ- and β-globin genes from a cloned library of human DNA. Cell 15:1157–1174, 1978.
21. Kaufman RE, Kretschmer PJ, Adams JW, Coon HC, Anderson WF, Nienhuis AW: Cloning and characterization of DNA sequences surrounding the human γ, δ and β globin genes. Proc Natl Acad Sci USA 77:4229–4233, 1980.
22. Enquist LW, Vande Woude GF, Wagner M, Smiley JR, and Summers WC: Construction and characterization of a recombinant plasmid encoding the gene for the thymidine kinase of herpes simplex type 1 virus. Gene 7:335–342, 1979.
23. Summers, WC, Summers WP: [^{125}I]deoxycytidine used in a rapid, sensitive, and specific assay for herpes simplex virus type 1 thymidine kinase. J Virol 24:314–318, 1977.
24. Southern EM: Detection of specific sequences among DNA fragments separated by gel electrophoresis. J Mol Biol 98:503–517, 1975.
25. Benz EJ Jr, Kretschmer PJ, Geist CE, Kantor JA, Turner PH, Nienhuis AW: Hemoglobin switching in sheep. Synthesis, cloning, and characterization of DNA sequences coding for the β^B-, β^C-, and γ-globin mRNAs. J Biol Chem 254:6880–6888, 1979.
26. Willing MC, Nienhuis AW, Anderson WF: Selective activation of human β- but not γ-globin gene in human fibroblast × mouse erythroleukemia cell hybrids. Nature 277:534–538, 1979.
27. Wilson JT, Wilson LB, deRiel JK, Villa-Komaroff L, Efstratiadis A, Forget BG, Weissman SM: Insertion of synthetic copies of human genes into bacterial plasmids. Nucleic Acids Res 5:563–581, 1978.
28. Wold B, Wigler M, Lacy E, Maniatis T, Silverstein S, Axel R: Introduction and expression of a rabbit β-globin gene in mouse fibroblasts. Proc Natl Acad Sci USA 76:5684–5688, 1979.

Organization and Expression of Globin Genes, pages 313–334

Gene Therapy Techniques: Use of Drug Resistance Selections in Intact Animals to Insert Human Globin Genes Into Bone-Marrow Cells of Living Mice

Winston Salser, Benton D. Tong, Howard D. Stang, Jeffrey K. Browne, Karen Mercola, Menashe Bar-Eli, and Martin J. Cline

Molecular Biology Institute (J.K.B.) and Department of Biology (W.S., B.D.T.), Department of Medicine, Center for the Health Sciences (H.D.S., K.M., M.B.-E., M.J.C.), University of California at Los Angeles

INTRODUCTION

We have developed techniques for the insertion of new genes into the bone marrow cells of living mice. The results show that bone-marrow stem cells (CFU-S) are among the cell types which we were able to genetically alter. By the use of selective procedures we can insure that modified stem-cells, carrying the active inserted genes, predominate in the bone marrow. In the case of the genes coding for dihydrofolate reductase (DHFR) or herpes simplex virus thymidine kinase (HSV-tk), we have shown that the altered bone marrow cells produce high levels of the active enzymes specified by these genes. We have also demonstrated that globin genes, which confer no selective advantage, can be introduced into living animals by cotransformation with a selective marker. Human β-globin genes carried in the plasmid pBR322 appear to be capable of replication in mouse cells for prolonged periods as autonomous circular molecules.

These findings may ultimately lead to the cure of life-threatening hemoglobinopathies by gene-therapy techniques. These techniques may be equally important, from a theoretical point of view, in helping to determine what nucleotide sequence "signals" are involved in gene-switching or other processes of cell differentiation. Monitoring the behavior of precisely altered genes in the cells of intact animals may provide information which would be difficult or impossible to obtain in artificial model systems (eg, Friend cells) in which the experimenter attempts to mimic normal differentiation in cell culture.

In initiating these experiments we decided to focus on gene-therapy procedures which might have clinical relevance for the treatment of sickle cell disease and the beta thalassemias. This placed significant constraints on our approach, since these diseases require procedures which can insert functioning globin genes into the great majority of bone marrow stem cells. This is especially true in sickle cell disease, where a successful gene-therapy procedure would have to produce significant amounts of normal β-globin in all or most of the patient's red blood cells in order to avoid the vascular occlusion and consequent tissue damage from the cells which sickle.

Originally, the idea of gene therapy suggested to us protocols in which a mammalian virus vector would be used to carry the desired genes into the cells of the target tissue. Although very promising work is being carried out in this area, [1, 2] it seemed difficult to imagine a viral vector which could establish itself in all of the cells of a particular tissue but was sufficiently disarmed to avoid harmful effects. We feared that such an approach might be inefficient and mutagenic or oncogenic. In addition there are severe limitations on the quantity of DNA which can be packaged in viral vectors, and it may ultimately be necessary to incorporate the large stretches of DNA surrounding globin genes in order to achieve proper regulation.

Therefore, we decided to utilize the technique of cell transformation by calcium precipitates of DNA as developed by Bachetti and Graham [3] and modified by Wigler et al [4]. This procedure allows the insertion of even quite large DNA sequences without special modification (eg, the mouse DHFR gene which we have inserted is about 42 kilobases (kb) in size [5]) and avoids the possibly toxic effects of viral vectors). Unfortunately, this approach has the disadvantage of low efficiencies of gene insertion. In the experiments of Pellicer et al [6] and Wigler et al [4] the efficiency of transformation of the HSV-tk gene was about 10^{-5} in fibroblasts. This efficiency varies dramatically with different cell lines (eg, the efficiency with Friend cells is much lower than with fibroblasts; [unpublished results of J.B.]) and with the conditions of growth (eg, adherent Friend cells show higher transformation efficiencies than do suspension cultures; [T. Maniatis, personal communication]) or the conditions of transformation (eg, treatment with DMSO or glycerol can cause dramatic increases in the efficiency of DNA uptake in some cell lines [7]).

USE OF SELECTION TECHNIQUES

An essential part of our strategy has been to develop ways to use selective procedures in living animals. In this way even a small number of cells with the desired gene insertions could be selectively induced to proliferate so that they would come to predominate in the bone marrow. Many selective strategies could theoretically be employed. We have chosen two which permit cells to grow in the presence of the chemotherapy agent methotrexate (MTX). Methotextrate binds to the enzyme DHFR and therefore inhibits the synthesis of thymidine monophosphate (TMP) and DNA synthesis. Our first strategy involved the insertion of altered DHFR genes which produce an enzyme resistant to MTX and thus can directly overcome the metabolic block. Our second strategy involved the insertion of HSV-tk genes which code for the enzyme that catalyzes the conversion of thymidine to TMP and enable the cells to grow in the presence of MTX by utilizing the salvage pathway.

THE DHFR MADE IN 3T6-R1 CELLS IS HIGHLY RESISTANT TO MTX

We have used DNA isolated from mouse fibroblast Swiss 3T6 cells containing reiterated DHFR structural genes [8] in these experiments. These cells, designated 3T6-R1, have been shown by Kellums et al [8] to contain about 30 copies of the

DHFR structural gene and to produce correspondingly elevated levels of the enzyme. Kellums et al did not report any difference between the properties of the enzyme made in 3T6-R1 or normal cells; however, we find that the 3T6-R1 enzyme is strikingly resistant to MTX. Figure 1 shows a comparison of the DHFR activities of enzymes from normal mouse cells, 3T6-R1 cells, and the $Pro^{-}3$ MTX^{RIII} line of Chinese hamster ovary (CHO) cells. The $Pro^{-}3$ MTX^{RIII} CHO cell line is one which was isolated by Siminovitch et al [9] and shown to contain an altered DHFR which is about tenfold more resistant to MTX than the normal CHO DHFR. In Figure 1 the enzyme activities as a function of increasing levels of MTX are shown and the CHO enzyme is seen to be about tenfold more resistant than the wild-type mouse enzymes. The 3T6-R1 cells appear to contain two DHFR activities. One which comprises 25% of the DHFR activity is sensitive to MTX. The other, which represents 75% of the total activity, appears to be totally resistant to MTX.

HSV-TK GENES CAN BE USED AS A SELECTIVE MARKER EVEN IN CELLS WITH ENDOGENOUS THYMIDINE KINASE ACTIVITY

Cell lines lacking thymidine kinase (tk^{-}) are readily transformed by appropriate DNA to a tk^{+} status [4, 6]. The one cell in 10^5 or 10^6 which integrates and functionally expresses the DNA specifying thymidine kinase has a proliferative advantage over tk^{-} cells because it can utilize exogenous thymidine to synthesize DNA when DNA synthesis in the parent cell line is blocked by the presence of a folate antagonist. Since there are no known strains of animals lacking the normal cellular thymidine kinase activity, such techniques would not be applicable to intact animals. The application of such techniques in cell culture has heretofore required the time-consuming creation of a nonreverting tk^{-} variant of every cell line studied. We felt that it might be possible to establish selective conditions in which the amount of endogenous tk activity was limiting for cell growth so that cells with additional tk genes (which we denote tk^{+++}) would have a selective advantage over wild-type tk^{+} cells. In addition we noted that the HSV-tk enzyme has a greater affinity for its substrate than does the normal mammalian enzyme and reasoned that this might confer an additional selective advantage. The K_m for the virus enzyme is estimated to be 0.4 μM thymidine [10] whereas that for the normal mouse enzyme is estimated as 9 μM thymidine [11].

First we tested the validity of this approach by using it to select HSV-tk transformants in tissue culture [12] and showed that this can be accomplished in a straightforward manner (Table I). Subsequently, we have used these techniques to insert HSV-tk genes into the bone marrow cells of living animals. In preliminary experiments we used the technique of Wigler et al [4] to insert the HSV-tk gene into tk^{-} mouse L cells (Ltk^{-}). The anticipated results for transformation of Ltk^{-} cells to tk^{+} status were observed, and transformation efficiencies were approximately the same as that reported by Wigler et al [4]. The thymidine kinase specific activities of different tk^{+} clones obtained in this way varied from 1.8 to 4.2 times the specific activity of wild-type (tk^{+}) cells (eg, clone 207 in Table I) when assayed with carrier-free thymidine (10^{-8} M added ^{3}H thymidine). When the specific pyrimidine kinase assay is used, with ^{125}I dC as substrate, the HSV-tk transformants show activities manyfold over the background level of the assay in normal (tk^{+}) cells (data not shown).

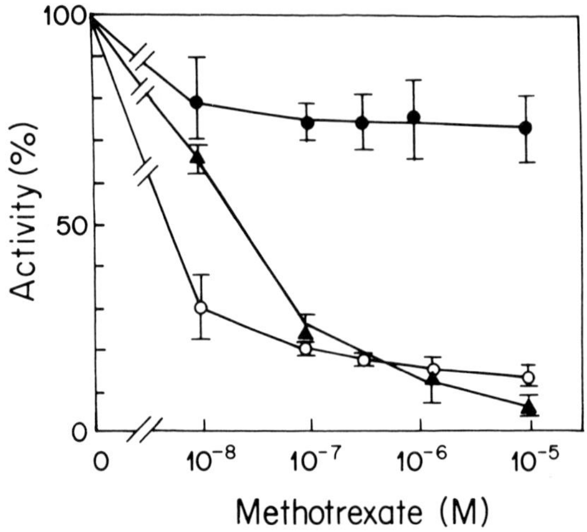

Fig. 1. The percentage activity of dihydrofolate reductase from 3T6-R1 (●——●), 3T6WT (○——○), and CHO (▲——▲) in the presence of increasing concentration of methotrexate. The results with 3T6-R1 and 3T6 (wild type) cells are the mean of three experiments, while those with enzyme from the MTX^R CHO cell line [9] are the mean of two experiments. Standard deviations are shown. The dihydrofolate reductase activity was assayed by using a sensitive fluorometric method [35] to determine the oxidation of NADP H → $NADP^+$. Cell lines have been maintained as monolayers in Dulbecco's Minimal Essential media (DME) supplemented with 10% dialyzed fetal calf serum. The 3T6-R1 and the CHO cell lines were grown in the presence of 4×10^{-4} M methotrexate (MTX). Cells to be assayed for DHFR activity were split into T-flasks (750 ml) and fed for three days with DME-MTX free medium in order to ensure complete washout of MTX bound to dihydrofolate reductase. The cells were then gently treated with trypsin, suspended in tris-HCl buffer 0.15 M pH 7.5 media at 0° C and brought to a concentration of 2×10^7 cells/ml. Lysates of the cells were prepared by sonication (biosorik IV, Bronwill) for ten seconds at a power output of 70 W. Chemical compounds were obtained from Sigma (St. Louis, Mo.). The activity of the enzyme was assayed in a final reaction mixture volume of 1 ml as follows: 0.7 ml tris-HCl buffer 0.15 M pH 7.5, 50 μl KCl 2.25 M, 50 μl NADPH 1.0 mM, 50 μl 2-mercaptoethanol 0.3 M, and 50 μl dihydrofolic acid 2.0 mM. These reaction mixtures were mixed and allowed to stand for five minutes at room temperature (RT) before adding the extract samples (100 microliters in tris-HCl buffer). Blanks consisted of a standard assay mixture minus extract. For the assay of activity in the presence of MTX extracts were preincubated at RT in the standard assay mixture containing MTX minus dihydrofolate. The reaction was initiated by the addition of dihydrofolate. The reaction mixtures were incubated at RT for 15 minutes, then 0.5 ml of 0.6N HCl was added to each tube for three to five minutes to destroy any residual NADPH. Samples of 100 μl were taken and mixed vigorously with 10 N NaOH (150 microliters) in 10 × 75 mm borosilicate glass tubes (diSPo culture tubes; Scientific Products, McGaw Park, Ill.) and allowed to stand for at least 60 minutes at RT to allow the strongly fluorescent alkali reaction products of $NADP^+$ to form. Each sample was taken, then diluted 6.5-fold with distilled water before reading on an Aminco fluorocolorimeter equipped with a Klett-Summersom 44 secondary filter. This secondary filter "peaked" at 420 nm, showed 1% of maximum transmission at 400 nm and 472 nm, and passed 57% of maximum transmission at 460 nm. A quinine-sulfate solution in 0.1 N H_2SO_4 served as working fluorometer standard solution. The instrument was calibrated so that 0.5 μg per ml of quinine sulfate gave 100 relative intensity units at the lowest sensitivity setting of the multiplier. Standard curves were prepared by adding known amounts of $NADP^+$ (1–20 nmoles) in 50 μl of control incubation mixtures. Standard curves were linear between 1 and 20 nmoles $NADP^+$ added. One enzyme unit was defined as the activity which produced 1 nmole $NADP^+$ per 15 minutes at 25°C.

TABLE I. Thymidine Kinase Specific Activity of Mouse L Cells Transformed in Vitro With HSV-TK DNA[1]

Clone	Cell lineage	Specific activity (cpm/gm)
	Untransformed	
	Ltk$^-$	0.08
	Ltk$^+$ (wild type)	4.9 to 5.5
	Transformed with HSV-tk plasmid	
207	Ltk$^-$	9.9
205	Ltk$^+$	10.3
214	Ltk$^+$	12.3
	Sal cut and ligated	
214-3	Ltk$^+$	22

[1]Clones were derived from the parent tk$^-$ cell line or tk$^+$ wild-type cells (NCTC) by transformation with calcium-precipitated HSV-tk DNA as described earlier [3, 4, 12], with 4 to 20 micrograms of tk DNA per 5×10^6 cells. The tk DNA was either in circular form in plasmid pBR322 (a clone of the 3.5 kb Bam HI HSV-tk gene carried in pBR322 was the kind gift of L. Enquist) or in plasmid that had been cut with Sal I and ligated [23]. Clones of Ltk$^-$ lineage were selected in HAT medium. Cells of NCTC (wild type) lineage were selected for in modified HAT medium containing 10^{-4} M methotrexate and 0.5 μg/ml thymidine.

HSV-tk-transformed clones were then used to determine conditions which would make tk activity limiting for cell growth. We explored a range of methotrexate and thymidine concentrations that would inhibit normal cells with a single copy of the tk gene but allow cells with the HSV-tk gene to grow. The conditions finally chosen involve higher folate antagonist concentrations (methotrexate, 10^{-4} M) and lower thymidine levels (0.5 μg/ml) than used in conventional HAT medium [12].

This selection was then used to isolate tk^{+++} cells obtained by transforming wild-type tk$^+$ cells with HSV-tk genes. Wild-type (tk$^+$) L cells were exposed to herpes virus tk DNA under transforming conditions and then cultured under selective conditions. A number of transformed clones were isolated and several were grown in quantity to permit measurement of thymidine kinase specific activity (clones 205, 214, and 214-3 in Table I), as well as for analysis of tk-specific DNA sequences by Southern blot hybridization [13] (data not shown). The efficiency of transformation of tk$^+$—tk^{+++} varied between 0.2 and five colonies per 10^6 cells when 4 μg of HSV-tk DNA was used. Spontaneous growth of tk$^+$ cells in the selective medium was less than one per 10^8. The success of this selection procedure encouraged us to undertake parallel experiments in mice, as described below. This selection procedure should also be useful for experiments in tissue culture cell lines since HSV-tk genes now provide a convenient selective marker in wild-type cells. This makes it possible to easily introduce globin genes into a variety of cell lines, either by cotransduction or by direct linkage of the globin genes to HSV-tk genes, without having to laboriously derive nonreverting tk$^-$ clones from each line.

TRANSFORMATION OF MOUSE BONE MARROW WITH DHFR GENES

We first established an escalating schedule of MTX treatment which would select for drug-resistant hematopoietic cells without killing control animals [14]. We also wished to be able to carry out the experiments with convenient internal controls and for this purpose we chose to use a syngeneic pair of mouse strains (CBA/Ca and CBA/T6T6), one of which carries the distinctive T6T6 chromosomal marker (Fig. 2). This enabled us to follow two cell types independently and, for instance, to determine that T6T6 marked cells, which have been transformed with MTX^R DNA, are able to successfully compete with Ca cells which have been transformed with a control DNA sample. In a successful transformation the cell type receiving the methotrexate-resistant (MTX^R) DNA should show a marked growth advantage and come to predominate in recipient animals treated with MTX.

The experiment shown in Table II illustrates the type of results obtained. The T6T6 marked bone marrow cells transformed with MTX^R DHFR genes showed 74%–96% predominance of the T6T6 marker in the bone marrow of recipient animals after treatment with MTX for 33–65 days. Control experiments show that these results are not due to any inherent advantage of the T6T6 cells over Ca cells.

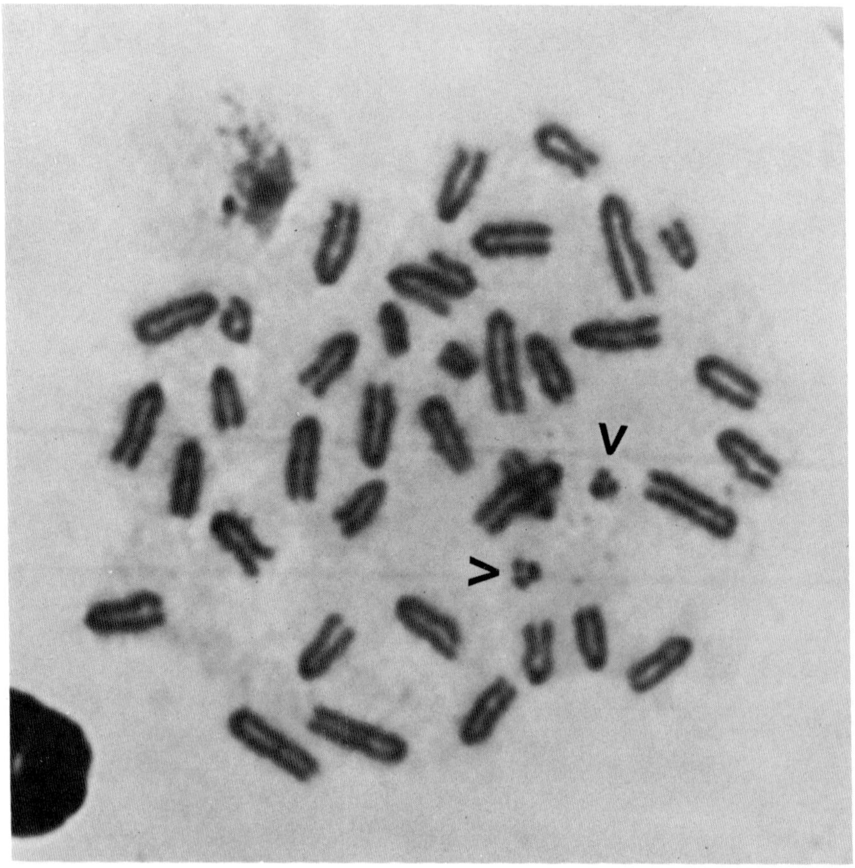

Fig. 2. Karyotype of CBA/T6T6 cell showing the distinctive chromosomal marker (arrows).

When the experimental design is reversed, so that the MTX^R DNA is used to transform Ca cells, it is the MTX^R Ca cells which successfully compete against mock-transformed T6T6 cells and become predominant in the MTX-treated recipient animals (CBA/T6T6 mice were used as the recipients in such experiments). Moreover, we find that treatment of the recipient animals with MTX is essential if the cell type transformed with MTX^R DNA is to become predominant. When MTX treatment is omitted the MTX^R DNA transformed cell karyotypes are typically found at levels of only 33%–40% [14], with the lower than 50% values reflecting the contribution of surviving cells from the marrow of the irradiated recipients. Finally, in an additional set of control experiments, irradiated CBA/Ca mice received an

TABLE II. Karyotype Analysis of Marrow Cells of CBA/Ca Mice Receiving a 1:1 Mixture of Control Ca and Transformed T6 Marrow Cells[1]

Recipient	Period with MTX (days)	Period without MTX (days)	Karyotype (% T6)
Primary 1	0–33	—	79
Primary 2	0–40	—	75
Primary 3	0–47	—	74
Primary 3	0–47	48–68	83
Secondary 3	47–61	—	88, 88, 100*
Primary 4	0–54	—	75
Secondary 4	54–72	—	83
Primary 5	0–65	—	96
Primary 5	0–65	66–113	63

[1]Donor T6T6 and Ca mice were pretreated with 3–4 mg per kg vinblastine three days before in vitro transformation by the method of Pellicer et al [6]. The resulting mitotic inhibition is followed by a burst of proliferation [33] resulting in a roughly threefold increase in pluripotent CFU-S [14]. On the day of transformation single cell suspensions in McCoy's 5A medium with 15% fetal calf serum (FCS) were obtained from femurs and tibias. Cells from Ca and T6T6 animals were placed in separate pools. Cell suspensions of 150×10^6 in 10 ml complete medium (CM) were incubated with 1.0 ml Ca-precipitated DNA containing 20 μg/ml DNA as described by Wigler et al [4] for four hours at 37° C in 5% CO_2 in tissue culture flasks. (In other experiments the DNA inputs have varied from 10 to 40 μg/ml DNA.) In the case of T6T6 cells the DNA was from 3T6 R1 MTX-resistant cells, while Ca marrow cells were incubated with control DNA preparations from MTX-sensitive cells. Thereafter, loosely adherent cells were collected and centrifuged at 150 gm for ten minutes and resuspended in DNA-free CM. After careful cell counts Ca and T6T6 cells were combined in a ratio of 1:1 and $1\text{–}2 \times 10^7$ of the combined cells were injected intravenously into recipient CBA/Ca mice in 0.3–0.4 ml of McCoy's medium with FCS. These recipients had received 850 rad from a cobalt source 24 hours previously, to eradicate endogenous hematopoiesis. This dose had low lethality but greatly reduced endogenous CFU-S. Between 48 and 96 hours after injection the recipient animals began treatment with a schedule of thrice-weekly MTX treatments at a level of 0.5 mg/kg for four doses, 2 mg/kg for four doses, and then 4 mg/kg for all subsequent treatments. With the radiation doses used in these experiments the surviving cells of the recipient animal made a substantial contribution to the repopulation of the marrow, as shown by karyotype analysis of marrow cells of CBA/Ca animals given a 1:1 mixture of mock transformed Ca and mock transformed T6T6 cells. In such recipients the karyotype analysis showed only 32% T6T6 marked cells (%T6T6 = 38 ± 12; mean ± 2 s.d.). Note that some of the details of the transformation conditions above differ from those given by Cline et al [14]. This is due to an oversight in the original paper which caused transformation conditions used in HSV-tk experiments to be listed for the DHFR experiments.

*Three secondary recipients.

equal mixture of mock-transformed Ca and mock-transformed T6T6 cells. These animals were followed up to two months with or without MTX treatment and both groups showed the result expected if neither karyotype had any selective advantage in the absence of gene therapy and MTX selection: The %Ca karyotype was about 68% in both groups, as expected from the contribution of infused Ca cells plus endogenous Ca cells (%Ca = 68 ± 12; mean ± 2 s.d.).

In addition to karyotype analyses (Table II), the marrow samples obtained periodically by limb amputation were analyzed for cellularity and CFU-S content and were also injected into secondary irradiated CBA/Ca recipients. This allowed us to assess the stability of the transformed cell population either with or without continued MTX treatments. Primary 3 (Table II) reached 74% T6T6 predominance after 47 days of MTX treatment, at which time marrow was transferred to three secondary recipients and the primary recipient was taken off MTX treatment. The level of T6T6 predominance continued to climb with continued MTX therapy, reaching 88%, 88%, and 100% in the three MTX-treated secondary recipients by 61 days and remaining stable or climbing slightly, from 74% to 83%, during the three-week-period in which MTX was withheld from the primary animal.

On the other hand, primary 5 (Table II) reached 96% T6T6 predominance after 65 days with MTX treatment but declined to 63% T6T6 during a period of 48 additional days without MTX treatment. In later studies the stability of the transformed karyotype in the absence of drug selection was variable (data not shown). This does not provide enough data to form conclusions but does suggest that any clinical use of a gene therapy procedure based on DHFR as a selective marker might require continued or periodic treatment with MTX. In this regard it should be stressed that use of MTX in a gene therapy procedure would seem to be entirely different from its use in cancer therapy. In cancer therapy there are clear reasons for employing a dose near the maximum which the patient can survive. Reductions in bone-marrow cellularity to 15% to 30% of the normal values are common in standard chemotherapy routines (intermittent treatments), with all of the attendant side-effects that this implies. In a gene therapy protocol the appropriate goal would be to achieve a moderate inhibition of the growth rate of MTX-sensitive cells so that the bone marrow cellularity would fall only 10–15% and the desired MTX^R cells would be stimulated to maximal growth rates. It is well understood from the mathematics of population biology and ecology that when two cell types occupy the same ecological niche, as seems certainly the case here, conferring even a slight growth advantage on one population will cause it to entirely supplant the other population over the course of time. Once the MTX^R cell population had been established this same maintenance dose of MTX would be expected to maintain the predominance of MTX^R cells with no detectable effects on bone-marrow cellularity. There is already extensive data about long-term administration of low levels of MTX since chronic treatments have been used extensively for other purposes, such as treatment of psoriasis [15].

TRANSFORMATION OF PLURIPOTENT STEM CELLS

We presumed that the long-term predominance of T6T6-marked cells indicated the successful implantation of MTX^R genes in pluripotent stem cells. To directly test this hypothesis, marrow was taken from selected primary recipient animals and 5×10^4 cells were injected into irradiated recipient CBA/Ca mice to assay for

CFU-S [16]. In such an assay CFU-S from the bone marrow may be detected because they give rise to individual colonies of hematopoietic cells in the spleens of irradiated recipient animals. Ten days later the secondary recipients were killed and individual spleen colonies were removed for karyotype analysis. As seen in Table III, the extent of T6T6 karyotype predomination in the pluripotent marrow stem cell population was similar to that seen in other bone marrow cells. Thus the results are as expected if there is efficient implantation of MTX^R genes in the stem-cell population.

HEMATOLOGICAL STATUS AND DHFR LEVELS IN MICE RECEIVING TRANSFORMED BONE MARROW

When it became apparent that cells transformed with MTX^R DNA predominated in the bone marrow of drug-treated recipient animals, simple tests were performed to assess the animal's hematological status. Animals receiving transformed bone marrow and MTX have tibial cellularity values averaging 9.0×10^6 cells per tibia which are indistinguishable from the values for saline-treated control animals (average 8.7×10^6 cells per tibia) and more than twice as high as the tibial cellularity in animals receiving MTX but no "gene therapy" with MTX^R DNA. Furthermore, they have high levels of CFU-S similar to control animals that did not receive radiation or bone marrow cells. Similar results were seen in several independent experiments. Thus, the hematological status of mice receiving transformed bone marrow treated with MTX^R DNA in vitro returned toward normal despite persistent treatment with MTX. Autopsies were performed on these animals at intervals up to 150 days after transformation, and sections of visceral organs were taken for microscopic study. No abnormalities were detected and the animals were clinically well at the time of sacrifice.

DHFR SPECIFIC ACTIVITY IN TRANSFORMED CELLS

To assess whether drug resistance was related to higher enzyme levels, spleens were removed from primary and secondary recipients of transformed marrow from four independent experiments and assayed for DHFR. Methotextrate treatment was terminated five to seven days before collection of tissue from these mice for the DHFR assays. Spleens were collected from 12 control animals that were either irradiated or not, and untreated or treated with MTX for periods of up to six weeks. In each case, conditions for control animals were chosen to match those of the experimental group. A radiometric assay for DHFR was performed on sonicated cell-free extracts of spleen [17, 18]. DHFR-specific activity was two to fourfold greater in the spleen extracts of animals receiving transformed bone marrow than in controls (Table IV). These were considered to be underestimates of the elevations of DHFR in the hematopoietic cells of animals receiving transformed marrow, as the extracts of spleen also included stromal and capsular tissues and nonproliferating lymphoid cells.

TRANSFORMATION OF MOUSE BONE MARROW WITH HSV-tk GENES

As outlined in an earlier section, we were successful in selecting for transformation of Ltk^+ cells to Ltk^{+++} HSV-tk transformants in tissue culture. On the basis of these results we undertook parallel experiments in mice [12]. An early result is

TABLE III. Karyotype Analysis of Bone Marrow and Pluripotent Stem Cells From CBA/Ca Mice Receiving a 1:1 Mixture of Control Ca and Transformed T6T6 Bone Marrow Cells[1]

Recipient	Duration of MTX (days)	Bone marrow karyotype T6T6 (%)	Karyotype of spleen colonies	
			T6T6 (%)	Ca (%)
Primary 1	0–24	57	50 (50)	50
Primary 2	0–40	75	74 (57)	26
Primary 3	0–47	74	92 (58)	8

[1]Individual spleen colonies were removed ten days after inoculation of the irradiated recipient with bone marrow cells. A single cell suspension was made from each colony and cells were incubated with colcemid 3 μg/ml for 90 minutes before treatment with hypotonic KCl and fixation with acetic acid/ethanol for chromosomal spreads [34]. In addition to colonies with all T6T6 or all Ca cells, colonies with a mixture of T6T6 and Ca cells were observed. Such a mixed colony can result from growth of a T6T6 CFU-S colony on a background of endogenous hematopoeisis in the Ca animals. It is much less likely that it results from overlap of a Ca and a T6T6 marked colony. Therefore those colonies listed as originating from T6T6 CFU-S included the sum of pure T6T6 and mixed colonies. The numbers in parentheses indicate the number of pure T6T6 colonies in each case.

TABLE IV. Dihydrofolate Reductase Activity in Spleen and Karyotype Analysis of Bone Marrow[1]

Recipient	Duration of MTX (days)	Bone marrow karyotype predominance (%)	DHFR specific activity (% control)
Primary	0–42	49	168
Secondary	39–67	93	375
Secondary	54–72	83	183
Secondary	65–107	—	208
Tertiary	53–80	—	224
Primary	0–59	62	375

[1]CBA/Ca recipient mice received a 1:1 mixture of control Ca and transformed T6T6 bone marrow cells. Spleen fragments were sonicated for 30 seconds in the cold and assayed for DHFR by the method of Hayman et al [18]. Results are expressed as the percentage of DHFR in spleens of simultaneously killed control mice. In the last animal listed the inverse procedure was followed, with transformed Ca cells being injected along with control T6T6 cells into a T6T6 recipient and the karyotype analysis showed 62% predominance of T6T6 cells.

shown in Table V. The strategy of the experiments was similar to that employed for selection of expression of the DHFR gene. Mouse bone marrow cells with the T6T6 chromosomal marker were treated, in vitro, with a calcium microprecipitate of the HSV-tk gene. The treated marrow, presumed to contain a few stem cells transformed by the viral gene, was mixed in a ratio of 1:1.5 with "mock"-transformed CBA/Ca marrow cells lacking the chromosomal marker. As in the experiments with MTX^R DHFR genes, marrow samples were obtained by limb amputations or by sacrificing the animal, and karyotype analysis was used to determine whether there was a predominance of T6T6-marked cells indicating successful gene therapy. As shown in the Table, high levels of the T6T6 karyotype were frequently observed in

TABLE V. Sizes of Restriction Fragments Hybridizing to HSV-tk Gene Sequences[1]

Enzyme	Animal					Control plasmid
	KMB2-1	MB2-8	KMB8-2A	KMB8-1A	KMB8-3B	
Eco RI	3.0 kb 2.5 kb 1.1 kb	8.2 kb 5.0 kb 2.5 kb 1.1 kb	0.5 kb	2.5 kb	4.9 kb 2.5 kb	4.5 2.5 1.0
Pro II	ND	ND	—	8.8 kb 2.5 kb 1.9 kb	3.5 kb 2.0 kb	2.9 2.7 2.0
Bam HI	ND	ND	ND	3.5 kb	ND	3.5
Xba 1	ND	ND	ND	ND	ND	*

[1] Spleens were excised from mice which had received bone marrow transformed with the HSV-tk gene. The transforming DNA included intact plasmid pBR322 carrying the tk gene in every case. In addition, ligated 3.5 kb Bam HI fragments containing the HSV-tk gene were included in the DNA used to transform KMB2-1 and KMB2-3 while the same Bam HI fragments in unligated form were included in the DNA used to transform KMB8-2A, KMB8-1A, and KMB8-3B. DNA extracted from spleens of animals was cleaved with the indicated restriction enzymes and Southern blots were probed with radioactive probe prepared by nick translation of the 3.5 kb HSV-tk Bam HI fragment. Fragment sizes are indicated in kilobase pairs (kb). The fragment sizes seen in a control digestion of intact plasmid containing the HSV-tk gene sequence are also shown. The percentages of karyotype predominance at the times of sacrifice were as follows: KMB2-1, 74%; KMB2-8, sample lost; KMB8-2A, 70%; KMB8-1A, 62%; KMB8-3B, 87%. KMB8-3B was a secondary recipient. All others were primary recipients.

*There is no cleavage site for Xba 1 in the plasmid pBR322 containing the HSV-tk fragment.

the treated animals but never in controls which had received mixtures of Ca and T6T6-marked cells in which both cell types had been mock transformed. In selected cases where T6T6 predominance was observed DNA was extracted, then digested with appropriate restriction enzymes and subjected to Southern blot analysis in order to test for the presence of HSV-tk gene sequences. HSV-tk gene sequences were shown to be present, confirming the success of the transformation procedure.

In analyzing the results of further experiments we noted that HSV-tk transformation in intact animals shows important differences from MTX^R DHFR transformations. First, gene therapy with DHFR genes shows a much higher success rate (15 out of 18) than gene therapy with HSV-tk genes (27 successes out of 70 mice as estimated from the frequency of karyotype predominance). The second surprise came in our examination of the control group of animals which had received T6T6 cells transformed with HSV-tk DNA but which were not treated with MTX. Some of these animals also developed T6T6 karyotype predominance, although usually at a somewhat later time than in MTX-treated animals. In fact, at late times the fraction of gene therapy-treated animals showing T6T6 predominance with and without MTX treatment appears to be quite similar.

The appearance of T6T6 karyotype predominance constitutes a rather indirect way of demonstrating successful introduction of HSV-tk genes. The possibility of artifact is effectively ruled out by direct demonstrations that the animals showing

T6T6 predominance contain HSV-tk gene sequences and that these are fully functional. This was accomplished by: (1) Southern blot analysis [13, 19], and (2) pyrimidine kinase assays showing that the spleen extracts from transformed animals contain the expected unique enzyme activity which can convert ^{125}I dC to ^{125}I dCMP—an activity not seen in extracts similarly prepared from normal mice. We therefore concluded that the HSV-tk gene confers a proliferative advantage on mouse bone marrow cells even in the absence of selective drug therapy with MTX. This may indicate that the salvage pathway for DNA synthesis is somewhat limiting to the growth of normal bone marrow so that cells with an enhanced salvage pathway capacity have an appreciable advantage during repopulation of the irradiated bone marrow.

COTRANSFORMATION OF HUMAN β-GLOBIN GENES WITH THE DHFR OR HSV-tk SELECTIVE MARKERS

Recently Wigler et al [20] described the cotransformation of thymidine kinase-deficient (Ltk$^-$) cells with both a selectable gene (HSV-tk) and molecules providing no selective advantage (linear ØX174 DNA). Although the tk genes and ØX174 sequences were not physically linked, cells which took up and expressed the tk gene were found to have integrated and maintained multiple copies of the ØX174 DNA. Other examples of cotransformation with, and presumed integration of, nonlinked [21, 22] or linked [23, 24] genes have been reported. Huttner et al, [25] cotransformed Ltk$^-$ cells with HSV-tk DNA and a plasmid containing human chromosomal β-globin gene sequences and suggested that the recombinant plasmid was not always integrated into the chromosomes of the recipient cells but could possibly exist in circular form.

We have used a similar approach to introduce human β-globin genes into mouse bone marrow cells. Figures 3 and 4 outline results from two such experiments, one involving cotransformation with the HSV-tk selective strategy, the other using MTX^R DHFR genes as the selective agent. In the experiments shown in Figure 3, HSV-tk transformed T6T6 cells were injected into three irradiated animals (primaries) as described earlier. At day 82 bone marrow cells from primary number 2 were injected into irradiated secondary CBA/Ca mice (secondaries 2a, 2b, 2c, Fig. 3). Finally, at day 150, bone marrow cells from secondary 2b were used to transfuse five irradiated tertiary recipients (2b1 through 2b5). This experiment is instructive because it illustrates the variety of responses seen in other experiments using HSV-tk genes and because a variety of analyses illustrate in detail the results of the procedures at a molecular level. Primary 1 was treated with MTX and showed levels of T6T6 karyotypes which started low (35% on the 28th day) and increased (62% on the 70th day). Southern blot analysis showed that both HSV-tk genes and human β-globin genes were present in DNA from this animal.

Primary animals 2 and 3 both showed a predominance of T6T6 karyotypes even though no MTX treatment was used, and the fact that HSV-tk genes were present and functioning at high level was shown by strong signals in the pyrimidine kinase assay. Primary recipient 3 was shown by Southern blot hybridization to have incorporated human β-globin genes along with HSV-tk genes. The DNA from this animal (denoted mouse A) has been studied in detail and results will be presented in a subsequent section, indicating that the human globin plasmid can be maintained as an autonomously replicating plasmid in mouse cells. As explained in the legend to Figure 3, the presence of HSV-tk enzyme activity in these experiments was estimated by measuring the pyrimidine kinase activity of this enzyme. Activities severalfold over background were observed even though hematopoietic cells are a small fraction of the tissue present in the spleens of these animals. In an animal treated with phenylhydrazine, in which such cells are expected to hypertrophy and form the majority of tissue mass, the pyrimidine kinase activity seen was 31-fold over the background even when tetrahydrouridine was not used to block an extraneous background activity seen in phenylhydrazine-treated animals (see legend to Fig. 3).

The T6T6 karyotype was stably maintained at a high level in primary 2 but decreased over time in primary 3. In the case of animal 2 the karyotype predominance was 80% at 82 days and continued high when bone marrow was transferred into secondary recipients (87%, 70%, and 71% in three secondary recipients at 150 days). We very seldom see successful colonization of tertiary recipients. Presumably this is because the pluripotent stem cell lines, like other normal cells not "transformed" for continuous growth in culture, have a limited proliferation potential. In part this may also be due to the fact that our protocol leaves a substantial contingent of surviving endogenous cells in the recipient animal which can compete with the injected cells. Therefore it is remarkable that in this experiment two tertiary recipients showed 50% T6T6 karyotypes when the expected result in a control experiment would have been 0–5%.

Figure 4 shows a similar experiment carried out using MTX^R DHFR genes as the selective marker. One of the animals was unusual in showing a failure to stably transform with the DHFR genes: T6T6 predominance was only 57% at day 57 in the primary recipient and only about 30% at day 115 in the secondary recipient. The other primary recipient showed 69% T6T6 predominance at day 47, and the secondary recipient had reached 80% T6T6 predominance by day 115. Southern blot analysis showed that human β-globin gene sequences were present in both the primary and secondary recipients of the latter animal. The DNA from the secondary recipient (denoted mouse B) has been analyzed in detail with results indicating that it, like the DNA from primary recipient 3 in Figure 3, contains human β-globin sequences in circular plasmid form.

RESTRICTION MAPPING OF HUMAN β-GLOBIN SEQUENCES CARRIED IN INTACT MICE

The spleen is a major site of hematopoeisis in the mouse and served as a convenient source of DNA for investigation of the status of DNA sequences inserted by gene therapy procedures. DNA isolated from the spleens of these mice was digested to completion with restriction endonucleases Eco RI, Bam HI, Hind III and Xba I, subjected to agarose gel electrophoresis, and transferred to nitrocellulose filters by the procedure of Southern [13] with minor modifications [12]. Plasmid $H_{\beta}1$, cut with appropriate restriction endonucleases and/or wild-type λ bacteriophage digested with Hind III plus ØX174 digested with Hae III were run in parallel as size markers. The plasmids were hybridized [26] with ^{32}P-labeled plasmid pBR322, or plasmid $H_{\beta}1$, or the fragment containing the human β-globin gene excised from $H_{\beta}1$.

Co-transformation of human beta globin genes using HSV-tk selection

Transforming DNA and donor cells

HSV-tk DNA and Hbg DNA used to transform CBA/T6T6 cells — 1.0

Salmon sperm DNA used to transform CBA/Ca cells — 1.5

→ CBA/Ca recipients irradiated with 850 rad

Primary recipient:	1	2	3
MTX treatment given:	yes	no	no
Bone marrow sampled at day:	28→70	82	28→70
%T6T6 marker observed:	35→62%	80%	70→33%
HSV-tk gene sequences observed (Southern blot):	yes	no	ND
HSV-tk enzyme observed by pyrimidine kinase assay (fold over control animal)	ND	6.6X	9.9X
Human globin gene sequence observed (Southern blot)	yes	no	yes

Secondary recipient (at day 82):	2a	2b	2c
MTX treatment given:	no	no	no
Bone marrow sampled at day:	150	150	150
%T6T6 marker observed:	87%	70%	71%
HSV-tk gene sequences observed (Southern blot):	yes	no	ND
HSV-tk enzyme observed by pyrimidine kinase assay (fold over control animal)	ND	31X	ND
Human globin gene sequence observed (Southern blot):	no	no	ND

Tertiary recipient:	2b1	2b2	2b3	2b4	2b5
%T6T6 marker observed at day 206:	50%	5%	50%	0%	11%

Figure 3

The sizes and relative intensities of bands seen in the autoradiographs of these gels suggest that most of the $H_{\beta}1$-derived sequences detected in the DNA from the spleens of these mice are in free plasmid which is not integrated into the mouse chromosomal DNA. Huttner, Scangos, and Ruddle [25] have also suggested the existence of circular plasmids in mouse tissue culture cells and Anderson and his colleagues report related findings in this volume.

Figure 6a shows an example of the results we have obtained. When $H_{\beta}1$ plasmid is digested with Bam H1 four fragments of 0.7, 1.8, 2.8, and 3.4 kb in size are obtained, and these same fragment sizes are seen when DNA from the spleens of mice A and B are digested with Bam H1 (Fig. 6a). In this experiment the labeled probe was made by excising the 4.4 kb human chromosomal DNA fragment from $H_{\beta}1$ (by Pst cleavage) and nick translating it. The relative intensities of the bands are as would be expected for this probe: the 3.4 kb fragment contains only a few hundred bp of human DNA and so gives a signal of low intensity in the autoradiograph. When nick-translated pBR322 probe was used instead only the 3.4 and 2.8 kb fragments were labeled [19]; labeling of the 1.0 and 0.7 kb fragments of $H_{\beta}1$ is not expected since they contain only human sequences. As shown by lane A, the probe shows no cross reactivity with a control DNA sample from an untreated mouse. Cleavage of $H_{\beta}1$ with Eco RI yields two fragments, 1.5 and 7.3 kb in size, each of

---

Fig. 3. The results of a detailed analysis of animals transformed with HSV-tk genes and human β-globin DNA are outlined. The technique used for transformation was similar to that described by Mercola et al [12] and is the same as that given in the legend to Table II except for the following: Cell suspensions of 5×10^7 cells in 10 ml complete medium were incubated with 1.0 ml Ca-precipitated DNA containing 40 μg/ml DNA. In the case of the CBA/T6T6 cells this DNA consisted of 4 μg of plasmid pBR322 containing the 3.5 kb HSV-tk gene fragment (a gift of L. Enquist) [36] in supercoiled form, 4 μg of this same plasmid digested to completion with Bam HI (BRL), 4 μg of $H_{\beta}1$ and 30 μm of salmon sperm DNA. The CBA/Ca control cells were mixed in a ratio of 1.0:1.5 and infused into irradiated CBA/Ca recipients. Autoradiographic analysis for HSV-tk or human globin gene sequences by the Southern blot technique was carried out as described earlier [12, 19]. The pyrimidine kinase assays were carried out as described in Figure 5. At day 70 primary recipient number 3 was sacrificed and DNA was extracted from its spleen for a detailed restriction enzyme digestion analysis of the inserted human globin gene sequences (animal A, Fig. 5–6). The assays for pyrimidine kinase were carried out according to a modification of the procedure of Cremer et al [37], as follows: The assay mixture contained 0.01 M ATP, 0.008 M creatine phosphate, 1 unit/ml creatine phosphokinase, 0.025 M NaF, 0.01 M Mg Ac_2, 0.1 M Na H_2PO_4, pH 6.0 and 5.0 μCi ^{125}I dC in a volume of 0.1 ml. This reaction mixture also contained 200 μg protein obtained by mincing a tissue sample with sterile forceps, sonicating two $\times$ 30 seconds on an ice/salt bath, and taking the supernate after centrifugation at 12,000 rpm in an eppendorf microfuge for two to five minutes. Incubation was at 37°C. Animal 2b was treated with phenylhydrazine to induce hypertrophy of hematopoieses. The increased ratio of hematopoietic to other spleen cells probably accounts for the increased enzyme activity seen. More recently we have found it desirable to add 30 μg tetrahydrouridine (THU) to the assay to suppress an endogenous background activity seen in phenylhydrazine-treated control animals (we thank Dr. Summers for suggesting this and for the kind gift of samples of THU). Had THU been used in these assays we would have expected the assays for animal 2b to show an even greater signal to background ratio. Samples were adsorbed on DEAE paper and washed twice with 5 ml 0.04 M NH_4HCO_3, 0.02 M KI, and once with 5 ml distilled H_2O at 37°C before counting radioactivity. Note that no TK gene sequences were demonstrated in the DNA from animals 2 and 2b despite the marked increase in pyrimidine kinase activity. We suggest that the position of integration of the TK gene sequences varies in the mixed transformed cell population, giving several different restriction endonuclease fragments on digestion, none of which are present in sufficient concentration to be detected and scored.

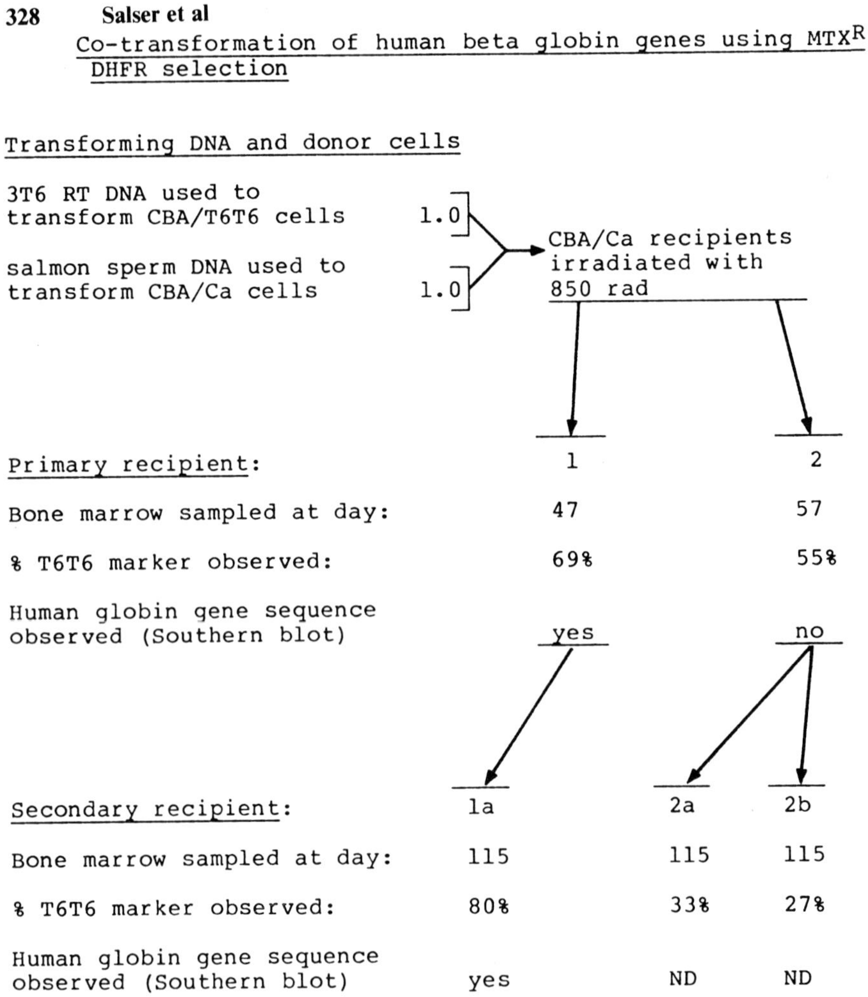

Fig. 4. Recipient CBA/Ca mice irradiated with 850 rads were transfused with a mixture of transformed CBA/Ca cells according to the protocol in the footnote to Table II except that the transforming DNA was a mixture of 32 μg MTX^R 3T6-R1 DNA and 4 μm Hβ1 plasmid DNA containing the human β-globin gene sequences. Hβ1 [38] was a gift from T. Maniatis and contains a 4.4 kb Pst fragment of chromosomal DNA with the human β-globin gene inserted into the Pst I site of plasmid pBR322. Human globin gene sequences were detected faintly in primary recipient 2 and more strongly in secondary recipient 2a. Both primary and secondary recipients were treated with MTX. Secondary recipient 1 was sacrificed on day 115 and DNA was extracted from its spleen for a detailed restriction enzyme digestion analysis of the status of human globin gene sequences in the mouse cells as shown in Figures 5–6 (animal B).

which contains both human and pBR322 sequences (Fig. 5b, lane A). Similar results are obtained with DNA from mouse B (tracks D and F) or mouse A (tracks C and E) regardless of whether the probe used was nick-translated pBR322 (lanes C and D) or human chromosomal sequences (lanes E and F).

INTEGRATED VERSUS CIRCULAR DNA MOLECULES

In addition to these major bands at 1.5 and 7.3 kb which could come from intact plasmids in the mouse cells, additional faint bands can be visualized in some experiments. These can be seen most clearly in Figure 5b, lanes D, E, and F, where

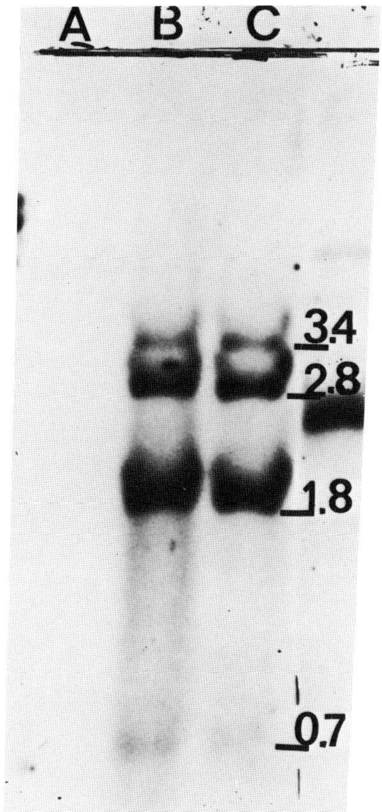

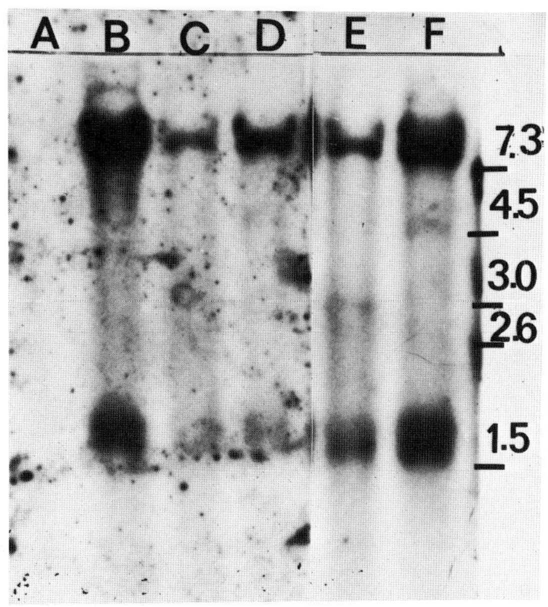

Fig. 5. Detection of human β-globin and pBR322 sequences in DNA isolated from the spleens of mice that had received bone marrow cotransformed with 3T6-R1 or HSV-tk DNA and Hβ1 DNA. DNA was purified, digested, and analyzed as described by Stang et al [19]. a. DNA digested by Bam H1 and hybridized to nick-translated human chromosomal Pst I fragment from Hβ1. Lane A: DNA from a control Ca mouse. Lane B: DNA from mouse A. Lane C: DNA from mouse B. Exposure time is seven days. b. DNA digested by Eco RI and hybridized to nick-translated pBR322 (lanes A–D) or human chromosomal Pst I fragment from Hβ1, 1.4×10^{-4} μg (4.5×10^{-5} μg Hβ1 represents single copy frequency in a 15 μg sample of genomic DNA). Lanes C and E: DNA from mouse A. Lanes D and F: DNA from mouse B. Exposure time is three days.

faint bands occur at about 2.6, 2.8, 3.1, 3.4 (lane F), and 4.5 kb (lanes D and F) for mouse B and 2.0, 2.7, and 3.0 kb in the case of mouse A (lane E).

Since such bands were never observed in multiple control mouse samples (data not shown) they may represent fragments of chromosomally integrated Hβ1 which coexist in the same cells with unintegrated plasmid or they may represent separate populations of transformed cells. This ambiguity of interpretation arises because, unlike the situation with tissue culture cell lines, it is impossible to work with pure clones of transformed cells in gene therapy procedures in living animals. The transformed cells in a particular animal may arise from many independent transformation events. Because of the limited proliferative capacity of normal cells it may be difficult or impossible to reproducibly obtain either primary or secondary recipients in which all of the transformed cells are descended from a single trans-

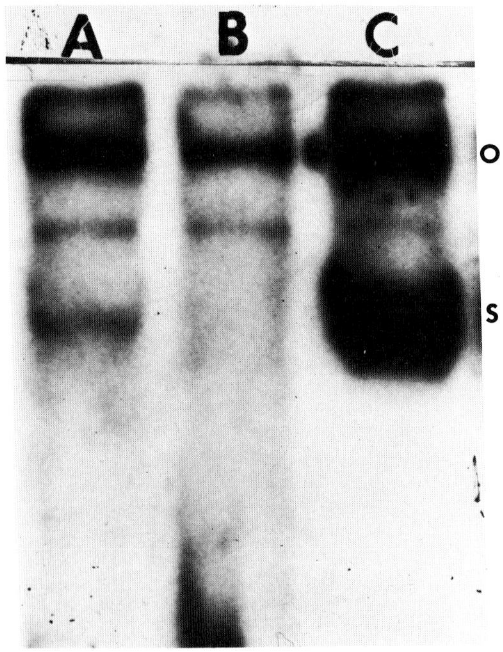

a

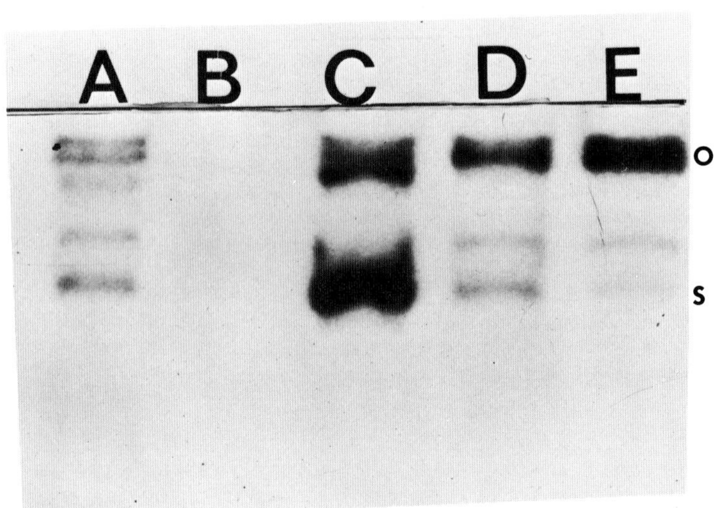

b

Fig. 6. Detection of open and supercoiled circular DNAs. Methods are as in Figure 5 and reference 14. a. DNA digested by Xba I and hybridized to nick-translated Hβ1. Lane A: DNA from mouse B. Lane B: DNA from mouse A. Lane C: Hβ1, 1.4×10^{-4} μg, supercoiled fraction. Exposure time is three days. S and O mark the position of supercoiled and open circles of Hβ1. b. Undigested DNA was electrophoresed and analyzed by filter hybridization using nick-translated Hβ1 as probe. Lane A: (15 μg) from mouse A, undigested. Lane B: control mouse DNA, unidigested. Lane C: Hβ1, supercoiled fraction (1.4×10^{-4} μg). Lane D: DNA from mouse B (15 μg) digested with Xba I; Lane E: Hβ1, supercoiled fraction, DNAse digested. Exposure time is 24 hours. S and O mark the positions of supercoiled and open circles of Hβ1.

formed stem cell. Each integration of a plasmid sequence into the host DNA should result in some novel restriction fragments and these should be different in each case. If an animal contained tens or hundreds of different kinds of cells with different integration events then there would be tens or hundreds of "novel" bands, but if these were present at equal levels then each would be too faint to visualize in the autoradiograph of the Southern blot. This situation could arise either if there were many independent transformations of different stem-cells or if one stem cell were transformed but carried the plasmid in an unintegrated form for several generations with many different integration events subsequently occurring in different CFU-S descendants of the initial transformant.

EVIDENCE OF HSV-tk GENES INTEGRATED INTO MOUSE DNA

We have presented evidence consistent with the idea that plasmid pBR322 containing the HSV-tk gene sequence may also be maintained autonomously (Fig. 1 of Mercola et al [12]). It should be emphasized that we have more frequently (in other animals) seen evidence for HSV-tk genes integrated in mouse chromosomal DNA. Table V shows such results in which cleavage of mouse DNA from transformed animals reveals fragments of expected sizes (eg, the 25 kb Eco RI fragment) but also fragments of novel sizes which hybridize to radioactive HSV-tk gene sequence probes. Since such bands are never seen in the DNA of control animals we presume that they result from integration of the HSV-tk gene into a variety of different sites in different animals. It should be emphasized that the bands seen are usually faint, as might be expected if the tissue from which DNA was obtained was derived from many cell lineages in which different integration events had occurred.

DETECTION OF OPEN AND SUPERCOILED CIRCLES

Digestion of the DNA from mouse A and from mouse B with Hind III, which cuts $H_{\beta}1$ at a single site, gave 8.8 kb fragments which hybridize to radioactive $H_{\beta}1$ probe [14]. Again this is the result expected if circular plasmid molecules were present in the cells of these mice, but these data and that shown in Figure 5 are not adequate to prove this hypothesis. Alternatively, similar results might be obtained if the cells contained integrated concatomers of the $H_{\beta}1$ plasmid sequence. This caveat applies also to the experiments of (Huttner et al [25]). To distinguish between these alternative hypotheses we performed a series of experiments designed to directly detect circular molecules. These involved electrophoretic analysis of either DNA which had been cleaved with Xba I or undigested DNA from mouse A or mouse B. As seen from track C of Figure 6a, plasmid $H_{\beta}1$ is not cleaved by Xba I. The lower of the two intense bands is due to supercoiled circular plasmid while the upper band is due to open (nicked) circles. When the DNA from mouse B was similarly digested bands were also seen in the positions of open and supercoiled circles. With the DNA from mouse A only the band in the position of open circles is seen, as expected if the DNA were extensively nicked during preparation. The Xba I digestion should have given very different results (ie, material electrophoresing in the position of very large or undigested DNA) if the $H_{\beta}1$ plasmid sequences had been present as integrated concatomers.

The presence of free plasmid circles in the mouse cells is further indicated by electrophoresis of undigested DNA samples (Fig. 6b). The lower bands in tracks A, C, and D are in the position expected for supercoiled $H_{\beta}1$. Authentic $H_{\beta}1$ supercoils (track C) are converted to open circles when $H_{\beta}1$ is nicked by partial digestion with DNAse I (track E). Undigested and Xba I digested mouse B DNA, shown in tracks A and D, both show substantial amounts of material in the position of supercoiled plasmid. Interestingly, two faint bands in track A are eliminated by Xba I digestion as if they were $H_{\beta}1$ circles which had incorporated other DNA sequences with Xba I sites. This may be similar to the alterations in $H_{\beta}1$ plasmids carried in tissue culture cells as reported in Dr. Anderson's paper in this volume.

From the intensities of the bands of Bam HI, Eco RI, and Xba I cleaved spleen DNA, relative to those of known amounts of similarly digested $H_{\beta}1$ DNA run in parallel, we estimate that the spleen of mouse A contained an average of 0.5 copy and spleen B contained an average of one copy of $H_{\beta}1$ per haploid genome. Since the spleen cell population also contains untransformed bone marrow cells, lymphocytes and nonhemotopietic spleen cells (eg, stromal cells) we estimate that the number of $H_{\beta}1$ copies in transformed cells may exceed five copies per cell in animal A and ten copies per cell in animal B.

DISCUSSION

We have shown the practicality of selective procedures which permit the effective insertion of genes into the bone marrow cells of living mice. Because of our ability to confer a proliferative advantage on those cells which have been altered in a desired way, we can frequently convert a great majority of the cells in the bone marrow in the desired manner. We have also shown that these procedures can be used to introduce a gene such as that coding for human β-globin, which by itself would confer no selective advantage on the recipient cells. The presence of human globin genes in mouse reticulocytes does not result in the production of human globin in quantities comparable to that from the mouse globin genes. We are currently carrying out experiments designed to test for the presence of human globin mRNA synthesis or the small amounts of human globin which may result if human globin mRNA or protein is unstable in mouse cells. In assessing the ultimate likelihood of being able to effectively manipulate genes in such gene therapy procedures it is encouraging to note that even a very complex gene like DHFR, which includes five intervening sequences and which extends for at least 42 kb [32], appears to function effectively after insertion using our approach.

Like others working in different systems (tissue culture cells), we have seen evidence that plasmids can be maintained as free circular molecules and have demonstrated supercoiled forms of these molecules for the first time. From transformation studies in yeast [27–29] it appears unlikely that the origin of replication from a bacterial plasmid would be active in eukaryotic cells. Therefore, it is interesting that Velinek et al [30] and Rubin et al [31] recently demonstrated a nucleotide sequence homology between a 30-residue portion of a family of interspersed repeated sequences in human genomic DNA and a sequence at or near the origin of DNA replication in BK virus, Simian virus 40, and polyoma virus. A similar sequence is found in $H_{\beta}1$ at the junction between the large intervening sequence and the codon for amino acid 105 in the human β-globin gene [30].

SUMMARY

This article reports the successful implantation of DNA containing functional genes into the bone marrow cells of living mice.

We have shown that drug resistance genes such as those coding for a methotrexate-resistant dihydrofolate reductase (DHFR) or the thymidine kinase from herpes simplex virus can be used to confer a proliferative advantage on cells having desired genetic alterations. As a result of this proliferative advantage, all or most of the bone marrow cells come to have the desired characteristics. Such gene therapy techniques may ultimately lead to the cure of life-threatening globinopathies such as sickle cell disease or the β thalassemias. As a step in this direction we have shown that these selective techniques can be used to insert human globin genes into the bone marrow cells of living mice. Efficient gene expression was obtained for both the thymidine kinase and DHFR genes inserted into mouse bone marrow but has not yet been shown for the human globin genes. The fact that a very large gene like DHFR, with many intervening sequences extending over at least 40 kb, is efficiently expressed, leads us to believe that this technique will be generally useful for a variety of mammalian genes.

ACKNOWLEDGMENTS

We thank D. Robertson, D. Morse, J. Lipeles, G. Kato, C. LeFevre, and S. Pitts for valuable technical assistance and L. Morse, R. Ruprecht, and A. Lusis for help and advice. This research was supported by USPHS grants GM18586, HL21831, CA15619, CA27682-01; PHS National Research Award (to J.B.); fellowships to H.D.S. and B.D.T. from the Leukemia Society of America; and an endowment from the Ambrose and Gladys Bowyer Foundation to M.J.C.

REFERENCES

1. Mulligan RC, Howard BH, Berg P: Synthesis of rabbit β-globin in cultured monkey kidney cells following infection with a SV40 β-globin recombinant genome. Nature 277:108–114, 1979.
2. Hamer DH, Leder P: Expression of the chromosomal mouse β^{maj}-globin gene cloned in SV40 Nature 281:35–40, 1979.
3. Bachetti S, Graham F: Transfer of the gene for thymidine kinase to thymidine kinase-deficient human cells by purified herpes simplex viral DNA. Proc Natl Acad Sci USA 74:1590–1594, 1977.
4. Wigler M, Pellicer A, Silverstein S, Axel R: Biochemical transfer of single-copy eucaryotic genes using total cellular DNA as donor. Cell 14:725–731, 1978.
5. Nunberg JH, Kaufman RJ, Chang ACY, Cohen SN, Schimke RT: Structure and genomic organization of the mouse dihydrofolate reductase gene Cell 19:355–364, 1980.
6. Pellicer A, Wigler M, Axel R, Silverstein S: The transfer and stable integration of the HSV thymidine kinase gene into mouse cells. Cell 14:133–141, 1978.
7. Copeland NG, Cooper GM: Transfection by exogenous and endogenous murine retrovirus DNAs. Cell 16:347–356, 1979.
8. Kellums R, Morhenn V, Pfendt E, Alt F, Schimke R: Polyoma virus and cyclic AMP-mediated control of dihydrofolate reductase on RNA abundance in methotrexate-resistant mouse fibroblasts. J Biol Chem 254:309–318, 1979.
9. Flintoff WF, Davidson SV, Siminovitch L: Isolation and partial characterization of three methotrexate-resistant phenotypes from chinese hamster ovary cells. Somatic Cell Genet 2:245–261, 1976.

10. Klemperer HG, Haynes GR: A virus-specific thymidine kinase in BHNZ1 cells infected with herpes simplex virus. Virology 31:120–128, 1967.
11. Chang Yang-Chi, Prusoff WH: Mouse ascites sarcoma 280 deoxythymidine kinase. General properties and inhibition studies. Biochemistry 13:1179–1185, 1974.
12. Mercola KE, Stang HD, Browne J, Salser W, Cline MJ: Insertion of a new gene of viral origin into bone marrow cells of mice. Science 208:1033–1035, 1980.
13. Southern E: Detection of specific sequences among DNA fragments spearated by gel electrophoresis. J Mol Biol 98:503–517, 1975.
14. Cline MJ, Stang H, Mercola K, Morse L, Ruprecht R, Browne J, Salser W: Gene transfer in intact animals. Nature 284:422–425, 1980.
15. "Physicians Desk Reference," Oradell, NJ: Medical Economics Company, 1979.
16. Till JE, McCullock EA: A direct measurement of the radiation sensitivity of normal mouse bone marrow cells. Radiat Res 14:213–222, 1961.
17. Hillcoat BL, Sweet V, Bertino JR: Increase of dihydrofolate reductase activity in cultured mammalian cells after exposure to methotrexate. Proc Natl Acad Sci USA 58:1632–1637, 1967.
18. Hayman R, McCready R, Van Der Weyden M: A rapid radiometric assay for dihydrofolate reductase. Analy Biochem 87:460–465, 1968.
19. Stang HD, Cline MJ, Mercola K, Browne J, Tong B, Lusis A, Salser W: Insertion of human β-globin gene sequences in intact mice. Manuscript in preparation.
20. Wigler M, Sweet R, Sim GK, Wold B, Pellicer A, Lacy E, Maniatis T, Silverstein S, Axel R: Transformation of mammalian cells with genes from procaryotes and eucaryotes. Cell 16:777–785, 1979.
21. Wold B, Wigler M, Lacy E, Maniatis T, Silverstein S, Axel R: Introduction and expression of a rabbit β-globin gene in mouse fibroblasts. Proc Natl Acad Sci USA 76:5684–5688, 1979.
22. Pellicer A, Wagner EF, Kareh AEl, Dewey MJ, Reuser AJ, Silverstein S, Axel R, Mintz B: Introduction of a viral thymidine kinase gene and the human β-globin gene into developmentally multipotential mouse teratocarcinoma cells. Proc Natl Acad Sci USA 77:2098–2102, 1980.
23. Mantei N, Boll W, Weissman C: Rabbit beta-globin mRNA production in mouse L cells transformed with cloned rabbit β-globin chromosomal DNA. Nature 281:40–46, 1979.
24. Breathnach R, Mantei H, Chambon P: Correct splicing of a chicken ovalbumin gene transcript in mouse L cells. Proc Natl Acad Sci USA 77:740–744, 1980.
25. Huttner KM, Scangos GA, Ruddle FH: DNA-mediated gene transfer of a circular plasmid into murine cells. Proc Natl Acad Sci USA 76:5820–5824, 1979.
26. Wahl GM, Stern M, Stark GR: Efficient transfer of large DNA fragments from agarose gels to diazobenzyloxymethyl-paper and rapid hybridization by using dextran sulfate. Proc Natl Acad Sci USA 76:3683–3687, 1979.
27. Struhl K, Stinchcomb D, Scherer S, Davis R: High-frequency transformation of yeast: Autonomous replication of hybrid DNA molecules. Proc Natl Acad Sci USA 76:1035–1039, 1979.
28. Beggs J: Transformation of yeast by a replicating hybrid plasmid. Nature 275:104–109, 1978.
29. Gerbaud C, Fournier P, Blanc H, Aigle M, Heslot H, Guerineau M: High frequency of yeast transformation by plasmids carrying part or entire 2-μm yeast plasmid. Gene 5:233–253, 1979.
30. Jelinek W, Toomey T, Leinwand L, Duncan C, Biro P, Chaudary P, Weissman S, Rubin C, Houck C, Deininger P, Schmid C: Ubiquitous, interspersed repeated sequences in mammalian genomes. Proc Natl Acad Sci USA 77:1398–1402, 1980.
31. Rubin C, Houck C, Deininger P, Friedmann T, Schmid C: Partial nucleotide sequence of the 300-nucleotide interspersed repeated human DNA sequences. Nature 284:372–374, 1980.
32. Nunberg JH, Kaufman RJ, Chang ACY, Cohen SN, Schimke RT: Structure and genomic organization of the mouse dihydrofolate reductase gene. Cell 19:355–364, 1980.
33. Smith WW, Wilson SM, Fred SS: Kinetics of stem cell depletion and proliferation: Effects of vinblastine and vincristine in normal and irradiated mice. J Natl Cancer Inst 40:847–854, 1968.
34. Metcalf D, Moore MAS: "Haemopoietic Cells," Amsterdam: North Holland, pp 44–47, 1971.
35. Lowry OH, Roberts NR, Kapphan JI: The fluorometric measurement of pyrimidine nucleotides. J Biol Chem 224:1047–1064, 1957.
36. Enquist LW, Madden MJ, Schiop-Stansly P, VandeWoude GF: Cloning of herpes simplex type 1 DNA fragments in a bacteriophage lambda vector. Science 203:541–544, 1979.
37. Cremer K, Bodemer M, Summers WC: Characterization of the mRNA for herpes simplex virus thymidine kinase by cell-free synthesis of active enzyme. Nucleic Acids Res 5:2333–2344, 1978.
38. Lawn R, Fritsch E, Parker R, Blake G, Maniatis, T: The isolation and characterization of linked β and β-globin genes from a cloned library of human DNA. Cell 15:1157–1174, 1978.

Author Index

Subject Index